T0291796

CAMBRIDGE LIBRARY COLLECTION

Books of enduring scholarly value

Astronomy

From ancient times, humans have tried to understand the workings of the world around them. The roots of modern physical science go back to the very earliest mechanical devices such as levers and rollers, the mixing of paints and dyes, and the importance of the heavenly bodies in early religious observance and navigation. The physical sciences as we know them today began to emerge as independent academic subjects during the early modern period, in the work of Newton and other 'natural philosophers', and numerous sub-disciplines developed during the centuries that followed. This part of the Cambridge Library Collection is devoted to landmark publications in this area which will be of interest to historians of science concerned with individual scientists, particular discoveries, and advances in scientific method, or with the establishment and development of scientific institutions around the world.

On the Determination of the Distance of a Comet from the Earth

Even while professionally engaged in banking, Sir John William Lubbock (1803–65) applied his formidable mind to scientific questions. Several of his early writings on astronomy – his particular sphere of interest – are gathered together in this reissue, notably *On the Determination of the Distance of a Comet from the Earth, and the Elements of its Orbit* (1832), *On the Theory of the Moon and on the Perturbations of the Planets* (1833), and *An Elementary Treatise on the Computation of Eclipses and Occultations* (1835). Lubbock received a Royal Society medal for tidal research in 1834, and herein is his *Elementary Treatise on the Tides* (1839). Also included is Lubbock's *On the Heat of Vapours and on Astronomical Refractions* (1840), in which he relates celestial observations to Gay-Lussac's gas expansion law. The collection closes with *On the Discovery of the Planet Neptune* (1861), Lubbock's lecture discussing how John Couch Adams first predicted the planet's existence.

Cambridge University Press has long been a pioneer in the reissuing of out-of-print titles from its own backlist, producing digital reprints of books that are still sought after by scholars and students but could not be reprinted economically using traditional technology. The Cambridge Library Collection extends this activity to a wider range of books which are still of importance to researchers and professionals, either for the source material they contain, or as landmarks in the history of their academic discipline.

Drawing from the world-renowned collections in the Cambridge University Library and other partner libraries, and guided by the advice of experts in each subject area, Cambridge University Press is using state-of-the-art scanning machines in its own Printing House to capture the content of each book selected for inclusion. The files are processed to give a consistently clear, crisp image, and the books finished to the high quality standard for which the Press is recognised around the world. The latest print-on-demand technology ensures that the books will remain available indefinitely, and that orders for single or multiple copies can quickly be supplied.

The Cambridge Library Collection brings back to life books of enduring scholarly value (including out-of-copyright works originally issued by other publishers) across a wide range of disciplines in the humanities and social sciences and in science and technology.

On the Determination of the Distance of a Comet from the Earth

And Other Works

J.W. Lubbock

CAMBRIDGE
UNIVERSITY PRESS

University Printing House, Cambridge, CB2 8BS, United Kingdom

Cambridge University Press is part of the University of Cambridge.
It furthers the University's mission by disseminating knowledge in the pursuit of education, learning and research at the highest international levels of excellence.

www.cambridge.org
Information on this title: www.cambridge.org/9781108068628

This edition first published 1832–61
This digitally printed version 2014

ISBN 978-1-108-06862-8 Paperback

ON

THE DETERMINATION

OF

THE DISTANCE OF A COMET

FROM

THE EARTH,

AND THE ELEMENTS OF ITS ORBIT.

BY J. W. LUBBOCK, ESQ. F.R.S.

LONDON:

PRINTED FOR C. KNIGHT, PALL MALL EAST; AND
J. AND J. J. DEIGHTON, CAMBRIDGE.

1832.

PRINTED BY RICHARD TAYLOR,
RED LION COURT, FLEET STREET.

PREFACE.

THE following pages are intended to convey an exposition of different methods which have been proposed for the solution of one of the most important problems in Physical Astronomy. The inconvenience of various notation is nowhere more apparent than in the methods which have been given by different authors for the solution of this problem, their mutual connexion being thereby obscured.

Some time since, in a paper which was printed in the Memoirs of the Astronomical Society, I endeavoured to present the different methods which deserve attention in a uniform system of notation; and I also proposed a new method, which results from a combination of the equations of Legendre and Lagrange, by which the distance of the comet from the sun is eliminated, and the question of the determination of the distance of a comet from the earth is reduced to the solution of a quadratic equation. It is needless for me to point out the advantage which results from the employment of a method by which the necessity of having recourse to repeated trials is superseded.

In the paper in question, I presented those equations in the form in which they had been given previously, in which, however, a term is neglected, sensible in both, but more so in the equation of Lagrange than in that of Legendre. I have now given details which could not enter into the plan of that paper, and also a new method of obtaining the perturbations of a comet, which consists in determining directly the perturbations of the rectangular coordinates.

23 St. James's Place,
14th July, 1832.

CONTENTS.

ON THE DETERMINATION OF THE DISTANCE OF A COMET FROM THE EARTH, AND THE ELEMENTS OF ITS ORBIT.

TYCHO BRAHE concluded, from their small parallax, that comets were not to be confounded with meteors at the earth's surface; but to Kepler we are indebted for the first important step in the determination of the distances of the heavenly bodies. He discovered that the observations of the planets were satisfied by supposing them to move in ellipses having the sun in one of the foci, and such that their periodic times were dependent in a certain ratio on their mean distances from the sun; hence, their periodic times being known from observation, it was easy to infer their mean distances from the sun in terms of the earth's mean distance. Doerfell, a clergyman of Plaven in Saxony, made the analogous step for comets, by showing that the observations of a comet, which appeared in 1680, were satisfied by supposing it to move in a parabola, having the sun in the focus. This conclusion, like those of Kepler, was empirical. The complete solution of this difficult problem was first achieved by Newton, to whom it afforded an important confirmation of the truth of the law of gravitation. It is, in fact, to the mechanical conditions of the system resulting from that law, that we are indebted for the means of determining the motions of the heavenly bodies with remarkable precision.

A comet is observed successively at three different times, separated by short intervals; and with only the data furnished by these observations, the mathematician is enabled, after the calculation of a few hours, to assign its present position in space, and its future path in the heavens, during the same apparition. It is not so easy to determine one of the elements, upon which the time of its return depends, the semi-axis major; but as comets are found to move in very eccentric ellipses or hyperbolas, such that the portion of their orbit in which they appear coincides sensibly with a parabola, the reciprocal of the semi-axis major may be neglected, and the parabolic elements being known, the comet may be recognised if it should reappear. Its reappearance furnishes the means of determining its period, upon which alone, when the perturbations are neglected, its semi-axis major, or mean distance from the sun, depends. More than one hundred comets have been observed, but up to the present time the periods of three only have been accurately ascertained. These bodies acquire importance from the probability which exists, that they will instruct us, not only with respect to the mass of the Georgium Sidus *, and perhaps of the other planets, but also with respect to an æther or fluid of small density pervading space, which may, after the lapse of ages, affect the stability of the planetary system. Already the existence of such a fluid appears scarcely doubtful, through Professor Encke's most interesting and laborious researches. See the *Astronomische Nachrichten*, No. CCX. and CCXI., or the translation by Professor Airy. Should such a fluid exist, we may be able, after a few years, to establish the existence of changes in the orbit of the comet which has been the subject of Professor Encke's investigations, analogous to those which hereafter, in that case, will be recognised in our own.

Halley was the first who applied extensively the method of determining the orbit of a comet given by Newton in the Principia. Having calculated the elements of twenty-four comets, "he suspected, from the like situation of their planes and perihelions, that the comets which appeared in the years 1531, 1607, and 1682, were one and the same comet that had made three revolutions in its elliptic orbit." He conjectured, therefore, that it would return about 1758. Afterwards Clairaut, with the assistance of Lalande, calculated how much the planets Jupiter and Saturn would disturb its course, and predicted that it would return to the perihelion about the middle of April 1759, stating, however, that owing to the intricacy of the calculations and the quantities neglected, there might be the error of a month in this determination. The event fully justified his prediction, for the comet did pass the perihelion

* *Mécanique Céleste*, vol. iv. p. 215.

on the 13th March. It was first seen, in December 1758, by a farmer named Palitzch, near Dresden, with an 8-feet telescope. Messier discovered it on January 21st, and observed it until February 14th; again from March 31st until April 16th, and also from May 1st until June 3rd; but although Pingré, Lalande, Lemonnier, La Caille, and all the astronomers at Paris, were anxiously expecting it, as likely to afford to the Newtonian theory of gravitation a more remarkable confirmation than any it had yet received, Delisle did not permit Messier to give notice of its appearance until April, when he could no longer retain the secret. It is to be hoped that such an occurrence as this concealment will never again sully the annals of astronomy. In May the comet was visible to the naked eye, but very faint. When Messier first saw it in January, it appeared, according to his description *, as a very faint light, equally extended round a luminous point which was the nucleus. The perturbations of this comet by the planets in its present revolution have been calculated by MM. de Pontécoulant and Damoiseau, according to whom it will return to the perihelion in the beginning of November 1835.

In February 1826 M. Biela at Josephstadt in Bohemia discovered a comet appearing in the form of a small nebula. It was soon found that the elements of this comet bore great resemblance to those of the comets of 1772 and 1806, and MM. Clausen and Gambart separately found ellipses which satisfied all the observations, the comet having a period of six years and three quarters. M. Damoiseau has calculated the perturbations caused by the planets, and has found that its next perihelion passage will take place on November 27th of the present year (1832). See *Mém. de l'Académie*, vol. viii. The time of the present revolution is diminished nearly ten days by the action of the planets.

This comet, called the Comet of Biela, was seen for the first time by Montaigne, at Limoges, in March 1772. When Messier observed it shortly afterwards, it appeared as a nebula without nucleus or tail. On one occasion (3rd April), he saw a small telescopic star shining through the comet, and mistook it for the nucleus.

The third comet whose period is known, is called the Comet of Encke. † " It was first seen by Mechain and Mes-

* " Elle ne paroissoit alors que comme une lumière extrêmement foible, également étendue en rond autour d'un point lumineux qui en étoit le noyau, sans être terminé, sa lumière étoit assez vive et blanchâtre."—*Mém. de l'Académie*, 1760, p. 390.

† See the translation by Professor Airy, before referred to, from which the passage within inverted commas is taken.

"sier in 1786; but they observed it only twice, and were
"therefore unable to determine the elements of its orbit. Miss
"Herschel discovered it in 1795, and it was observed by
"several European astronomers. In 1805 Pons, Huth and
"Bouvard, discovered it on the same day. In 1819 Pons dis-
"covered it again. Hitherto it was supposed that the four
"comets were different; but Encke (Bode's *Astron. Jahrb.*
"1822) not only pointed out their identity, but showed that an
"elliptic orbit agreed better with each set of observations than
"a parabola. In Bode's *Astron. Jahrb.* 1823 (published in
"1820), Encke gave new calculations of the perturbations, &c.
"and as there still appeared to be some unknown cause of un-
"certainty, he gave two ephemerides for its appearance in 1822.
"This was observed by Rumker in New South Wales: and
"Encke, after discussing his observations in the *Astron. Jahrb.*
"1826, concluded that the supposition of a resisting medium
"was necessary to reconcile all the observations. The comet
"was again generally observed in Europe in 1825 and 1828:
"and the circumstances of the last appearance were particu-
"larly favourable for determining the influence of Jupiter's
"mass and the absolute amount of the retardation, which the
"other observations had left undetermined." The period of this comet is about three years and one third. It passed the perihelion on May 3 of the present year, but has not (I believe) been seen, owing to unfavourable circumstances. M. Struve has given a drawing of this comet, in the *Astronomische Nachrichten*, No. 154., as it appeared in 1828 with his large refracting telescope. It was an ill-defined luminous patch, about 4′ diameter, without any well-defined nucleus. Some part of it, not in the centre, was brighter than the rest. It is easy to imagine that these bodies, of which the density is so inconsiderable, must be very sensible to the action of a resisting medium, if such exists.

As no comet has hitherto sensibly disturbed the motions of any of the planets, their mass is probably extremely small. The comet of 1770 appears to have traversed the system of the satellites of Jupiter without producing the slightest alteration; and Laplace says that its mass must have been less than the $\frac{1}{5000}$th that of the earth, or it would have produced a sensible effect on the length of our year. As these bodies do not move in the plane of the ecliptic, but in all directions in space, and as their mass is so small, all fears with respect to the disastrous consequences which might arise from our collision with one of them, appear quite groundless.

The equations which serve to determine the distance of a

comet from the earth, must of course be independent of its elements; since we are supposed to be ignorant of the value of those quantities. Such are the differential equations of motion which are furnished by dynamics and the law of gravitation.

In a method of approximation, it is obvious that, instead of the differential coefficients, their values in terms of finite differences may be substituted. Thus $\frac{x_1 + x_3 - 2x_2}{t^2}$* may be put for the value of $\frac{d^2 x}{dt^2}$ corresponding to x_2. If x_1, x_2, x_3, are the coordinates of the comet referred to the axis x, the origin being at the sun at the first, second and third observation, and r_2 its distance from the sun at the second observation, the interval t between the first and second observation being equal to that between the second and third with a contrary sign; then by the equations of motion,

$$\frac{x_1 + x_3 - 2x_2}{t^2} + \frac{\mu x_2}{r_2^3} = 0$$

$$\frac{y_1 + y_3 - 2y_2}{t^2} + \frac{\mu y_2}{r_2^3} = 0 \qquad (1)$$

$$\frac{z_1 + z_3 - 2z_2}{t^2} + \frac{\mu z_2}{r_2^3} = 0.$$

If the curtate distance of the comet from the earth be
called .... ϱ
distance of the sun from the earth .... R
longitude of the sun .... $\odot$
geocentric longitude of the comet .... λ
geocentric latitude .... β

$$x = \varrho \cos\lambda - R\cos\odot \quad y = \varrho \sin\lambda - R\sin\odot \quad z = \varrho\tan\beta \quad (2)$$

$$r^2 = \frac{\varrho^2}{\cos^2\beta} - 2\varrho R\cos(\odot - \lambda) + R^2$$

$$\varrho_1 \cos\lambda_1 + \varrho_3 \cos\lambda_3 - 2\varrho_2 \cos\lambda_2 \left\{1 - \frac{\mu t^2}{2r_2^3}\right\}$$

$$- R_1 \cos\odot_1 - R_3 \cos\odot_3 + 2R_2 \cos\odot_2 \left\{1 - \frac{\mu t^2}{2r_2^3}\right\} = 0$$

μ = mass of the sun + mass of comet.

Neglecting the difference of the masses of the earth and

---

* $\frac{d^n x}{dt^n} = \{\log. (1 + \Delta x)\}^n$. See Lacroix, *Traité du Calcul Intégral*, vol. iii. p. 352.

comet in comparison with the mass of the sun, this constant may be considered as the same for both, and the equations of the earth's motion give to the same degree of approximation,

$$R_1 \cos \odot_1 + R_3 \cos \odot_3 - 2 R_2 \cos \odot_2 \left\{1 - \frac{\mu t^2}{2 R_2^3}\right\} = 0$$

therefore

$$\varrho_1 \cos \lambda_1 + \varrho_3 \cos \lambda_3 - 2 \varrho_2 \cos \lambda_2 \left\{1 - \frac{\mu t^2}{2 r_2^3}\right\}$$

$$- \mu t^2 R_2 \cos \odot_2 \left\{\frac{1}{R_2^3} - \frac{1}{r_2^3}\right\} = 0$$

similarly

$$\varrho_1 \sin \lambda_1 + \varrho_3 \sin \lambda_3 - 2 \varrho_2 \sin \lambda_2 \left\{1 - \frac{\mu t^2}{2 r_2^3}\right\}$$

$$- \mu t^2 R_2 \sin \odot_2 \left\{\frac{1}{R_2^3} - \frac{1}{r_2^3}\right\} = 0 \quad (3)$$

$$\varrho_1 \tan \beta_1 + \varrho_3 \tan \beta_3 - 2 \varrho_2 \tan \beta_2 \left\{1 - \frac{\mu t^2}{2 r_2^3}\right\} = 0.$$

By elimination from these equations an expression may be found for either of the quantities ϱ_1, ϱ_2, ϱ_3, or either may be found in terms of another; thus,

$$\varrho_1 = A \gamma_1 \varrho_2 \qquad \varrho_3 = A \gamma_2 \varrho_2 \qquad (4)$$

where $A = 1 - \frac{\mu t^2}{2 r_2^3} = 1$ nearly, and γ_1, γ_2, are functions of known quantities.

These are the only equations which would furnish the means of determining the distance of a comet, were it not that the integrals in which the reciprocal of the semi-axis major is introduced may also be employed, rejecting that quantity. Thus in the parabola

$$\frac{\mathrm{d} x^2 + \mathrm{d} y^2 + \mathrm{d} z^2}{\mathrm{d} t^2} = \frac{2 \mu}{r} \qquad (5)$$

See *The Mechanism of the Heavens*, by Mrs. Somerville, p. 187.

$$\frac{\mathrm{d} x}{\mathrm{d} t} = \frac{x_3 - x_1}{2 t}, \qquad \frac{\mathrm{d} y}{\mathrm{d} t} = \frac{y_3 - y_1}{2 t}, \qquad \frac{\mathrm{d} z}{\mathrm{d} t} = \frac{z_3 - z_1}{2 t}$$

substituting these values in equation (5),

$$(x_1 - x_3)^2 + (y_1 - y_3)^2 + (z_1 - z_3)^2 = \frac{8 \mu t^2}{r_2} \qquad (6)$$

By substituting for x, y, z, their values in terms of ϱ, λ, R, $\odot$, from equations (2), an equation is found which contains ϱ_1, ϱ_3, and by again substituting for these quantities their values $A\gamma_1\varrho_2$, $A\gamma_2\varrho_3$, and for r_2 its value from the equation

$$r_2^2 = \frac{\varrho_2^2}{\cos^2\beta_2} - 2\,\varrho_2 R_2 \cos(\odot_2 - \lambda_2) + R_2^2 \qquad (7)$$

an equation is obtained in which ϱ_2 is the only unknown quantity. As the form of this equation is inconvenient, it is necessary in practice to obtain by repeated trials values of r_2 and ϱ_2, which satisfy the equations (6) and (7) simultaneously. When the quantity $\frac{\mu t^2}{2 r_2^3}$ is neglected in the factor A, which then becomes equal to unity, this method coincides with that proposed by M. Legendre, *Nouvelles Méthodes pour la Détermination des Orbites des Comètes*, p. 32, and which is explained by M. de Pontécoulant, *Théorie Analytique du Système du Monde*, vol. ii. p. 30. The equation of M. Legendre, although identical with that described above, is not given in the same form.

Another transformation of the equations of motion which obtains in the parabola may also be used.

$$\frac{d^2 . r^2}{2\,d\,t^2} = \frac{\mu}{r} \qquad (8)$$

Mechanism of the Heavens, p. 305.

This equation gives

$$r_1^2 + r_3^2 - 2\,r_2^2 = \frac{2\,\mu\,t^2}{r_2} \qquad (9)$$

and by similar substitutions as those above in equation (6), an equation may be obtained which may be combined with equation (7), and a value of ϱ_2 found by trials, which satisfies both. When the quantity $\frac{\mu t^2}{2 r_2^3}$ is neglected in the factor A, this method coincides with that proposed by Lagrange, and which is given by him in the *Mécanique Analytique*, vol. ii. p. 33, but which has not been often employed. This method is given by Delambre, *Astronomie Théorique et Pratique*, vol. iii., who says, "On entrevoit la peine qu'aura le calculateur, sans entrevoir quel sera le succès." In fact, however, this method is equally exact with that of Legendre; and as the quantities involved are nearly the same, the labour of the calculator is not increased by forming the equations which are used in both *.

* It may be worth while here to notice a mistake which occurs in the *Mécanique Anal.* vol. ii. p. 34, and which continues for some pages:

For $\frac{ds}{dt} = g\left\{\frac{1}{a} - \frac{1}{r}\right\}$ read $\frac{ds}{dt} = g\left\{\frac{1}{r} - \frac{1}{a}\right\}$

g is the quantity which I have called μ after Laplace and other writers.

The influence of the quantity $\frac{\mu t^2}{2 r_2^3}$ in the factor A is much more sensible in the equation of Lagrange than in that of Legendre, and should not be neglected in either.

The method of Laplace was first given in the *Mémoires de l'Académie* for 1780, and is also to be found in the first volume of the *Mécanique Céleste.*

The equations of motion give

$$\frac{d^2 \rho \cos \lambda + \sin \lambda \, d \rho \, d \lambda - \rho \cos \lambda \, d \lambda^2 + \rho \sin \lambda \, d^2 \lambda}{d t^2}$$

$$+ \mu \left\{ \frac{\cos \odot}{R^2} - \frac{R \cos \odot - \rho \cos \lambda}{r^3} \right\} = 0$$

$$\frac{d^2 \rho \sin \lambda + 2 \cos \lambda \, d \rho \, d \lambda - \rho \sin \lambda \, d \lambda^2 + \rho \cos \lambda \, d^2 \lambda}{d t^2}$$

$$+ \mu \left\{ \frac{\sin \odot}{R^2} - \frac{R \sin \odot - \rho \sin \lambda}{r^3} \right\} = 0$$

$$\frac{d^2 \varrho \tan \beta + \varrho \frac{d^2 \beta}{\cos^2 \beta} + \frac{2 \, d \rho \, d \beta}{\cos^2 \beta} + \frac{2 \, d \rho \sin \beta}{\cos^2 \beta} d^2 \beta}{d t^2}$$

$$+ \mu \rho \frac{\tan \beta}{r^3} = 0.$$

By elimination between these equations, either of the quantities $\frac{d^2 \varrho}{d t^2}$, $\frac{d \varrho}{d t}$, or ϱ, may be found in terms of the differential coefficients $\frac{d^2 \beta}{d t^2}$, $\frac{d \beta}{d t}$, $\frac{d^2 \lambda}{d t^2}$, $\frac{d \lambda}{d t}$, and known quantities. These differential coefficients may easily be obtained in terms of the finite differences of the quantities λ, β, which are given by observation. The method would be rigorous if the observations were indefinitely numerous and exact: practically the errors of the observations and the additional labour required destroy the advantages which would accrue from employing more than three observations.

By substitutions for $\frac{d \varrho}{d t}$ and $\frac{d^2 \varrho}{d t^2}$, which are implicitly contained in equations (6) and (8), equations may be found for the determination of ϱ, which are substantially identical with those of Legendre and Lagrange before explained, but which differ by containing the quantities $\frac{d \lambda}{d t}$, $\frac{d \beta}{d t}$, &c., or the finite differences of the quantities λ, β, instead of the finite differences of the trigonometrical lines $\sin \lambda$, $\cos \lambda$, $\sin \beta$, $\cos \beta$, &c. This method, though of great analytical beauty, does not present the equations in a convenient form for numerical calculation.

The method of Olbers is founded on a very elegant theorem due to Euler, in which the interval $(2\,t)$ between the first and third observations is given in terms of the radii vectores r_1, r_3, and the chord k.

$$t = \frac{\{r_1 + r_3 + k\}^{\frac{3}{2}} \pm \{r_1 + r_3 - k\}^{\frac{3}{2}}}{12\sqrt{\mu}} \qquad (10)$$

which must be combined with the equations

$\rho_3 = M^* \rho_1$, M being a function of known quantities.

$$\begin{aligned} k^2 = {} & \frac{\rho_1^2}{\cos^2\beta_1} + \frac{\rho_3^2}{\cos^2\beta_3} + 2\,\rho_1\,\rho_3\,\{\cos(\lambda_1 - \lambda_3) + \tan\beta_3\tan\beta_1\} \\ & -2\,\{\rho_1\,R_1\cos(\odot_1 - \lambda_1) + \rho_3\,R_3\cos(\odot_3 - \lambda_3) \\ & \quad - \rho_1\,R_3\cos(\odot_3 - \lambda_1) - \rho_3\,R_1\cos(\odot_1 - \lambda_3)\} \qquad (11) \\ & + R_1^2 + R_3^2 - 2\,R_1\,R_3\cos(\odot_1 - \odot_3) \end{aligned}$$

$$r_1^2 = \frac{\rho_1^2}{\cos^2\beta_1} - 2\,\rho_1\,R_1\cos(\odot_1 - \lambda_1) + R_1^2$$

$$r_3^2 = \frac{\rho_3^2}{\cos^2\beta_3} - 2\,\rho_3\,R_3\cos(\odot_3 - \lambda_3) + R_3^2$$

Simultaneous values of the unknown quantities r_1, r_3, ρ_1, ρ_3 and k, which satisfy all these equations, can only be found by repeated trials and approximations. Although the theorem furnished by the equation (10) is a most elegant property of the parabola, nothing I apprehend is gained by making use of the equation in this form. It appears better to have recourse to the value of k or k^2 in a series proceeding according to ascending powers of t, which may easily be deduced from equation (10) by the reversion of series as is explained in the Berlin Ephemeris for 1833, p. 267.

$$k^2 = \frac{16\,\mu\,t^2}{r_1 + r_3}\left\{1 + \frac{\mu\,t^2}{3\,(r_1 + r_3)^2}\right\} \qquad (12)$$

This method, however, then becomes very similar to that already described, p. 6, and is identical with that proposed by Mr. Ivory, " On a new method of deducing a first approximation to the orbit of a Comet from three Geocentric Observations," *Phil. Trans.* 1814.

The orbit of the comet being in a plane passing through the sun, this condition furnishes immediately the equation

---

* $M = \frac{\gamma_2}{\gamma_1}$.

$$z_1(x_2y_3-y_2x_3)+z_2(x_3y_1-x_1y_3)+z_3(y_2x_1-x_2y_1)=0 \quad (13)$$

By substituting $\gamma_1\rho_2$ for ρ_1, and $\gamma_2\rho_2$ for ρ_3 in this equation, a quadratic is obtained for the determination of ρ_2. In practice, however, this equation is not applicable, because the coefficients of the unknown quantity are small, and liable to considerable alteration from the inevitable errors of the observations *.

Besides the methods which have already been explained, any combination of equations (6) and (8) may be employed. Thus an equation may be obtained by eliminating $\frac{1}{a}$ between them, and which might, perhaps, be advantageously used in some cases. The combination, however, which appears to me to be the most preferable, and to offer the best method of determining the distance of a comet from the earth, is that which is obtained by eliminating $\frac{1}{r_2}$; by this means an equation arises, which, by neglecting at first the quantity $\frac{\mu t^2}{2r_2^3}$ in the factor A, (that is, taking A equal to unity,) may be treated as a quadratic, and furnishes an approximate value of ρ_2. This value of ρ_2 gives r_2, which being substituted in the term $\frac{\mu t^2}{2r_2^3}$ before neglected, the accurate value of ρ_2 may now be found from the same quadratic equation. By this method the necessity is obviated of having recourse to repeated trials; for the small error produced by the substitution of the value of ρ_2, given by the first approximation, in the term $\frac{\mu t^2}{r_2^3}$, is quite insensible.

The elements of the orbit can easily be determined from the heliocentric distances r_1, r_2, r_3, and the geocentric distances ρ_1, ρ_2, ρ_3, as will hereafter be shown. The elements thus found approximately from three observations of the comet, must afterwards be corrected. Let $\delta\lambda$ and $\delta\beta$ be the errors in geocentric longitude and latitude of a calculated place of the comet,

$$\delta\lambda=\frac{d\lambda}{d\iota}\delta\iota+\frac{d\lambda}{d\nu}\delta\nu+\frac{d\lambda}{d\varpi}\delta\varpi+\frac{d\lambda}{de}\delta e+\frac{d\lambda}{d\epsilon}\delta\epsilon \quad (14)$$

* "Die coefficienten wurden immer sehr klein, und deswegen hatten die geringsten Fehler der Beobachtungen immer einen ungemein grossen Einfluss auf den Werth der unbekannten Grösse."—*Abhandlung über die leichteste und bequemste Methode die Bahn eines Cometen zu berechnen.* von W. Olbers.

$$\delta\beta = \frac{d\beta}{d\iota}\delta\iota + \frac{d\beta}{d\nu}\delta\nu + \frac{d\beta}{d\varpi}\delta\varpi + \frac{d\beta}{de}\delta e + \frac{d\beta}{d\varepsilon}\delta\varepsilon \qquad (15)$$

the elements being denoted by ι, ν, ϖ, e, ε. The coefficients $\frac{d\lambda}{d\iota}$, $\frac{d\beta}{d\iota}$, &c. must be found by ascertaining the difference $(\delta\lambda)$, $(\delta\beta)$ in the place of the comet, which is produced by a small change in any element ι, the other elements remaining the same. Every observation of the comet furnishes two equations of the form (14) and (15), all of which, when great accuracy is required, must be combined by the method of least squares, and the errors $\delta\iota$, $\delta\nu$, &c. finally obtained by elimination.

Having thus given a short sketch of the different methods which have been proposed for finding the distance of a comet, I shall now proceed to examine the question more in detail, and to carry the approximation to terms which I have hitherto neglected.

2. Let the geocentric longitude of the comet be called... λ
——————— latitude.............................. β
——— curtate distance................................ ρ
——— distance of the comet from the sun .... r
——— longitude of the sun ....$\odot$
——— coordinates of the comet, the origin being at the sun, and the plane xy coinciding with the ecliptic .... x, y, z

and let $\frac{r\,dr}{dt} = s$

$$x = \rho\cos\lambda - R\cos\odot \quad y = \rho\sin\lambda - R\sin\odot \quad z = \rho\tan\beta \quad (16)$$

$$r^2 = \frac{\rho^2}{\cos^2\beta} - 2R\rho\cos(\odot - \lambda) + R^2$$

I shall first suppose the intervals between the observations to be the same, and equal t: this condition, which somewhat simplifies the expressions, may always be realized by interpolation; and then by Maclaurin's theorem,

$$\left.\begin{aligned} x_1 &= x_2 - \frac{dx}{dt}t + \frac{d^2x}{2\,dt^2}t^2 - \frac{d^3x}{2.3\,dt^3}t^3 + \&c. \\ x_3 &= x_2 + \frac{dx}{dt}t + \frac{d^2x}{2\,dt^2}t^2 + \frac{d^3x}{2.3\,dt^3}t^3 + \&c. \end{aligned}\right\} \quad (17)$$

The equations of motion give

$$\frac{d^2x}{dt^2} + \frac{\mu x}{r^3} = 0 \quad \frac{d^2y}{dt^2} + \frac{\mu y}{r^3} = 0 \quad \frac{d^2z}{dt^2} + \frac{\mu z}{r^3} = 0 \qquad (18)$$

$$\frac{d^3 x}{d t^3} = -\frac{\mu\, d x}{r^3 d t} + \frac{3 \mu x d r}{r^4 d t} = -\frac{\mu\, d x}{r^3 d t} + \frac{3 \mu x s}{r^5}$$

$$x_1 = x_2 \left\{1 - \frac{\mu t^2}{2 r_2^3} - \frac{\mu s t^3}{2 r_2^5}\right\} - \frac{d x}{d t} t \left\{1 - \frac{\mu t^2}{2 . 3 r_2^3}\right\}$$

$$x_3 = x_2 \left\{1 - \frac{\mu t^2}{2 r_2^3} + \frac{\mu s t^3}{2 r_2^5}\right\} + \frac{d x}{d t} t \left\{1 - \frac{\mu t^2}{2 . 3 r_2^3}\right\} \quad (19)$$

$$x_3 - x_1 = 2 t \frac{d x}{d t} \left\{1 - \frac{\mu t^2}{2 . 3 r_2^3}\right\} + \frac{\mu s x_2}{r_2^5} t^3 \quad (20)$$

And similarly,

$$y_3 - y_1 = 2 t \frac{d y}{d t} \left\{1 - \frac{\mu t^2}{2 . 3 r_2^3}\right\} + \frac{\mu s y_2 t^3}{r_2^5}$$

$$z_3 - z_1 = 2 t \frac{d z}{d t} \left\{1 - \frac{\mu t^2}{2 . 3 r_2^3}\right\} + \frac{\mu s z_2 t^3}{r_2^5}$$

since $\quad s = \dfrac{r\, d r}{d t} = \dfrac{x\, d x + y\, d y + z\, d z}{d t}$

neglecting t^6,

$$(x_3 - x_1)^2 + (y_3 - y_1)^2 + (z_3 - z_1)^2$$

$$= 4 t^2 \left\{\frac{d x^2 + d y^2 + d z^2}{d t^2}\right\} \left\{1 - \frac{\mu t^2}{2 . 3 r_2^3}\right\}^2 + \frac{4 \mu s^2 t^4}{r_2^5}$$

and since $\quad \dfrac{d x^2 + d y^2 + d z^2}{d t^2} = \dfrac{2 \mu}{r} - \dfrac{\mu}{a}$. *Mechanism of the Heavens*, p. 187.

$$(x_3 - x_1)^2 + (y_3 - y_1)^2 + (z_3 - z_1)^2$$

$$= 4 \mu t^2 \left\{\frac{2}{r_2} - \frac{1}{a}\right\} \left\{1 - \frac{\mu t^2}{3 r_2^3}\right\} + \frac{4 \mu s^2 t^4}{r_2^5} \quad (21)$$

By Maclaurin's theorem,

$$r_1 = r_2 - \frac{d r}{d t} t + \frac{d^2 r}{2 d t^2} t^2 + \&c.$$

$$r_3 = r_2 + \frac{d r}{d t} t + \frac{d^2 r}{2 d t^2} t^2 + \&c.$$

$$r_2 = \frac{r_1 + r_3}{2} - \frac{d^2 r}{2 d t^2} t^2 + \&c.$$

$$\frac{d^2 . r^2}{2 d t^2} = \mu \left\{ \frac{1}{r} - \frac{1}{a} \right\} \qquad (22)$$

Mechanism of the Heavens, p. 305.

whence $\dfrac{d^2 r}{d t^2} = \dfrac{\mu}{r^2} - \dfrac{s^2}{r^3} - \dfrac{\mu}{a r}$

therefore $r_2 = \dfrac{r_1 + r_3}{2} - \dfrac{1}{2} \left\{ \dfrac{\mu}{r_2^2} - \dfrac{s^2}{r_2^3} - \dfrac{\mu}{a r_2} \right\} t^2$

$$\frac{2}{r_2} = \frac{4}{r_1 + r_3} \left[1 + \left\{ \frac{4 \mu}{(r_1 + r_3)^3} - \frac{8 s^2}{(r_1 + r_3)^4} \right\} t^2 \right\}$$

nearly.

and substituting for r_2 in equation (21) its value from the preceding expression,

$$(x_3 - x_1)^2 + (y_3 - y_1)^2 + (z_3 - z_1)^2$$
$$= \frac{16 \mu t^2}{(r_1 + r_3)} \left\{ 1 + \frac{\mu t^2}{3 (r_1 + r_3)^2} - \frac{1}{4 a} \right\} \qquad (23)$$

If k be the chord of the arc described by the comet in the time $2 t$,

$$k^2 = (x_3 - x_1)^2 + (y_3 - y_1)^2 + (z_3 - z_1)^2$$
$$= 4 \mu t^2 \left\{ \frac{2}{r_2} - \frac{1}{a} \right\} \left\{ 1 - \frac{\mu t^2}{3 r_2^3} \right\} + \frac{4 \mu s^2 t^4}{r_2^5}$$
$$= \frac{16 \mu t^2}{(r_1 + r_3)} \left\{ 1 + \frac{\mu t^2}{3 (r_1 + r_3)^2} - \frac{1}{4 a} \right\}$$

neglecting $\dfrac{1}{a}$,

$$k^2 = \frac{8 \mu t^2}{r_2} \left\{ 1 - \frac{\mu t^2}{3 r_2^3} \right\} + \frac{4 \mu s^2 t^4}{r_2^5} \qquad (24)$$

$$= \frac{16 \mu t^2}{(r_1 + r_3)} \left\{ 1 + \frac{\mu t^2}{3 (r_1 + r_3)^2} \right\} \qquad (25)$$

Again, by Maclaurin's theorem,

$$r_1^2 = r_2^2 - \frac{d . r^2}{d t} t + \frac{d^2 . r^2}{2 d t^2} t^2 - \frac{d^3 . r^2}{2 . 3 d t^3} t^3 + \text{\&c.}$$

$$r_3^2 = r_2^2 + \frac{d . r^2}{d t} t + \frac{d^2 . r^2}{2 d t^2} t^2 + \frac{d^3 . r^2}{2 . 3 d t^3} t^3 + \text{\&c.}$$

$$r_1^2 + r_3^2 - 2\,r_2^2 = \frac{d^2.r^2}{d\,t^2}\,t^2 + \frac{d^4.r^2}{3.4\,d\,t^4}\,t^4 + \&c.$$

neglecting $\frac{1}{a}$,

$$\frac{d^3.r^2}{d\,t^3} = -\frac{2\,\mu\,d\,r}{r^2\,d\,t} = -\frac{2\,\mu\,s}{r^3} \qquad \frac{d^4.r^2}{d\,t^4} = \frac{2.3\,\mu\,s^2}{r^5} - \frac{2\,\mu^2}{r^4}$$

therefore

$$\Delta^2.r^2 = r_1^2 + r_3^2 - 2\,r_2^2$$

$$= 2\,\mu\,t^2\left\{\frac{1}{r_2} - \frac{1}{a} - \frac{\mu\,t^2}{12\,r_2^4}\right\} + \frac{\mu\,s^2\,t^4}{2\,r_2^5} \qquad (26)$$

$$4\,\Delta^2.r^2 - k^2 = -\frac{4\,\mu\,t^2}{a} + \frac{2\,\mu^2\,t^4}{r_2^4} - \frac{2\,\mu\,s^2\,t^4}{r_2^5} \qquad (27)$$

Resuming the equations

$$x_1 = x_2\left\{1 - \frac{\mu\,t^2}{2\,r_2^3} - \frac{\mu\,s\,t^3}{2\,r_2^5}\right\} - \frac{d\,x}{d\,t}\,t\left\{1 - \frac{\mu\,t^2}{2.3\,r_2^3}\right\}$$

$$x_3 = x_2\left\{1 - \frac{\mu\,t^2}{2\,r_2^3} + \frac{\mu\,s\,t^3}{2\,r_2^5}\right\} + \frac{d\,x}{d\,t}\,t\left\{1 - \frac{\mu\,t^2}{2.3\,r_2^3}\right\}$$

$$x_1 + x_3 - 2\,x_2\left\{1 - \frac{\mu\,t^2}{2\,r_2^3}\right\} = 0 \qquad (28)$$

which last equation, as before remarked, may be obtained at once from the equation $\frac{d^2\,x}{d\,t^2} + \frac{\mu\,x}{r^3} = 0$, by substituting $\frac{\Delta^2\,x}{t^2}$ for $\frac{d^2\,x}{d\,t^2}$, Δ being the characteristic of finite differences.

Similarly,

$$y_1 + y_3 - 2\,y_2\left\{1 - \frac{\mu\,t^2}{2\,r_2^3}\right\} = 0$$

$$z_1 + z_3 - 2\,z_2\left\{1 - \frac{\mu\,t^2}{2\,r_2^3}\right\} = 0$$

Substituting the geocentric coordinates of the comet,

$$\rho_1\cos\lambda_1 - R_1\cos\odot_1 + \rho_3\cos\lambda_3 - R_3\cos\odot_3$$

$$-\,2\left\{\rho_2\cos\lambda_2 - R_2\cos\odot_2\right\}\left\{1 - \frac{\mu\,t^2}{2\,r_2^3}\right\} = 0$$

By the equations of the earth's motion we obtain in a similar manner, to the same degree of approximation,

$$R_1 \cos \odot_1 + R_3 \cos \odot_3 - 2 R_2 \cos \odot_2 \left\{ 1 - \frac{\mu t^2}{2 R_2^3} \right\} = 0$$

and hence

$$\rho_1 \cos \lambda_1 + \rho_3 \cos \lambda_3 - 2 \rho_2 \cos \lambda_2 \left\{ 1 - \frac{\mu t^2}{2 r_2^3} \right\}$$
$$+ \mu t^2 R_2 \cos \odot_2 \left\{ \frac{1}{R_2^3} - \frac{1}{r_2^3} \right\} = 0$$

$$\rho_1 \sin \lambda_1 + \rho_3 \sin \lambda_3 - 2 \rho_2 \sin \lambda_2 \left\{ 1 - \frac{\mu t^2}{2 r_2^3} \right\}$$
$$+ \mu t^2 R_2 \sin \odot_2 \left\{ \frac{1}{R_2^3} - \frac{1}{r_2^3} \right\} = 0$$

$$\rho_1 \tan \beta_1 + \rho_3 \tan \beta_3 - 2 \rho_2 \tan \beta_2 \left\{ 1 - \frac{\mu t^2}{2 r_2^3} \right\} = 0$$

and by elimination,

$$\rho_1 = \frac{\mu t^2 R_2 \{ \tan \beta_2 \sin (\odot_2 - \lambda_3) - \tan \beta_3 \sin (\odot_2 - \lambda_2) \}}{\tan \beta_3 \sin (\lambda_1 - \lambda_2) + \tan \beta_2 \sin (\lambda_3 - \lambda_1) + \tan \beta_1 \sin (\lambda_2 - \lambda_3)} \left\{ \frac{1}{R_2^3} - \frac{1}{r_2^3} \right\}$$

$$\rho_2 \left\{ 1 - \frac{\mu t^2}{2 r_2^3} \right\}$$
$$= \frac{\mu t^2 R_2 \{ \tan \beta_1 \sin (\odot_2 - \lambda_3) - \tan \beta_3 \sin (\odot_2 - \lambda_1) \}}{2 \{ \tan \beta_3 \sin (\lambda_1 - \lambda_2) + \tan \beta_2 \sin (\lambda_3 - \lambda_1) + \tan \beta_1 \sin (\lambda_2 - \lambda_3) \}} \left\{ \frac{1}{R_2^3} - \frac{1}{r_2^3} \right\} \quad (29)$$

$$\rho_3 = \frac{\mu t^2 R_2 \{ \tan \beta_1 \sin (\odot_2 - \lambda_2) - \tan \beta_2 \sin (\odot_2 - \lambda_1) \}}{\tan \beta_3 \sin (\lambda_1 - \lambda_2) + \tan \beta_2 \sin (\lambda_3 - \lambda_1) + \tan \beta_1 \sin (\lambda_2 - \lambda_3)} \left\{ \frac{1}{R_2^3} - \frac{1}{r_2^3} \right\}$$

The curtate distance (ρ) of the comet from the earth being essentially positive, it is evident that $R_2 \lessgtr r_2$ according as the quantities

$$\tan \beta_1 \sin (\odot_2 - \lambda_3) - \tan \beta_3 \sin (\odot_2 - \lambda_1) \text{ and}$$
$$\tan \beta_3 \sin (\lambda_1 - \lambda_2) + \tan \beta_2 \sin (\lambda_3 - \lambda_1) + \tan \beta_1 \sin (\lambda_2 - \lambda_3)$$

have the same or contrary signs.

Either of the preceding equations would serve to determine the distance of the comet from the earth at the corresponding time, were it not that the quantity in the denominator is generally small, and greatly influenced by the errors of the observations.

If

$$\tan \beta_3 \sin (\lambda_1 - \lambda_2) + \tan \beta_2 \sin (\lambda_3 - \lambda_1) + \tan \beta_1 \sin (\lambda_2 - \lambda_3) = 0$$

the three places of the comet are in the same apparent great circle, and either

$$\frac{1}{R_2^3} - \frac{1}{r_2^3} = 0 \qquad (r_2 = R_2) \quad \text{or}$$

$$\tan\beta_1 \sin(\odot_2 - \lambda_3) - \tan\beta_3 \sin(\odot_2 - \lambda_1) = 0.$$

In the first case it is easy to show that

$$\left.\begin{aligned} \rho_1 &= 2\rho_2 \left\{1 - \frac{\mu t^2}{2 r_2^3}\right\} \frac{\sin(\lambda_3 - \lambda_2)}{\sin(\lambda_3 - \lambda_1)} \\ \rho_3 &= 2\rho_2 \left\{1 - \frac{\mu t^2}{2 r_2^3}\right\} \frac{\sin(\lambda_1 - \lambda_2)}{\sin(\lambda_1 - \lambda_3)} \end{aligned}\right\} \quad (30)$$

In the second case, if the places of the sun and comet be denoted by the letters C, S respectively, C_1, C_3, and S_2 are in the same great circle, and the method fails altogether. The only resource in this difficulty appears to be to take three other observations which are not subject to the same conditions.

If
$$\left.\begin{aligned} \gamma_1 &= 2\,\frac{\tan\beta_2 \sin(\odot_2 - \lambda_3) - \tan\beta_3 \sin(\odot_2 - \lambda_2)}{\tan\beta_1 \sin(\odot_2 - \lambda_3) - \tan\beta_3 \sin(\odot_2 - \lambda_1)} \\ \gamma_2 &= 2\,\frac{\tan\beta_1 \sin(\odot_2 - \lambda_2) - \tan\beta_2 \sin(\odot_2 - \lambda_1)}{\tan\beta_1 \sin(\odot_2 - \lambda_3) - \tan\beta_3 \sin(\odot_2 - \lambda_1)} \end{aligned}\right\} \quad (31)$$

$$A = 1 - \frac{\mu t^2}{2 r_2^3} = 1 \text{ nearly.} \quad (32)$$

then $\rho_1 = A\gamma_1\rho_2 \qquad \rho_3 = A\gamma_2\rho_2$

Putting

$$\begin{aligned} A\gamma_1\rho_2 \cos\lambda_1 - R_1 \cos\odot_1 &\quad \text{for } x_1, \\ \rho_2 \cos\lambda_2 - R_2 \cos\odot_2 &\quad \text{for } x_2, \\ A\gamma_2\rho_2 \cos\lambda_3 - R_3 \cos\odot_3 &\quad \text{for } x_3, \end{aligned}$$

equation (21) becomes

$$\begin{aligned} &\left\{\frac{\gamma_1^2}{\cos^2\beta_1} + \frac{\gamma_2^2}{\cos^2\beta_3} - 2\gamma_1\gamma_2\{\cos(\lambda_1 - \lambda_3) + \tan\beta_1 \tan\beta_3\}\right\} A^2\rho_2^2 \\ &+ 2\left\{\gamma_1 R_3 \cos(\odot_3 - \lambda_1) + \gamma_2 R_1 \cos(\odot_1 - \lambda_3)\right. \\ &\left.- \gamma_1 R_1 \cos(\odot_1 - \lambda_1) - \gamma_2 R_3 \cos(\odot_3 - \lambda_3)\right\} A\rho_2 \quad (33) \\ &+ R_1^2 + R_3^2 - 2R_1 R_3 \cos(\odot_1 - \odot_3) \\ &= 4\mu t^2 \left\{\frac{2}{r_2} - \frac{1}{a}\right\}\left\{1 - \frac{\mu t^2}{3 r_2^3}\right\} + \frac{4\mu s^2 t^4}{r_2^5} \end{aligned}$$

Similarly, equation (26) becomes

$$\left\{\frac{\gamma_1^2}{\cos^2\beta_1}+\frac{\gamma_2^2}{\cos^2\beta_3}-\frac{2}{A^2\cos^2\beta_2}\right\}A^2\rho_2^2$$

$$-2\left\{\gamma_1 R_1\cos(\odot_1-\lambda_1)+\gamma_2 R_3\cos(\odot_3-\lambda_3)\right.$$

$$\left.-\frac{2R_2}{A}\cos(\odot_2-\lambda_2)\right\}A\rho_2$$

$$+R_1^2+R_3^2-2R_2^2=2\mu t^2\left\{\frac{1}{r_2}-\frac{1}{a}-\frac{\mu t^2}{12 r_2^4}\right\}+\frac{\mu s^2 t^4}{2 r_2^5} \qquad (34)$$

If E be the eccentricity of the earth's orbit, and ϖ the longitude of the perihelion, equation (33) may be put in the form,

$$\left\{\frac{\gamma_1^2}{\cos^2\beta_1}+\frac{\gamma_2^2}{\cos^2\beta_3}-2\gamma_1\gamma_2\{\cos(\lambda_1-\lambda_3)+\tan\beta_1\tan\beta_3\}\right\}A^2\rho_2^2$$

$$+4t\sqrt{\mu}\left\{\{\gamma_2\sin(\odot_2-\lambda_3)-\gamma_1\sin(\odot_2-\lambda_1)\}\frac{\left(1-\frac{E^2}{2}\right)}{R_2}\right.$$

$$\left.+E\sin(\odot_2-\varpi)\{\gamma_2\cos(\odot_2-\lambda_3)-\gamma_1\cos(\odot_2-\lambda_1)\}\right\}A\rho_2$$

$$+4\mu t^2\left\{\frac{2}{R_2}-1\right\}\left\{1-\frac{\mu t^2}{3R_2^3}\right\}+\frac{4\mu^2}{R_2^3}\frac{d R^2}{d t^2}t^4$$

$$=4\mu t^2\left\{\frac{2}{r_2}-\frac{1}{a}\right\}\left\{1-\frac{\mu t^2}{3 r_2^3}\right\}+\frac{4\mu s^2 t^4}{r_2^5} \qquad (35)$$

Nothing, however, is gained in the numerical application of the equation by this change.

In the notation of M. Legendre (*Nouvelles Méthodes*, &c.),

$$\frac{2P}{C}=\gamma_1\sin(\odot_2-\lambda_1)-\gamma_2\sin(\odot_2-\lambda_3)$$

$$\frac{2Q}{C}=\gamma_1\cos(\odot_2-\lambda_1)-\gamma_2\cos(\odot_2-\lambda_3)$$

$$\frac{2H}{C}=\gamma_1\tan\beta_1-\gamma_2\tan\beta_3 \qquad \theta=t\sqrt{\mu}$$

Neglecting the quantity $\frac{\mu t^2}{2 r_2^3}$ in the factor A, &c., it is evident that the equation (35) may be put in the form,

$$\frac{2}{r_2} = \frac{2}{R_2} - 1 - \frac{2\rho_2 \cos\beta_2}{\theta C}\left\{\frac{\left(1-\frac{E^2}{2}\right)}{R_2} P + E\sin(\odot_2 - \varpi) Q\right\}$$

$$+ \frac{\rho_2^{\,2}\cos\beta_2^{\,2}}{\theta^2 C^2}\{P^2 + Q^2 + H^2\} \tag{36}$$

ρ_2 being the distance of the comet from the earth, (not the curtate distance as before,) which is the equation given by M. Legendre, *Nouvelles Méthodes*, &c., p. 38, and by M. de Pontécoulant, *Théor. Anal.* vol. ii. p. 30. The quantities, however, which are neglected in this form of the equation are by no means insensible, and should not be disregarded.

Such are the expressions which result when the fourth power of the time is neglected: it may however be worth while to obtain the expressions which result when the terms depending on this quantity are retained, although they are too complicated to be used with advantage in practice.

$$\frac{d^4 x}{d t^4} = -\frac{\mu\, d^2 x}{r^3\, d t^2} + \frac{3\mu s\, d x}{r^5\, d t} + \frac{3\mu s\, d x}{r^5\, d t} + \frac{3\mu x\, d s}{r^5\, d t} - \frac{3.5\,\mu x s^2}{r^7}$$

$$\frac{d s}{d t} = \frac{\mu}{r} \qquad \text{therefore, neglecting } \frac{\mu}{a},$$

$$\frac{d^4 x}{d t^4} = \frac{4\mu^2 x}{r^6} - \frac{3.5\,\mu x s^2}{r^7} + \frac{6\mu s\, d x}{r^5\, d t}$$

$$x_1 = x_2\left\{1 - \frac{\mu t^2}{2 r_2^{\,3}} - \frac{\mu s t^3}{2 r_2^{\,5}} + \frac{\mu^2 t^4}{6 r_2^{\,6}} - \frac{5\mu s^2 t^4}{8 r_2^{\,7}}\right\}$$

$$- t\frac{d x}{d t}\left\{1 - \frac{\mu t^2}{2.3 r_2^{\,3}} - \frac{\mu s t^3}{4 r^5}\right\}$$

$$x_3 = x_2\left\{1 - \frac{\mu t^2}{2 r_2^{\,3}} - \frac{\mu s t^3}{2 r_2^{\,5}} + \frac{\mu^2 t^4}{6 r_2^{\,6}} - \frac{5\mu s^2 t^4}{8 r_2^{\,7}}\right\}$$

$$+ t\frac{d x}{d t}\left\{1 - \frac{\mu t^2}{2.3 r_2^{\,3}} + \frac{\mu s t^3}{4 r_2^{\,5}}\right\}$$

$$x_1 + x_3 = 2 x_2\left\{1 - \frac{\mu t^2}{2 r_2^{\,3}} - \frac{\mu s t^3}{2 r_2^{\,5}} + \frac{\mu^2 t^4}{6 r_2^{\,6}} - \frac{5\mu s^2 t^4}{8 r_2^{\,7}}\right\}$$

$$+ \frac{d x}{d t}\frac{\mu s t^4}{2 r_2^{\,5}}$$

$$x_1\left\{1 + \frac{\mu s t^3}{4 r_2^{\,5}}\right\} + x_3\left\{1 - \frac{\mu s t^3}{4 r_2^{\,5}}\right\}$$

$$- 2\,x_2 \left\{1 - \frac{\mu\,t^2}{2\,r_2^{\,3}} - \frac{\mu\,s\,t^3}{2\,r_2^{\,5}} + \frac{\mu^2\,t^4}{6\,r_2^{\,6}} - \frac{5\,\mu\,s^2\,t^4}{8\,r_2^{\,7}}\right\} = 0$$

$$\left\{\rho_1 \cos\lambda_1 - R_1 \cos\odot_1\right\}\left\{1 + \frac{\mu\,s\,t^3}{4\,r_2^{\,5}}\right\}$$

$$+ \left\{\rho_3 \cos\lambda_3 - R_3 \cos\odot_3\right\}\left\{1 - \frac{\mu\,s\,t^3}{4\,r_2^{\,5}}\right\}$$

$$- 2\left\{\rho_2 \cos\lambda_2 - R_2 \cos\odot_2\right\}\left\{1 - \frac{\mu\,t^2}{2\,r_2^{\,3}} - \frac{\mu\,s\,t^3}{2\,r_2^{\,5}}\right.$$

$$\left. + \frac{\mu^2\,t^4}{6\,r_2^{\,6}} - \frac{5\,\mu\,s^2\,t^4}{8\,r_2^{\,7}}\right\} = 0$$

Neglecting $\frac{\mathrm{d}\,R}{\mathrm{d}\,t}$ in the terms multiplied by t^3, which is allowable, the earth moving nearly in a circle,

$$R_1 \cos\odot_1 + R_3 \cos\odot_3 - 2\,R_2 \cos\odot_2\left\{1 - \frac{\mu\,t^2}{2\,R_2^{\,3}} + \frac{\mu^2\,t^4}{6\,R_2^{\,6}}\right\} = 0$$

and therefore

$$\rho_1 \cos\lambda_1 \left\{1 + \frac{\mu\,s\,t^3}{4\,r_2^{\,5}}\right\} + \rho_3 \cos\lambda_3 \left\{1 - \frac{\mu\,s\,t^3}{4\,r_2^{\,5}}\right\}$$

$$- 2\,\rho_2 \cos\lambda_2 \left\{1 - \frac{\mu\,t^2}{2\,r_2^{\,3}} - \frac{\mu\,s\,t^3}{2\,r_2^{\,5}} + \frac{\mu^2\,t^4}{6\,r_2^{\,6}} - \frac{5\,\mu\,s^2\,t^4}{8\,r_2^{\,7}}\right\}$$

$$+ \mu\,t^2\,R_2 \cos\odot_2 \left\{\frac{1}{R_2^{\,3}} - \frac{1}{r_2^{\,3}} - \frac{\mu\,t^2}{3\,R_2^{\,6}} + \frac{\mu\,t^2}{3\,r_2^{\,6}} - \frac{5\,s^2\,t^2}{4\,r_2^{\,7}}\right\}$$

$$+ \left\{R_3 \cos\odot_3 - R_1 \cos\odot_1\right\}\frac{\mu\,s\,t^3}{4\,r_2^{\,5}} = 0$$

$$R_3 \cos\odot_3 - R_1 \cos\odot_1 = -\,2\,R_2 \sin\odot_2\,t\,\frac{\mathrm{d}\odot}{\mathrm{d}\,t}$$

$$= -\,\frac{2 \sin\odot_2\,t\sqrt{\mu}}{R_2} \quad \text{nearly ;}$$

therefore if

$$\tan\beta_2 \sin(\odot_2 - \lambda_3) - \tan\beta_3 \sin(\odot_2 - \lambda_2) = C_1$$
$$\tan\beta_1 \sin(\odot_2 - \lambda_2) - \tan\beta_3 \sin(\odot_2 - \lambda_1) = C_2$$
$$\tan\beta_1 \sin(\odot_2 - \lambda_2) - \tan\beta_2 \sin(\odot_2 - \lambda_1) = C_3$$
$$\tan\beta_3 \sin(\lambda_1 - \lambda_2) + \tan\beta_2 \sin(\lambda_3 - \lambda_1)$$
$$+ \tan\beta_1 \sin(\lambda_2 - \lambda_3) = D$$

$$1 - \frac{\mu t^2}{2 r^3} - \frac{\mu s t^3}{2 r_2^5} + \frac{\mu^2 t^4}{6 r_2^6} - \frac{5}{8} \frac{\mu s^2 t^4}{r_2^7} = A$$

by elimination,

$$\left\{1 + \frac{\mu s t^3}{4 r_2^5}\right\} D \rho_1 = \mu t^2 R_2 C_1 \left\{\frac{1}{R_2^3} - \frac{1}{r_2^3} - \frac{\mu t^2}{3 R_2^6} + \frac{\mu t^2}{3 r_2^6} - \frac{5 s^2 t^2}{4 r_2^7}\right\} + \frac{\mu^{\frac{3}{2}} s t^4 \, \mathrm{d} C_1}{2 r_2^5 R_2 \, \mathrm{d} \odot} *$$

$$2 A D \rho_2 = \mu t^2 R_2 C_2 \left\{\frac{1}{R_2^3} - \frac{1}{r_2^3} - \frac{\mu t^2}{3 R_2^6} + \frac{\mu t^2}{3 r_2^6} - \frac{5 s^2 t^2}{4 r_2^7}\right\} + \frac{\mu^{\frac{3}{2}} s t^4 \, \mathrm{d} C_2}{2 r_2^5 R_2 \, \mathrm{d} \odot}$$

$$\left\{1 - \frac{\mu s t^3}{4 r_2^5}\right\} D \rho_3 = \mu t^2 R_2 C_3 \left\{\frac{1}{R_2^3} - \frac{1}{r_2^3} - \frac{\mu t^2}{3 R_2^6} + \frac{\mu t^2}{3 r_2^6} - \frac{5 s^2 t^2}{4 r_2^7}\right\} + \frac{\mu^{\frac{3}{2}} s t^4 \, \mathrm{d} C_3}{2 r_2^5 R_2 \, \mathrm{d} \odot}$$

$$\rho_1 \left\{1 + \frac{\mu s t^3}{4 r_2^5}\right\} = 2 \left\{ \frac{C_1 + \frac{\mu^{\frac{1}{2}} s t^2}{2 r_2^5 R_2^5} \frac{\mathrm{d} C_1}{\mathrm{d} \odot} \left\{1 - \frac{R_2^3}{r_2^3}\right\}^{-1}}{C_2 + \frac{\mu^{\frac{1}{2}} s t^2 \, \mathrm{d} C_2}{2 r_2^5 R_2^5 \, \mathrm{d} \odot} \left\{1 - \frac{R_2^3}{r_2^3}\right\}^{-1}} \right\} A \rho_2$$

$$\rho_3 \left\{1 - \frac{\mu s t^3}{4 r_2^5}\right\} = 2 \left\{ \frac{C_3 + \frac{\mu^{\frac{1}{2}} s t^2}{2 r_2^5 R_2^5} \frac{\mathrm{d} C_3}{\mathrm{d} \odot} \left\{1 - \frac{R_2^3}{r_2^3}\right\}^{-1}}{C_2 + \frac{\mu^{\frac{1}{2}} s t^2 \, \mathrm{d} C_2}{r_2^5 R_2^5 \, \mathrm{d} \odot} \left\{1 - \frac{R_2^3}{r_2^3}\right\}^{-1}} \right\} A \rho_2$$

$s = 0$, if the comet is in its perihelion at the time of the second observation. If $C_1 = 0$, $C_2 = 0$, $C_3 = 0$, then

---

* $\frac{\mathrm{d} C_1}{\mathrm{d} \odot}$ is intended to denote

$$\tan \beta_2 \cos (\odot_2 - \lambda_3) - \tan \beta_3 \cos (\odot_2 - \lambda_2)$$

$$\rho_1 = 2\frac{\frac{dC_1}{d\odot}}{\frac{dC_2}{d\odot}} A\rho_2 \qquad \rho_3 = 2\frac{\frac{dC_3}{d\odot}}{\frac{dC_2}{d\odot}} A\rho_2$$

The difficulty, however, ensues when the term multiplied by $\frac{dC}{d\odot}$ becomes *comparable* to the other. It is evident that the nearer the comet is to the perihelion at the time of the second observation, the more nearly will the equations (31), (32), (33), (34), and also the methods of Olbers and Mr. Ivory, approach to accuracy.

If v be the true anomaly of the comet, reckoned from the perihelion,

$$r = \frac{a(1-e^2)}{1+e\cos v}$$

Let $a(1-e^2) = 2D \qquad r^2 dv = h dt$

$$dt = \frac{4D^2 dv}{h(1+e\cos v)^2} \qquad D = \frac{h^2}{2\mu} \qquad dt = \frac{(2D)^{\frac{3}{2}} dv}{\sqrt{\mu}(1+e\cos v)^2}$$

$$dt = \frac{(2D)^{\frac{3}{2}} dv}{\sqrt{\mu}(1+\cos v)^2}\left\{1 + 2(1-e)\frac{\cos v}{1+\cos v}\right.$$

$$\left. + 3(1-e)^2\frac{\cos v^2}{(1+\cos v)^2} + \&c.\right.$$

By integration,

$$t = \frac{D^{\frac{3}{2}}\sqrt{2}}{\sqrt{\mu}}\left\{\tan\frac{1}{2}v + \frac{1}{3}\tan^3\frac{1}{2}v\right.$$

$$\left. + (1-e)\tan\frac{1}{2}v\left\{\frac{1}{4} - \frac{1}{4}\tan^2\frac{1}{2}v - \frac{1}{5}\tan^4\frac{1}{2}v\right\} + \&c.\right.$$

t being reckoned from the time of the perihelion passage. See the *Mécanique Céleste*, vol. i. p. 186.

In the parabola

$$t = \frac{D^{\frac{3}{2}}\sqrt{2}}{\sqrt{\mu}}\left\{\tan\frac{1}{2}v + \frac{1}{3}\tan^3\frac{1}{2}v\right\} \tag{37}$$

See *Whewell's Dynamics*, new edition, p. 58.

From this equation a Table may be formed of values of v

and r corresponding to given values of $\frac{t\sqrt{\mu}}{D^{\frac{3}{2}}}$, from which by interpolation the heliocentric places of the comet may be found when the time is given from its perihelion passage. There is a Table of this kind in the *Astronomie Théorique et Pratique* of Delambre, vol. iii. p. 434. This interpolation, however, is troublesome when accuracy is required, and it is better to have recourse at once to the expression for $\tan\frac{v}{2}$, which is easily obtained by the reversion of series.

If $\tau = u + \mathrm{f}\,u$, then by Lagrange's theorem

$$u = \tau - \mathrm{f}\tau + \frac{\mathrm{d}\,(\mathrm{f}\tau)^2}{2\,\mathrm{d}\tau} - \frac{\mathrm{d}^2 . (\mathrm{f}\tau)^3}{2.3\,\mathrm{d}\,t^3} + \&\text{c.}$$

$$\text{If } u = \tan\frac{v}{2}, \quad \tau = \frac{t\sqrt{\mu}}{D^{\frac{3}{2}}\sqrt{2}} \quad \mathrm{f}\tau = \frac{\tau^3}{3}$$

$$\tan\frac{v}{2} = \tau - \frac{\tau^3}{3} + \frac{6}{2.9}\tau^5 - \frac{9.8}{2.3.27}\tau^7 + \frac{12.10.11}{2.3.4.81}\tau^9 + \&\text{c.}$$

$$= \tau - \frac{\tau^3}{3} + \frac{\tau^5}{3} - \frac{4}{9}\tau^7 + \frac{55}{81}\tau^9 + \&\text{c.} \qquad (38)$$

If v' be the true anomaly in a parabola whose perihelion distance is the same as in any ellipse, corresponding to the time t, v' may be found as above; and if $v' + x$ is the true anomaly in the ellipse in question,

$$\sin x = \frac{1}{10}(1-e)\tan\frac{v'}{2}\left\{4 - 3\cos^2\frac{v'}{2} - 6\cos^4\frac{v'}{2}\right\}\text{nearly.} \quad (39)$$

Méc. Cél. vol. i. p. 186.

The logarithm of the part of this expression which is independent of e is tabulated in Olbers's (*Abhandlung*, &c.) Table V. entitled, "Reduction der Parabel auf die Ellipse," and thus the correction x may be found approximately. In the case, however, even of the Comet of Halley, whose eccentricity is ·9676, I have found this method not sufficiently accurate, and it is better to proceed as follows: v' is to be found as before, either from such a Table as that in Delambre, or from the expression above, equation (38).

If u' be the eccentric anomaly which corresponds to v', $\tan\frac{u'}{2} = \sqrt{\frac{1-e}{1+e}}\tan\frac{v'}{2}$, the value of t corresponding to this value of v' is found from the equation $t = \frac{a^{\frac{3}{2}}}{\sqrt{\mu}}(v' - e \sin v')$

Let this value of t be called t' so that $t - t' = \delta t$; the corresponding value of v being v', $v - v' = \delta v'$

$$\frac{\delta v'}{\delta t} = \frac{d v}{d t} \text{ nearly}, \qquad \delta v' = \frac{\delta t \sqrt{\mu}}{a^{\frac{3}{2}}(1 - e \cos v')} \text{ nearly},$$

whence $\delta v'$, the correction to be applied to v', is known.

Neglecting the masses of the earth and comet in comparison with that of the sun, the constant μ may be considered as the same for both,

If P be the sidereal period of the earth,

$$P = \frac{2\pi}{\sqrt{\mu}} \qquad (\pi = 3{\cdot}14159)$$

If the time be expressed in days,

$$P = 365{\cdot}25638, \qquad \log. \sqrt{\mu} = 8{\cdot}2355821$$

If t be the time from the perihelion passage,

$$t = c_1 v - c_2 \sin v \tag{40}$$

c_1 = the periodic time of the comet, expressed in the same unit as t, divided by 360°, expressed in the same unit as v; and since the periodic times of the planets are as the major axes of their orbits to the power of $\frac{3}{2}$;

= the periodic time of the earth, expressed in the same unit as t, divided by 360°, expressed in the same unit as v, multiplied by the semi-axis major of the comet to the power of $\frac{3}{2}$, the mean distance of the earth from the sun, being, as usual, assumed for unity.

c_2 = the periodic time of the comet, expressed in the same unit as t, multiplied by the eccentricity and divided by $2 \times 3{\cdot}14159$.

= the periodic time of the earth, expressed in the same unit as t, multiplied by the eccentricity and by the semi-axis major of the comet to the power of $\frac{3}{2}$, and divided by $2 \times 3{\cdot}14159$.

$$\delta v' = \frac{\delta t}{c_1(1 - e \cos v')}$$

Resuming the equation,

$$t = \frac{D^{\frac{3}{2}}\sqrt{2}}{\sqrt{\mu}}\left\{\tan\frac{1}{2}\,v + \frac{1}{3}\,\tan^3\frac{1}{2}\,v\right\}$$

which is true in the parabola, D being the perihelion distance, if $2\,t$ be the interval between the first and third observation,

$$t = \frac{D^{\frac{3}{2}}}{\sqrt{2\,\mu}}\left[\tan\frac{1}{2}v_3 - \tan\frac{1}{2}v_1 + \frac{1}{3}\left\{\tan^3\frac{1}{2}v_3 - \tan^3\frac{1}{2}v_1\right\}\right]$$

If k is the chord of the arc described by the comet in the time $2\,t$,

$$k^2 = r_1^2 + r_3^2 - 2\,r_1\,r_3\cos(v_1 - v_3)$$

$$= (r_1 + r_3^2) - 4\,r_1\,r_3\cos^2\frac{1}{2}(v_1 - v_3)$$

$$2\cos\frac{1}{2}(v_1 - v_3)\sqrt{r_1\,r_3} = \pm\sqrt{(r_1 - r_3 + k)\,(r_1 + r_3 - k)}$$

Let $r_1 + r_3 + k = m^2, \quad r_1 + r_3 - k = n^2$

$$r_1 + r_3 = \frac{1}{2}(m^2 + n^2) \qquad 2\cos\frac{1}{2}(v_1 - v_3)\sqrt{r_1\,r_3} = \pm\,m\,n$$

$$\sin^2\frac{1}{2}(v_1 - v_3) = \cos^2\frac{1}{2}v_1 + \cos^2\frac{1}{2}v_3$$

$$-\,2\cos\frac{1}{2}(v_1 - v_3)\cos\frac{1}{2}v_1\cos\frac{1}{2}v_3$$

since $\cos^2\frac{1}{2}\,v = \frac{D}{r}$

$$\sin^2\frac{1}{2}(v_1 - v_3) = \frac{D}{r_1} + \frac{D}{r_3} - 2\cos\frac{1}{2}(v_1 - v_3)\frac{D}{\sqrt{r_1\,r_3}}$$

$$= \frac{D\left\{r_1 + r_3 - 2\cos\frac{1}{2}(v_1 - v_3)\sqrt{r_1\,r_3}\right\}}{r_1\,r_3}$$

$$= \frac{D\left\{\frac{m^2 + n^2}{2} \mp m\,n\right\}}{r_1\,r_3}$$

$$2\sin\frac{1}{2}(v_3 - v_1)\sqrt{r_1\,r_3} = (m \mp n)\sqrt{2\,D}$$

$$t = \frac{D^{\frac{3}{2}}}{\sqrt{2\mu}}\left[\tan\frac{1}{2}v_3 - \tan\frac{1}{2}v_1 + \frac{1}{3}\left\{\tan^3\frac{1}{2}v_3 - \tan^3\frac{1}{2}v_1\right\}\right]$$

$$= \frac{D^{\frac{3}{2}}}{\sqrt{2\mu}}\left\{\tan\frac{1}{2}v_3 - \tan\frac{1}{2}v_1\right\}\left\{1 + \frac{1}{3}\tan\frac{1}{2}v_1\right.$$

$$\left. + \frac{1}{3}\tan\frac{1}{2}v_1\tan\frac{1}{2}v_3 + \frac{1}{3}\tan^2\frac{1}{2}v_3\right\}$$

$$1 + \tan\frac{1}{2}v_1\tan\frac{1}{2}v_3 = \frac{\cos\frac{1}{2}(v_1 - v_3)}{\cos\frac{1}{2}v_1\cos\frac{1}{2}v_3} = \frac{\cos\frac{1}{2}(v_1 - v_3)\sqrt{r_1 r_3}}{D}$$

$$\tan\frac{1}{2}v_3 - \tan\frac{1}{2}v_1 = \frac{\sin\frac{1}{2}(v_3 - v_1)}{\cos\frac{1}{2}v_1\cos\frac{1}{2}v_3} = \frac{\sin\frac{1}{2}(v_3 - v_1)\sqrt{r_1 r_3}}{D}$$

$$t = \frac{D^{\frac{3}{2}}}{\sqrt{2\mu}}\left\{\frac{\sin\frac{1}{2}(v_3 - v_1)\sqrt{r_1 r_3}}{D}\right\}\left\{\frac{\cos\frac{1}{2}(v_1 - v_3)\sqrt{r_1 r_3}}{D}\right.$$

$$\left. + \frac{1}{3}\frac{\sin^2\frac{1}{2}(v_1 - v_3)\, r_1 r_3}{D^2}\right\}$$

$$t = \frac{D^{\frac{3}{2}}}{\sqrt{2\mu}}\left\{\frac{m \mp n}{\sqrt{2D}}\right\}\left\{\pm\frac{mn}{2D} + \frac{1}{3}\frac{(m \pm n)^2}{2D}\right\}$$

$$t\sqrt{\mu} = \frac{(m \mp n)^3 \pm 3mn(m \mp n)}{12}$$

$$= \frac{m^3 \mp n^3}{12}$$

$$= \frac{(r_1 + r_3 + k)^{\frac{3}{2}} \mp (r_1 + r_3 - k)^{\frac{3}{2}}}{12} \qquad (41)$$

This elegant theorem, often attributed to Lambert, was discovered by Euler*. The preceding proof is taken from the

* "Hac formula pro parabolâ quidem primo ab ill. Euler inventa esse videtur (*Miscell. Berolin*, t. vii. p. 20), qui tamen in posterum neglexit, neque etiam ad ellipsin et hyperbolam extendit, errant itaque qui formulam clar. Lambert tribuunt, etiamsi huic geometræ meritum, hanc expressionem oblivione sepultam proprio marte eruisse et ad reliquas sectiones conicas ampliavisse, non possit denegari."—Gauss *Theoria Motus Corporum Cœlestium*, p. 119.

Berliner Astronomisches Jahrbuch für 1833, p. 267. "Über "die Olbers'sche Methode zur Bestimmung der Cometen "bahnen." A somewhat different proof is given by Lagrange in the *Mécanique Analytique*, vol. ii. p. 30.

By expansion

$$12\,t\sqrt{\mu} = 3\,k\,(r_1 + r_3)^{\frac{1}{2}} - \frac{1\,.\,3}{4\,.\,6}k^3\,(r_1 + r_3)^{-\frac{3}{2}}$$

$$- \frac{1\,.\,3\,.\,3\,.\,5}{4\,.\,6\,.\,8\,.\,10}k^5\,(r_1 + r_3)^{-\frac{5}{2}} + \&\text{c.}$$

and by the reversion of series

$$k = \frac{4\,t\sqrt{\mu}}{(r_1 + r_3)^{\frac{1}{2}}}\left\{1 + \frac{\mu\,t^2}{6\,(r_1 + r_3)^2} + \&\text{c.}\right\} \tag{42}$$

$$k^2 = \frac{16\,\mu\,t^2}{r_1 + r_3}\left\{1 + \frac{\mu\,t^2}{3\,(r_1 + r_3)^2} + \&\text{c.}\right\} \tag{43}$$

which agrees with the equation (25), p. 13.

Referring to the annexed figure, which is that of Newton, *Principia*, lib. 3, lemma 9 and 10;

$$\text{A I} = \frac{\text{A C}}{2} = \frac{k}{2}$$

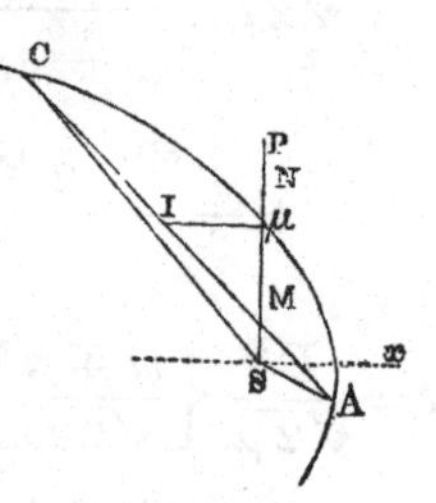

I μ is drawn parallel to the axis of the parabola. $\text{A C} = k$ $\quad \text{M}\,\mu = \text{I}\,\mu$.

Let x, y be the rectangular coordinates of any point of the parabola, the origin being at the focus as before, and the axis x coinciding with the axis of the curve.

$$x = r - 2\,D \qquad y^2 = 4\,D\,r - 4\,D^2$$

Let x_1, y_1 be the coordinates of the point A
x_3, y_3 C
x', y' .... μ

The coordinates of the point I are $\dfrac{x_1 + x_3}{2}$ and $\dfrac{y_1 + y_3}{2}$, therefore

$$\text{S}\,\mu = D + \frac{y'^2}{4\,D} = D + \frac{(y_1 + y_3)^2}{16\,D}$$

$$y_1 + y_3 = 2\,y_2\left\{1 - \frac{\mu\,t^2}{2\,r_2^{\,3}} - \frac{\mu\,s\,t^3}{2\,r_2^{\,5}}\right\}$$

$$S\mu = D + \frac{y_2^2}{4D}\left\{1 - \frac{\mu t^2}{r_2^3} - \frac{\mu s t^3}{r_2^5}\right\}$$

$$= D + \left\{r_2 - D\right\}\left\{1 - \frac{\mu t^2}{r_2^3} - \frac{\mu s t^3}{r_2^5}\right\}$$

$$\frac{r^2 \, d v^2}{d t^2} + \frac{d r^2}{d t^2} = \frac{2\mu}{r} - \frac{\mu}{a}$$

$$\frac{h^2}{r^2} + \frac{s^2}{r^2} = \frac{2\mu}{r} - \frac{\mu}{a}$$

$$a(1 - e^2) = 2 r_2 - \frac{r_2^2}{a} - \frac{s^2}{\mu} \tag{44}$$

In the parabola, $$D = r_2 - \frac{s^2}{2\mu} \tag{45}$$

therefore

$$S\mu = r_2 - \frac{s^2}{2\mu} + \frac{s^2}{2\mu}\left\{1 - \frac{\mu t^2}{r_2^3} - \frac{\mu s t^3}{r_2^5}\right\}$$

$$= r_2 - \frac{s^2 t^2}{2 r_2^3} - \frac{s^3 t^3}{2 r_2^5}$$

By a property of the parabola

$A I^2 = 4 S\mu \times I\mu$, whence $k^2 = 16 S\mu \times I\mu$

By equation (24), p. 13,

$$k^2 = \frac{8\mu t^2}{r_2}\left\{1 - \frac{\mu t^2}{3 r_2^3}\right\} + \frac{4\mu s^2 t^4}{r_2^5}$$

therefore

$$\frac{8\mu t^2}{r_2}\left\{1 - \frac{\mu t^2}{3 r_2^3}\right\} + \frac{4\mu s^2 t^4}{r_2^5} = 16\left\{r_2 - \frac{s^2 t^2}{2 r_2^3} - \frac{s^3 t^3}{2 r_2^5}\right\} I\mu$$

$$I\mu = \frac{\mu t^2}{2 r_2^2} \text{ nearly}, \qquad \frac{2}{3} I\mu = \frac{\mu t^2}{3 r_2^2}$$

By the equation

$$\frac{d x^2 + d y^2 + d z^2}{d t^2} = \frac{2\mu}{r}$$

if a body describing a parabola be at the distance

$$S\mu + \frac{2}{3} I\mu = r_2 - \frac{s^2 t^2}{2 r_2^3} + \frac{\mu t^2}{3 r_2^2}$$

$$\text{the square of the velocity} = \frac{2\mu}{r_2\left\{1+\frac{\mu t^2}{3 r_2^3}-\frac{s^2 t^2}{2 r_2^4}\right\}}$$

$$= \frac{2\mu}{r_2}\left\{1-\frac{\mu t^2}{3 r_2^3}\right\}+\frac{\mu s^2 t^2}{r_2^5}$$

$$= \frac{k^2}{4 t^2}$$

But $\frac{k^2}{4 t^2}$ is the square of the velocity of a body describing the chord A C uniformly in the time $2t$,

"Cometa igitur eâ cum velocitate, quam habet in altitudine "$S\mu + \frac{2}{3} I\mu$, eodem tempore describeret chordam A C "quam-proximè." *Principia*, Glasgow edition, p. 145.

I shall now consider the case when the intervals between the observations are unequal.

Let the time between the first and second observations $= t_1$

.... second and third ...... $= t_2$

$$\rho_1 \cos \lambda_1 - R_1 \cos \odot_1 = \left\{\rho_2 \cos \lambda_2 - R_2 \cos \odot_2\right\}\left\{1-\frac{\mu t_1^3}{2 r_2^3} + \frac{\mu s t_1^3}{2 r_2^5}\right\} + \frac{dx}{dt}\left\{t_1 - \frac{\mu t_1^3}{2 \,.\, 3 r_2^3}\right\}$$

$$R_1 \cos \odot_1 = R_2 \cos \odot_2 \left\{1 - \frac{\mu t_1^2}{2 R_2^3} + \frac{\mu S t_1^3}{2 R_2^5}\right\} + \frac{d \,.\, R \cos \odot}{dt}\left\{t_1 - \frac{\mu t^3}{2 \,.\, 3 r_1^3}\right\}$$

Neglecting the terms $\frac{\mu s t_1^3}{2 r_2^5}$, $\frac{\mu S t_1^3}{2 R_2^5}$, which are small,

$$\rho_1 \cos \lambda_1 = \rho_2 \cos \lambda_2 \left\{1-\frac{\mu t_1^2}{2 r_2^3}\right\} + \frac{\mu t_1^2}{2} R_2 \cos \odot_2 \left\{\frac{1}{r_2^3}-\frac{1}{R_2^3}\right\} + \frac{dx}{dt}\left\{t_1 - \frac{\mu t_1^3}{2 \,.\, 3 r_2^3}\right\} + \frac{d \,.\, R \cos \odot}{dt}\left\{t_1 - \frac{\mu t_1^3}{2 \,.\, 3 R_2^3}\right\}$$

Similarly

$$\rho_3 \cos \lambda_3 = \rho_2 \cos \lambda_2 \left\{1-\frac{\mu t_2^2}{2 r_2^3}\right\} + \frac{\mu t_2^2}{2} R_2 \cos \odot_2 \left\{\frac{1}{r_2^2}-\frac{1}{R_2^3}\right\}$$

$$+\frac{\mathrm{d}\,x}{\mathrm{d}\,t}\left\{t_2-\frac{\mu\,t_2^3}{2\,.\,3\,r_2^3}\right\}+\frac{\mathrm{d}\,.\,R\cos\odot}{\mathrm{d}\,t}\left\{t_2-\frac{\mu\,t_2^3}{2\,.\,3\,R_2^3}\right\}$$

$$\rho_1\cos\lambda_1=\rho_2\cos\lambda_2\left\{1-\frac{\mu\,t_1^2}{2\,r_2^3}\right\}+\frac{\mu\,t_1^2}{2}R_2\cos\odot_2\left\{\frac{1}{r_2^3}-\frac{1}{R_2^3}\right\}$$

$$+\frac{\mathrm{d}\,x}{\mathrm{d}\,t}t_1-\left\{\frac{\rho_1\cos\lambda_1-\rho_3\cos\lambda_3}{t_1-t_2}-\frac{\mathrm{d}\,.\,R\cos\odot}{\mathrm{d}\,t}\right\}\frac{t_1^3}{2\,.\,3\,r_2^3}$$

$$+\frac{\mathrm{d}\,.\,R\cos\odot}{\mathrm{d}\,t}\left\{t_1-\frac{\mu\,t_1^3}{2\,.\,3\,R_2^3}\right\}$$

$$\rho_3\cos\lambda_3=\rho_2\cos\lambda_2\left\{1-\frac{\mu\,t_2^2}{2\,r_2^3}\right\}+\frac{\mu\,t_2^2}{2}R_2\cos\odot_2\left\{\frac{1}{r_2^3}-\frac{1}{R_2^3}\right\}$$

$$+\frac{\mathrm{d}\,x}{\mathrm{d}\,t}t_2-\left\{\frac{\rho_1\cos\lambda_1-\rho_3\cos\lambda_3}{t_1-t_2}-\frac{\mathrm{d}\,.\,\mathrm{R}\cos\odot}{\mathrm{d}\,t}\right\}\frac{t_2^3}{2\,.\,3\,r_2^3}$$

$$+\frac{\mathrm{d}\,.\,R\cos\odot}{\mathrm{d}\,t}\left\{t_2-\frac{\mu\,t_2^3}{2\,.\,3\,R_2^3}\right\}$$

Eliminating $\frac{\mathrm{d}\,x}{\mathrm{d}\,t}$,

$$\rho_1\cos\lambda_1\,t_2\left\{1+\frac{\mu\,t_1\,(t_1+t_2)}{2\,.\,3\,r_2^3}\right\}-\rho_3\cos\lambda_3\,t_1\left\{1+\frac{u\,t_2\,(t_1+t_2)}{2\,.\,3\,r_2^3}\right\}$$

$$=\rho_2\cos\lambda_2\,(t_2-t_1)\left\{1+\frac{\mu\,t_2\,t_1}{2\,r_2^3}\right\}$$

$$-\frac{\mu\,t_1\,t_2\,(t_2-t_1)}{2}R_2\left\{\cos\odot_2\right.$$

$$\left.-\frac{t_1+t_2}{3}\sin\odot_2\frac{\mathrm{d}\,\odot}{\mathrm{d}\,t}\right\}\left\{\frac{1}{r_2^3}-\frac{1}{R_2^3}\right\}$$

$\cos\odot_2-\frac{t_1+t_2}{3}\sin\odot_2\frac{\mathrm{d}\,\odot}{\mathrm{d}\,t}$ is the cosine of the sun's longitude corresponding to the time $\frac{t_1+t_2}{3}$ from the second observation: if this longitude be called $\odot'$,

$$\rho_1\cos\lambda_1\,t_2\left\{1+\frac{\mu\,t_1\,(t_1+t_2)}{2\,.\,3\,r_2^3}\right\}-\rho_3\cos\lambda_3\,t_1\left\{1+\frac{\mu\,t_2(t_1+t_2)}{2\,.\,3\,r_2^3}\right\}$$

$$=\rho_2\cos\lambda_2\,(t_2-t_1)\left\{1+\frac{\mu\,t_2\,t_1}{2\,r_2^3}\right\}$$

$$-\frac{\mu t_1 t_2 (t_2 - t_1)}{2} R_2 \cos \odot \left\{\frac{1}{r_2^3} - \frac{1}{R_2^3}\right\}$$

similarly

$$\rho_1 \sin \lambda_1 t_2 \left\{1 + \frac{\mu t_1 (t_1 + t_2)}{2 \cdot 3 r_2^3}\right\} - \rho_3 \sin \lambda_3 t_1 \left\{1 + \frac{\mu t_2 (t_1 + t_2)}{2 \cdot 3 r_2^3}\right\}$$

$$= \rho_2 \sin \lambda_2 (t_2 - t_1) \left\{1 + \frac{\mu t_2 t_1}{2 r_2^3}\right\}$$

$$-\frac{\mu t_1 t_2 (t_2 - t_1)}{2} R_2 \sin \odot' \left\{\frac{1}{r_2^3} - \frac{1}{R_2^3}\right\}$$

$$\rho_1 \tan \beta_1 t_1 \left\{1 + \frac{\mu t_1 (t_1 + t_2)}{2 \cdot 3 r_2^3}\right\} - \rho_3 \tan \beta_3 t_2 \left\{1 + \frac{\mu t_2 (t_1 + t_2)}{2 \cdot 3 r_2^3}\right\}$$

$$= \rho_2 \tan \beta_2 (t_2 - t_1) \left\{1 + \frac{\mu t_2 t_1}{2 r_2^3}\right\}$$

and by elimination as before,

$$\rho_1 \left\{1 + \frac{\mu t_1 (t_1 + t_2)}{2 \cdot 3 r_2^3}\right\}$$

$$= \frac{\mu t_1 (t_2 - t_1) R_2 \{\tan \beta_3 \sin (\odot' - \lambda_2) - \tan \beta_2 \sin (\odot' - \lambda_3)\}}{2\{\tan \beta_3 \sin (\lambda_1 - \lambda_2) + \tan \beta_2 \sin (\lambda_3 - \lambda_1) + \tan \beta_1 \sin (\lambda_2 - \lambda_3)\}} \left\{\frac{1}{R_2^3} - \frac{1}{r_2^3}\right\}$$

$$\rho_2 \left\{1 + \frac{\mu t_1 t_2}{2 r_2^3}\right\}$$

$$= \frac{\mu t_1 t_2 R_2 \{\tan \beta_3 \sin (\odot' - \lambda_1) - \tan \beta_1 \sin (\odot' - \lambda_3)\}}{2\{\tan \beta_3 \sin (\lambda_1 - \lambda_2) + \tan \beta_2 \sin (\lambda_3 - \lambda_1) + \tan \beta_1 \sin (\lambda_2 - \lambda_3)\}} \left\{\frac{1}{R_2^3} - \frac{1}{r_2^3}\right\}$$

$$\rho_3 \left\{1 + \frac{\mu t_1 (t_1 + t_2)}{2 \cdot 3 r_2^3}\right\}$$

$$= \frac{\mu t_2 (t_2 - t_1) R_2 \{\tan \beta_1 \sin (\odot' - \lambda_2) - \tan \beta_2 \sin (\odot' - \lambda_1)\}}{2\{\tan \beta_3 \sin (\lambda_1 - \lambda_2) + \tan \beta_2 \sin (\lambda_3 - \lambda_1) + \tan \beta_1 \sin (\lambda_2 - \lambda_3)\}} \left\{\frac{1}{R_2^3} - \frac{1}{r_2^3}\right\}$$

Making $\rho_1 = A \gamma_1 \rho_2$, $\rho_3 = A \gamma_2 \rho_2$, and substituting in equations (1) and (2) as before, the right-hand side of each of these equations suffers no change.

$$k^3 = \mu (t_1 - t_2)^2 \left\{\frac{2}{r_2} - \frac{1}{a} - \frac{2 s (t_2 + t_1)}{r_2^3}\right\}$$

$$r_1^2 + r_3^2 - 2 r_2^2 = \mu (t_1^2 + t_2^2) \left\{\frac{1}{r_2} - \frac{1}{a}\right\} + \frac{\mu s (t_1^3 + t_2^3)}{r_2^3}$$

The equations

$$\frac{d^2 x}{d t^2} + \frac{\mu x}{r^3} = 0 \qquad \frac{d^2 y}{d t^2} + \frac{\mu y}{r^3} = 0 \qquad \frac{d^2 z}{d t^2} + \frac{\mu z}{r^3} = 0$$

give

$$\frac{d^2 \rho \cos\lambda - 2\sin\lambda\, d\rho\, d\lambda - \rho\cos\lambda\, d\lambda^2 + \rho\sin\lambda\, d^2\lambda}{d t^2} \tag{46}$$

$$+ \mu \left\{ \frac{\cos \odot}{R^2} - \frac{R\cos\odot - \rho\cos\lambda}{r^3} \right\} = 0$$

$$\frac{d^2 \rho \sin\lambda + 2\cos\lambda\, d\rho\, d\lambda - \rho\sin\lambda\, d\lambda^2 + \rho\cos\lambda\, d^2\lambda}{d t^2} \tag{47}$$

$$+ \mu \left\{ \frac{\sin \odot}{R^2} - \frac{R\sin\odot - \rho\sin\lambda}{r^3} \right\} = 0$$

$$\frac{d^2 \rho \tan\beta + \rho \dfrac{d^2\beta}{\cos^2\beta} + 2\dfrac{d\rho\, d\beta}{\cos^2\beta} + 2\rho\dfrac{\sin\beta}{\cos^2\beta} d\beta^2}{d t^2} \tag{48}$$

$$+ \mu \rho \frac{\tan\beta}{r^3} = 0$$

Whence if $\tan\beta = s$

$$\frac{d^2 \rho - \rho\, d\lambda^2}{d t^2} + \mu R \cos(\odot - \lambda) \left\{ \frac{1}{R^3} - \frac{1}{r^3} \right\} + \frac{\mu \rho}{r^3} = 0 \tag{49}$$

$$\frac{2\, d\rho\, d\lambda + \rho\, d^2\lambda}{d t^2} + \mu R \sin(\odot - \lambda) \left\{ \frac{1}{R^3} - \frac{1}{r^3} \right\} = 0 \tag{50}$$

$$\frac{s\, d^2 \rho + 2\, d\rho\, ds + \rho\, d^2 s}{d t^2} + \frac{\mu \rho s}{r^3} = 0 \tag{51}$$

Eliminating $\frac{d^2\rho}{d t^2}$ between (49) and (51)

$$\frac{2\, d\rho\, ds + \rho\, d^2 s + \rho s\, d\lambda^2}{d t^2} - \mu s R \cos(\odot - \lambda) \left\{ \frac{1}{R^3} - \frac{1}{r^3} \right\} = 0$$

Substituting in this equation the value of $\mu R \left\{ \frac{1}{R^3} - \frac{1}{r^3} \right\}$ in equation (50)

$$\frac{d\rho}{d t} = -\frac{\rho}{2} \, \frac{\left\{ s\cos(\odot - \lambda)\dfrac{d^2\lambda}{d t^2} + \sin(\odot - \lambda)\left\{ \dfrac{d^2 s}{d t^2} + \dfrac{s\, d\lambda^2}{d t^2} \right\} \right\}}{\dfrac{ds}{dt}\sin(\odot - \lambda) + \dfrac{d\lambda}{dt} s\cos(\odot - \lambda)} \tag{52}$$

and since

$$\frac{d\,s}{d\,t} = \frac{d\,\beta}{\cos^2\beta\,d\,t} \qquad \frac{d^2\,s}{d\,t^2} = \frac{d^2\,\beta}{\cos^2\beta\,d\,t^2} + \frac{2\sin\beta\,d\,\beta^2}{\cos^3\beta\,d\,t^2}$$

$$\frac{d\,\rho}{d\,t} = -\frac{\rho}{2}\left\{\frac{\frac{d^2\,\beta}{d\,t^2} + \sin\beta\cos\beta\frac{d\,\lambda^2}{d\,t^2} + 2\tan\beta\frac{d\,\beta^2}{d\,t^2} + \frac{\sin\beta\cos\beta\;d^2\,\lambda}{\tan(\odot - \lambda)\,d\,t^2}}{\frac{d\,\beta}{d\,t} + \frac{\sin\beta\cos\beta\;d\,\lambda}{\tan(\odot - \lambda)\,d\,t}}\right\}$$

Substituting this value of $\frac{d\,\rho}{d\,t}$ in the equation (50)

$$\rho = \frac{\sin(\odot - \lambda)\frac{d\,\beta}{d\,t} + \sin\beta\cos\beta\cos(\odot - \lambda)\frac{d\,\lambda}{d\,t}}{\frac{d^2\,\lambda\,d\,\beta}{d\,t^2\,d\,t} - \frac{d^2\,\beta\,d\,\lambda}{d\,t^2\,d\,t} - \sin\beta\cos\beta\frac{d\,\lambda^3}{d\,t^3} - 2\tan\beta\frac{d\,\beta^2\,d\,\lambda}{d\,t^2\,d\,t}}\left\{\frac{1}{r^3} - \frac{1}{R^3}\right\}.$$

If A be the longitude of the earth seen from the sun,

$$A = \odot - 180^\circ$$

$$\cos(A - \lambda) = \cos(\odot - \lambda - 180^\circ) = -\cos(\odot - \lambda)$$

$$\sin(A - \lambda) = \sin(\odot - \lambda - 180^\circ) = -\sin(\odot - \lambda)$$

and if

$$\mu' = \frac{\frac{d^2\,\beta\,d\,\lambda}{d\,t^2\,d\,t} - \frac{d^2\,\lambda\,d\,\beta}{d\,t^2\,d\,t} + \sin\beta\cos\beta\frac{d\,\lambda^3}{d\,t^3} + 2\tan\beta\frac{d\,\beta^2\,d\,\lambda}{d\,t^2\,d\,t}}{\sin\beta\cos\beta\cos(A - \lambda)\frac{d\,\lambda}{d\,t} + \sin(A - \lambda)\frac{d\,\beta}{d\,t}}$$

$$\rho = \frac{R}{\mu'}\left\{\frac{1}{r^3} - \frac{1}{R^3}\right\}$$

"The curtate distance ρ of the comet from the earth being "always positive, this equation shows that the distance r of "the comet from the sun is less or greater than the distance "R of the sun from the earth, according as μ' is positive or ne-"gative; these distances are equal if $\mu' = 0$. It is possible, "moreover, by the mere inspection of a celestial globe, to "determine the sign of μ', and consequently whether the "comet is nearer the sun than the earth, or not. For this "purpose let us conceive a great circle which passes through "two geocentric places of the comet indefinitely near to each "other. Let be the inclination of this circle to the ecliptic, "and ν the longitude of its ascending node, we have

$$\tan\ \sin(\lambda - \nu) = \tan\beta$$

" whence we get

$$d\beta \sin(\lambda - \nu) = d\lambda \sin\beta \cos\beta \cos(\lambda - \nu)$$

" and differentiating again, we have

$$0 = \frac{d\lambda}{dt}\frac{d^2\beta_{\prime}}{dt^2} - \frac{d\beta}{dt}\frac{d^2\lambda}{dt^2} + 2\tan\beta\frac{d\lambda}{dt}\frac{d\beta^2}{dt^2} + \sin\beta\cos\beta\frac{d\lambda^3}{dt^3}$$

" $\frac{d^2\beta_{\prime}}{dt^2}$ being the value of $\frac{d^2\beta}{dt^2}$, which would obtain, if the ap-
" parent motion of the comet continued in the great circle.
" The value of μ' thus becomes, by substituting for $\frac{d\beta}{dt}$ its va-
" lue $\frac{\sin\beta\cos\beta\cos(\lambda-\nu)}{\sin(\lambda-\nu)}\frac{d\lambda}{dt}$;

$$\mu' = \frac{\left\{\frac{d^2\beta}{dt^2} - \frac{d^2\beta_{\prime}}{dt^2}\right\}\sin(\lambda-\nu)}{\sin\beta\cos\beta\sin(A-\nu)}$$

" The quantity $\frac{\sin(\lambda-\nu)}{\sin\beta\cos\beta}$ is constantly positive, there-
" fore the value of μ' is positive or negative, according as
" $\frac{d^2\beta}{dt^2} - \frac{d^2\beta_{\prime}}{dt^2}$ has the same sign or a contrary sign to $\sin(A-\nu)$;
" but $A - \nu = 180° +$ the distance of the sun to the ascend-
" ing node of the great circle; whence it is easy to conclude
" that μ' will be positive or negative, according as in a third
" geocentric position of the comet indefinitely near the two
" former, the comet deviates from the great circle on the
" same side of the great circle as the sun, or not. Let us
" conceive a great circle of the sphere passing through two
" geocentric places of the comet indefinitely near to one other;
" if in a third place indefinitely near the former and consecu-
" tive, the comet deviates from this great circle, on the same
" side as the sun or on the opposite side, the comet will be
" nearer the sun than the earth, or not; it will be at the same
" distance if it continues to move in the same great circle."
Mécanique Céleste, vol. i. p. 208.

$$\frac{dx}{dt} = \frac{d\rho}{dt}\cos\lambda - \rho\sin\lambda\frac{d\lambda}{dt} - \cos\odot\frac{dR}{dt} + R\sin\odot\frac{d\odot}{dt}$$

$$\frac{d\,y}{d\,t} = \frac{d\,\rho}{d\,r}\sin\lambda + \rho\cos\lambda\frac{d\,\lambda}{d\,t} - \sin\odot\frac{d\,R}{d\,t} = R\cos\odot\frac{d\,\odot}{d\,t}$$

$$\frac{d\,z}{d\,t} = \frac{d\,\rho}{d\,r}\tan\beta + \rho\frac{d\,\beta}{\cos\beta^2\,d\,t}$$

Substituting these values of $\frac{d\,x}{d\,t}$, $\frac{d\,y}{d\,t}$ and $\frac{d\,z}{d\,t}$ in the equation

$$\frac{d\,x^2}{d\,t^2} + \frac{d\,y^2}{d\,t^2} + \frac{d\,z^2}{d\,t^2} - \frac{2\,\mu}{r} + \frac{\mu}{a} = 0$$

$$\frac{\rho^2\,d\,\lambda^2 + \frac{d\,\rho^2}{\cos\beta^2} + R^2\,d\,\odot^2 - 2\cos(\odot-\lambda)\{d\,\rho\,d\,R + R\,\rho\,d\,\lambda\,d\,\odot\}}{d\,t^2}$$

$$+\frac{2\sin(\odot-\lambda)\,\{R\,d\,\rho\,d\,\odot - \rho\,d\,R\,d\,\lambda\} + 2\,\rho\,d\,\rho\frac{d\,\beta}{\cos\beta^2} + \rho^2\frac{d\,\beta^2}{\cos\beta^4}}{d\,t^2}$$

$$-\frac{2\,\mu}{r} + \frac{\mu}{a} = 0 \qquad (53)$$

Substituting in this equation for $\frac{d\,\rho}{d\,t}$ its value given by equation (52), an equation is obtained which may be considered as identical with equation (33).

The method of Laplace differs from others essentially in employing as data in the equations which determine the distance of the comet from the earth, not the trigonometrical lines which refer to the geocentric places of the comet ($\sin\beta$, $\sin\lambda$, &c.) but the values of the differential coefficients $\frac{d\,\lambda}{d\,t}$, $\frac{d\,\beta}{d\,t}$, $\frac{d^2\,\lambda}{d\,t^2}$, $\frac{d^2\,\beta}{d\,t^2}$ in terms of the finite differences of the quantities λ_1, λ_2, β_1, β_2, &c.

Thus $\frac{d\,\lambda}{d\,t} = \frac{\lambda_3 - \lambda_1}{2\,t}$ $\qquad \frac{d\,\beta}{d\,t} = \frac{\beta_3 - \beta_1}{2}$ nearly.

When Laplace first gave this method, he intended that the differential coefficients $\frac{d\,\lambda}{d\,t}$, $\frac{d\,\beta}{d\,t}$, $\frac{d^2\,\lambda}{d\,t^2}$, $\frac{d^2\,\beta}{d\,t^2}$ should be determined by the concurrence of more than three observations; very great care is then required in applying the right signs to the quantities involved, and the labour is much increased. Besides, Legendre has pointed out that the influence of the errors of the observations on the differential coefficients, when the observations by which they are determined are multiplied, is such as in great measure to destroy the advantages which this method promised. Laplace has since admitted the force of this objection. "Independamment de la longueur du calcul, "les erreurs des observations nuisent à l'exactitude que l'on "peut attendre de la multiplicité des observations; et il me "parait préférable de n'en employer que trois, en fixant "l'epoque à l'observation intermediaire, et en prenant les ob-"servations extrêmes assez peu distantes entre elles, pour que "dans l'intervalle qui les separe, les données précédentes "puissent être supposées à fort peu près les mêmes." *Méc. Cél.* vol. 5. p. 342.

The geocentric coordinates of the comet being represented as before by x, y, z, the plane $x\,y$ coinciding with the plane of the ecliptic, let $x = a\rho$, $y = b\rho$, $z = c\rho$; if ρ be the distance of the comet from the earth, and not the curtate distance as before, a, b, c are the cosines of the angles which ρ makes with the coordinate axes; and if $A = \cos \odot$, $B = \sin \odot$, the equations of motion give

$$\frac{\rho\, d^2 a + 2\, da\, d\rho + a\, d^2\rho - d^2 R A}{d t^2} + \frac{\mu (a\rho - AR)}{r^3} = 0$$

$$\frac{\rho\, d^2 a + 2\, da\, d\rho + a\, d^2\rho}{d t^2} + \frac{\mu a \rho}{r^3} + \mu A R \left\{ \frac{1}{R^3} - \frac{1}{r^3} \right\} = 0$$

Similarly

$$\frac{\rho\, d^2 b + 2\, db\, d\rho + b\, d^2\rho}{d t^2} + \frac{\mu b \rho}{r^3} + \mu B R \left\{ \frac{1}{R^3} - \frac{1}{r^3} \right\} = 0$$

$$\frac{\rho\, d^2 c + 2\, dc\, d\rho + c\, d^2\rho}{d t^2} + \frac{\mu c \rho}{r^3} = 0$$

and by elimination

$$\rho = \frac{\mu R \{c\,(A\,db - B\,da) + a\,B\,dc - b\,A\,dc\}}{\{b\,(dc\,d^2a - da\,d^2c) + a\,(db\,d^2c - dc\,d^2b) + c\,(da\,d^2b - db\,d^2a)\}} \left\{ \frac{1}{r^3} - \frac{1}{R^3} \right\}$$

If Π be the pole of a great circle touching the apparent orbit of the comet;

φ be the distance from the pole of the osculating circle to its circumference, so that the radius of this circle $= \sin \varphi$;

δ be the distance upon the celestial sphere from the apparent place of the sun to the pole Π,

and if $\frac{d\,\sigma^2}{d\,t^2} = \frac{d\,a^2 + d\,b^2 + d\,c^2}{d\,t^2}$.

M. Binet has proved, in a paper in the *Journal de l'Ecole Polytechnique*, vol. xiii. cah. 20, that the preceding expression may be put in the form

$$\rho \frac{d\,\sigma^2}{d\,t^2} = \mu\, R \cos\delta \tan\phi \left\{\frac{1}{r^3} - \frac{1}{R^3}\right\} \quad (54)$$

this proof, however, rests upon some geometrical theorems which are not, to my knowledge, to be found in any work in the English language, and which to attempt to prove would lead me too far from the present subject.

3. The elements of the orbit can easily be found when the curtate distance ρ_2 has been obtained.

$$r_1^{\,2} = \frac{A^2 \gamma_1^{\,2} \rho_2^{\,2}}{\cos^2 \beta_1} - 2\,A\,\gamma_1 \rho_2 R_1 \cos(\odot_1 - \lambda_1) + R_1^{\,2}$$

$$r_2^{\,2} = \frac{\rho_2^{\,2}}{\cos^2 \beta_2} - 2\,\rho_2 R_2 \cos(\odot_2 - \lambda_2) + R_1^{\,2}$$

$$r_3^{\,2} = \frac{A^2 \gamma_2^{\,2} \rho_2^{\,2}}{\cos^2 \beta_3} - 2\,A\,\gamma_2 \rho_2 R_3 \cos(\odot_3 - \lambda_3) + R_3^{\,2}$$

When the intervals between the observations are equal,

$$r_1^{\,2} = r_2^{\,2} - \frac{d\,.\,r^2}{d\,t}\,t + \frac{d^2\,.\,r^2}{2\,d\,t^2}\,t^2 - \frac{d^3\,.\,r^2}{2\,.\,3\,d\,t^3}\,t^3 + \frac{d^4\,.\,r^2}{2\,.\,3\,.\,4\,d\,t^4}\,t^4 + \&c.$$

$$r_3^{\,2} = r_2^{\,2} + \frac{d\,.\,r^2}{d\,t}\,t + \frac{d^2\,.\,r^2}{2\,d\,t^2}\,t^2 + \frac{d^3\,.\,r^2}{2\,.\,3\,d\,t^3}\,t^3 + \frac{d^4\,.\,r^2}{2\,.\,3\,.\,4\,d\,t^4}\,t^4 + \&c.$$

$$r_3^{\,2} - r^{\,2} = \frac{2\,d\,.\,r^2}{d\,t}\,t + \frac{d^3\,.\,r^2}{3\,d\,t^3}\,t^3 + \&c.$$

Neglecting $\frac{1}{a}$,

$$\frac{d^2\,.\,r^2}{d\,t^2} = \frac{2\,\mu}{r} \qquad \frac{d^3\,.\,r^2}{d\,t^3} = -\frac{2\,\mu\,s}{r^3}$$

$$r_3^{\,2} - r_1^{\,2} = 4\,s\,t - \frac{2\,\mu\,s\,t^3}{3\,r_2^{\,3}}$$

$$s = \frac{r_3{}^2 - r_1{}^2}{4\,t}\left\{1 + \frac{\mu\,t^2}{6\,r_2{}^3}\right\} \tag{55}$$

The equation $\dfrac{d\,x^2 + d\,y^2 + d\,z^2}{d\,t^2} = \dfrac{2\,\mu}{r} - \dfrac{\mu}{a}$

gives $\dfrac{s^2}{r^2} + \dfrac{\mu\,a\,(1 - e^2)}{r^2} = \dfrac{2\,\mu}{r} - \dfrac{\mu}{a}$

$$a\,(1 - e^2) = 2\,r_2 - \frac{r_2{}^2}{a} - \frac{s^2}{2\,\mu} \tag{56}$$

From which equation the eccentricity e may be found when the semi-major axis a is known.

If the eccentric anomaly be called v,

$$\cos v = \frac{a - r}{a\,e}, \qquad \cos v_2 = \frac{a - r_2}{a\,e} \tag{57}$$

If t be the time from the perihelion passage

$$t = c_1\,v - c_2 \sin v. \qquad \text{See p. 23.} \tag{58}$$

In the parabola $D = r_2 - \dfrac{s^2}{2\,\mu}$ (59)

D being the perihelion distance.

$$\cos^2 \frac{1}{2}\,v_1 = \frac{D}{r_1}, \quad \cos^2 \frac{1}{2}\,v_2 = \frac{D}{r_2}, \quad \cos^2 \frac{1}{2}\,v_3 = \frac{D}{r_3} \tag{60}$$

from which equations v_1, v_2, and v_3 may be found.

The time of the passage of the perihelion is given by the equation

$$t = \frac{D^{\frac{3}{2}}\,\sqrt{2}}{\sqrt{\mu}}\left\{\tan \frac{1}{2}\,v + \frac{1}{3}\tan^3 \frac{1}{2}\,v\right\} \tag{61}$$

in which either v_1, v_2, or v_3 may be used.

The heliocentric longitudes and latitudes λ', β', are found by the equations

$$\sin \beta' = \frac{\rho \tan \beta}{r} \qquad \sin(\lambda' - \odot) = \frac{\rho \sin(\lambda - \odot)}{r \cos \beta'} \tag{62}$$

The equation to the plane of the orbit, the origin being at the sun and the plane $x\,y$ coinciding with the ecliptic, is,

$$x\,(z_1\,y_3 - y_1\,z_3) + y\,(x_1\,z_3 - z_1\,x_3) + z\,(x_3\,y_1 - y_3\,z_1) = 0$$

The equations to the intersection of the comet's orbit with the ecliptic or the line of nodes, are

$$x\,(z_1\,y_3 - y_1\,z_3) = y\,(z_1\,x_3 - x_1\,z_3) \qquad z = 0$$

and if the axis x coincide with the line from which the longitudes are reckoned, and ν is the longitude of the node reckoned from the same line,

$$\tan \nu = \frac{z_1\,y_3 - y_1\,x_3}{z_1\,x_3 - x_1\,z_3} = \frac{\tan \beta_1' \sin \lambda_3' - \sin \lambda_1' \tan \beta_3'}{\tan \beta_1' \cos \lambda_3 - \cos \lambda_1' \tan \beta_3'} \quad (63)$$

If the inclination of the orbit of the comet be called ι, by spherical trigonometry

$$\tan \iota = \frac{\tan \beta'}{\sin (\lambda' - \nu)} \quad (64)$$

in which equation either (λ_1', β_1'), (λ_2', β_2') or (λ_3', β_3') may be used.

Let S be the sun, ☊ S ☋ the line of nodes, C the place of the comet at either observation, S D the perihelion distance. $v =$ C S D. If C' be the projection of C on the ecliptic the angle C' S ☊ $= \nu - \lambda'$

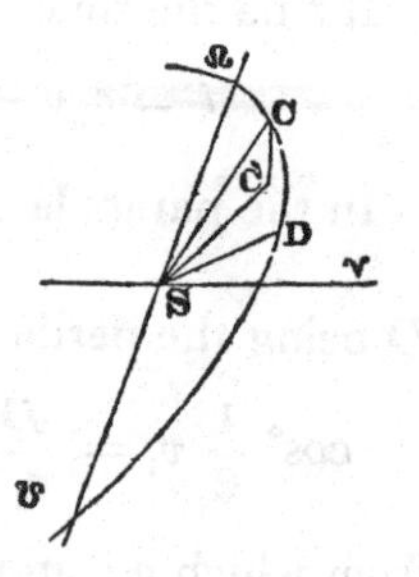

$$\tan \mathrm{C\,S}\,☊ = \frac{\tan \mathrm{C'\,S}\,☊}{\cos \iota} = \frac{\tan (\nu - \lambda')}{\cos \iota}$$

from which equation C S ☊ may be found, and hence D S ☊ or D S ☋ is known

ϖ, *the longitude of the perihelion*, is the angle ♈ S ☋ + *D* S ☋, which is thus determined.

4. When great accuracy is required, as in the case of any of the comets whose period has been ascertained, and when the observations employed are very precise, it is necessary to take into account the perturbations which the comet experiences even during the time of its apparition. The following appears to me to be the simplest method which can be proposed for that purpose.

Let $x_{,}$, $y_{,}$, $z_{,}$ be the coordinates of the disturbing planet, $m_{,}$ its mass, $\rho_{,}$ its distance from the comet,

$$R = m_{,} \left\{ \frac{x\,x_{,} + y\,y_{,} + z\,z_{,}}{r_{,}^3} - \frac{1}{\rho_{,}} \right\}$$

$$A = m_{,} \left\{ \frac{x_{,}}{r_{,}^3} - \frac{x_{,} - x}{\rho_{,}^3} \right\} = \frac{\mathrm{d}R}{\mathrm{d}x}$$

$$B = m_{\prime}\left\{\frac{y_{\prime}}{r_{\prime}^{3}} - \frac{y_{\prime} - y}{\rho_{\prime}^{3}}\right\} = \frac{dR}{dy}$$

$$C = m_{\prime}\left\{\frac{z_{\prime}}{r_{\prime}^{3}} - \frac{z_{\prime} - z}{\rho_{\prime}^{3}}\right\} = \frac{dR}{dz}$$

Then by the equations of motion

$$\frac{d^2 x}{dt^2} + \frac{\mu x}{r^3} + A = 0$$

$$\frac{d^2 y}{dt^2} + \frac{\mu y}{r^3} + B = 0 \qquad (65)$$

$$\frac{d^2 z}{dt^2} + \frac{\mu z}{r^3} + C = 0$$

hence

$$\frac{d^3 x}{dt^3} = -\frac{\mu\, dx}{r^3\, dt} + \frac{3\mu x\, dr}{r^4} - \frac{dA}{dt}$$

$$\frac{d^4 x}{dt^4} = -\frac{\mu\, d^2 x}{r^3\, dt^2} + \frac{3\mu x\, d^2 r}{r^4} - \frac{d^2 A}{dt^2} + \&c.$$

$$r\, dr = x\, dx + y\, dy + z\, dz$$

$$r\, d^2 r + dr^2 = dx^2 + dy^2 + dz^2 + x\, d^2 x + y\, d^2 y + z\, d^2 z$$

$$= \frac{\mu}{r} - \frac{\mu}{a} - Ax - By - Cz$$

therefore

$$\frac{d^4 x}{dt^4} = \frac{\mu A}{r^3} - \frac{3(Ax + By + Cz)\mu x}{r^5} - \frac{d^2 A}{dt^2} + \&c.$$

By Maclaurin's theorem

$$x = x_0 + \left(\frac{dx}{dt}\right)_0 t + \left(\frac{d^2 x}{2\, dt^2}\right)_0 t^2 + \&c.$$

$$= x_0\left\{1 - \frac{\mu t^2}{2 r_0^3} + \&c.\right\} + \left(\frac{dx}{dt}\right)_0 t\left\{1 - \frac{\mu t^2}{6 r_0^3} - \&c.\right\}$$

$$- \frac{A t^2}{2} - \frac{dA}{dt}\frac{t^3}{2.3}$$

$$+ \left\{\frac{\mu A}{r^3} - \frac{3(Ax + By + Cz)\mu x}{r^5} - \frac{d^2 A}{dt^2}\right\}\frac{t^4}{2.3.4}$$

x_0, r_0, $\left(\frac{dx}{dt}\right)_0$ &c. being the values of those variables when

$t = 0$. The expression for x consists of two parts; the one arises from the development according to powers of t of its value in the osculating ellipse, that is, the ellipse in which the values of x_0, y_0, z_0, $\left(\frac{dx}{dt}\right)_0$, $\left(\frac{dy}{dt}\right)_0$, $\left(\frac{dz}{dt}\right)_0$ coincide with the actual values of those quantities. The other part of the expression is the perturbation of x or the difference between the real value of x after the time t and that which would obtain in the ellipse in question: if this difference be called δx,

$$\delta x = -\frac{t^2}{2}\left\{A + \frac{dA}{3\,dt}t + \left\{\frac{d^2A}{dt^2} - \frac{\mu A}{r^3} + \frac{3(Ax + By + Cz)\mu x}{r^5}\right\}\frac{t^2}{3.4} + \&c.\right\}$$

similarly

$$\delta y = -\frac{t^2}{2}\left\{B + \frac{dB}{3\,dt}t + \left\{\frac{d^2B}{dt^2} - \frac{\mu B}{r^3} + \frac{3(Ax + By + Cz)\mu y}{r^5}\right\}\frac{t^2}{3.4} + \&c.\right\} \quad (66)$$

$$\delta z = -\frac{t^2}{2}\left\{C + \frac{dC}{3\,dt}t + \left\{\frac{d^2C}{dt^2} - \frac{\mu C}{r^3} + \frac{3(Ax + By + Cz)\mu z}{r^5}\right\}\frac{t^2}{3.4} + \&c.\right\}$$

A, $\frac{dA}{dt}$, $\frac{d^2A}{dt^2}$, &c. are properly the values of those quantities when $t = 0$, the index at foot having been suppressed for convenience; they may be obtained by differentiation, but more simply from the finite differences of the quantity A. From δx, δy, δz; $(\delta\lambda)$, $(\delta\beta)$ the perturbations in geocentric longitude and latitude are easily deduced.

$$\delta x = \delta\rho\cos\lambda - \rho\sin\lambda\,(\delta\lambda)$$

$$\delta y = \delta\rho\sin\lambda + \rho\cos\lambda\,(\delta\lambda)$$

$$\delta z = \delta\rho\tan\beta + \frac{\rho\,(\delta\beta)}{\cos^2\beta}$$

$$\rho\,(\delta\lambda) = -\sin\lambda\,\delta x + \cos\lambda\,\delta y \quad (67)$$

$$\rho\,(\delta\beta) = \{\cos\lambda\,\delta x + \sin\lambda\,\delta y\}\sin\beta\cos\beta - \cos^2\beta\,\delta z$$

If $\delta\lambda$, $\delta\beta$, as before, are the differences between the observed geocentric longitudes and latitudes of the comet and those calculated with approximate values (ι), (ν), (ϖ), (e), (ε) of the constants in the osculating ellipse,

$$\delta\lambda - (\delta\lambda) = \frac{d\lambda}{d\iota}\delta\iota + \frac{d\lambda}{d\iota}\delta\nu + \frac{d\lambda}{d\varpi}\delta\varpi + \frac{d\lambda}{de}\delta e + \frac{d\lambda}{d\varepsilon}\delta\varepsilon \quad (68)$$

$$\delta\beta - (\delta\beta) = \frac{d\beta}{d\iota}\delta\iota + \frac{d\lambda}{d\nu}\delta\nu + \frac{d\beta}{d\varpi}\delta\varpi + \frac{d\beta}{de}\delta e + \frac{d\beta}{d\varepsilon}\delta\varepsilon \quad (69)$$

Each observation furnishes two equations of this kind, and the quantities $\delta\iota$, $\delta\nu$, &c. being obtained by elimination, the resulting values $(\iota) + \delta\iota$, $(\nu) + \delta\nu$, &c. are more accurate values the constants in the osculating ellipse.

5. I shall now proceed to give some numerical examples, first making use of places calculated on purpose in a known orbit, in order that the magnitude of the quantities neglected may be seen, and the errors of method may not be mixed up with the errors of observation.

In the following example the geocentric places of the comet are calculated from the following elements.

Time of perihelion passage, Nov. 29. $12^h\ 42^m\ 46^s$
or 29·5297, 1781.

Longitude of the perihelion . . .	$=16^\circ\ 3'\ 7''$
Longitude of the node	$=77\ 22\ 55$
Inclination of orbit . . . . . . : . . .	$=27\ 12\ 4$
Perihelion distance	$=\cdot9609951$

Motion retrograde.

These elements are given for the second comet of 1781. *Mémoires de l'Académie*, 1780, p. 71.

Date.		
Nov. 14·3535	$\lambda_1 = 307^\circ\ 18'\ 41''$	$\beta_1 = 55^\circ\ 15'\ 24''$ North.
	$\odot_1 = 232\ 54\ 2$	log. $R_1 = 9\cdot9948640$
19·3535	$\lambda_2 = 306\ 52\ 58$	$\beta_2 = 39\ 13\ 45$ ....
	$\odot_2 = 237\ 57\ 4$	log $R_2 = 9\cdot9944260$
24·3535	$\lambda_3 = 306\ 43\ 31$	$\beta_3 = 31\ 6\ 25$ ....
	$\odot_3 = 243\ 0\ 41$	log. $R_3 = 9\cdot9940280$

$\odot_2 = 237^\circ\ 57'\ 4''$	$\odot_2 = 237^\circ\ 57'\ 4''$	$\odot_2 = 237^\circ\ 57'\ 4''$
$\lambda_2 = 306\ 52\ 58$	$\lambda_3 = 306\ 43\ 31$	$\lambda_1 = 307\ 18\ 41$
$\odot_2 - \lambda_2 = -\ 68\ 55\ 54$	$\odot_2 - \lambda_3 = -\ 68\ 46\ 27$	$\odot_2 - \lambda_1 = -\ 69\ 21\ 37$
$\odot_3 = 243\ 0\ 41$	$\odot_1 = 232\ 54\ 2$	$\odot_1 = 232\ 54\ 2$
$\lambda_1 = 307\ 18\ 41$	$\lambda_3 = 306\ 43\ 31$	$\lambda_1 = 307\ 18\ 41$
$\odot_3 - \lambda_1 = -\ 64\ 18\ 0$	$\odot_1 - \lambda_3 = -\ 73\ 49\ 29$	$\odot_1 - \lambda_1 = -\ 74\ 24\ 39$

$$\odot_3 = 243^\circ\ 0'\ 41''\qquad \lambda_1 = 307^\circ\ 18'\ 41''\qquad \odot_3 = 243^\circ\ 0'\ 41''$$
$$\lambda_3 = 306\ 43\ 31\qquad \lambda_3 = 306\ 43\ 31\qquad \odot_1 = 232\ 54\ 2$$
$$\odot_3 - \lambda_3 = -63\ 42\ 50\qquad \lambda_1 - \lambda_3 = 35\ 10\qquad \odot_3 - \odot_1 = 10\ 6\ 39$$

$$\gamma_1 = 2\frac{\tan\beta_2 \sin(\odot_2 - \lambda_3) - \tan\beta_3 \sin(\odot_2 - \lambda_2)}{\tan\beta_1 \sin(\odot_2 - \lambda_3) - \tan\beta_3 \sin(\odot_2 - \lambda_1)}$$ See p. 16.

$$\gamma_2 = 2\frac{\tan\beta_1 \sin(\odot_2 - \lambda_2) - \tan\beta_2 \sin(\odot_2 - \lambda_1)}{\tan\beta_1 \sin(\odot_2 - \lambda_3) - \tan\beta_3 \sin(\odot_2 - \lambda_1)}$$

log. tan β_2 = 9·9119180
log. sin $(\odot_2 - \lambda_3)$ = 9·9694908
9·8814088
nearest log. = 9·8814075
·761042

log. tan β_3 = 9·7806081
log. sin $(\odot_2 - \lambda_2)$ = 9·9699526
9·7505607
nearest log. = 9·7505547
·563068

log. tan β_1 = 0·1589208
log. sin $(\odot_2 - \lambda_2)$ = 9·9699526
0·1288734
nearest log. = 0·1288514
1·34547

log. tan β_2 = 9·9119180
log. sin $(\odot_2 - \lambda_1)$ = 9·9711902
9·8831082
nearest log. = 9·8831047
·764026

log. tan β_1 = 0·1589208
log. sin $(\odot_2 - \lambda_3)$ = 9·9694908
0·1284116
nearest log. = 0·1283993
1·34404

log. tan β_3 = 9·7806081
log. sin $(\odot_2 - \lambda_1)$ = 9·9711902
9·7517983
nearest log. = 9·7517947
·56467

1·34404
·56467
log. ·77937 = 9·8917437
log. 2 = 0·3010300
9·5907137

·761042
·563068
log. ·197974 = 9·2966082
9·5907137
log. γ_1 = 9·7058945

1·34547
·76403
log. ·58144 = 9·7645049
9·5907137
log. γ_2 = 0·1737912

$$\left\{\frac{\gamma_1^2}{\cos^2\beta_1} + \frac{\gamma_2^2}{\cos^2\beta_3} - 2\gamma_1\gamma_3\{\cos(\lambda_1-\lambda_3)+\tan\beta_1\tan\beta_3\}\right\}A^2\rho_2^2$$

$$+ 2 \{\gamma_1 R_3 \cos(\odot_3 - \lambda_1) + \gamma_2 R_1 \cos(\odot_1 - \lambda_3)$$
$$- \gamma_1 R_1 \cos(\odot_1 - \lambda_1) - \gamma_2 R_3 \cos(\odot_3 - \lambda_3)\} A \rho_2 \quad (1)$$
$$+ R_1^2 + R_2^2 - 2 R_1 R_3 \cos(\odot_1 - \odot_3)$$
$$= \frac{8 \mu t^2}{r_2}\left\{1 - \frac{\mu t^2}{3 r_2^3}\right\} + \frac{4 \mu s^2 t^4}{r_2^5}$$

$$\left\{\frac{\gamma_1^2}{\cos^2\beta_1} + \frac{\gamma_2^2}{\cos^2\beta_3} - \frac{2}{A^2\cos^2\beta_2}\right\} A^2 \rho_2^2$$
$$- 2\left\{\gamma_1 R_1 \cos(\odot_1 - \lambda_1) + \gamma_2 R_3 \cos(\odot_3 - \lambda_3)\right.$$
$$\left. - \frac{2 R_2}{A}\cos(\odot_2 - \lambda_2)\right\} A \rho_2 \quad (2)$$
$$+ R_1^2 + R_3^2 - 2 R_2^2 = \frac{2 \mu t^2}{r_2}\left\{1 - \frac{\mu t^2}{12 r_2^3}\right\} + \frac{\mu s^2 t^4}{2 r_2^5}$$

$A = 1 - \dfrac{\mu t^2}{2 r_2^3} = 1$ nearly. When the time is expressed in days, log. $\mu = 6{\cdot}4711642$

log. γ_1 =	9·7058945	log. γ_2 =	0·1737912
log. $\cos\beta_1$ =	9·7557995	log. $\cos\beta_3$ =	9·9325775
	9·9500950		0·2412137
	2		2
	9·9001800		0·4824274
nearest log. =	9·9001868	nearest log. =	0·4824162
	·79467		3·03688

log. γ_1 =	9·7058945	log. γ_1 =	9·7058945
log. γ_2 =	0·1737912	log. γ_2 =	0·1737912
log. $\tan\beta_1$ =	0·1589208	log. $\cos(\lambda_1 - \lambda_3)$ =	9·9999773
log. $\tan\beta_3$ =	9·7806081		9·8796630
	9·8192146	nearest log. =	9·8796635
nearest log. =	9·8192148		·75799
	·65950		·65950
			1·41749
			2
			2·83498
		·79467 + 3·03688 =	3·83155

$$\frac{\gamma_1^2}{\cos^2\beta_1} + \frac{\gamma_2^2}{\cos^2\beta_3} - 2\gamma_1\gamma_3\{\cos(\lambda_1 - \lambda_3) + \tan\beta_1\tan\beta_3\} = {\cdot}99657$$

$$\log.\cos\beta_2 = 9{\cdot}8890902$$
2
$9{\cdot}7781804$
$\log. 2 = 0{\cdot}3010300$
$0{\cdot}5228496$
nearest log. $= 0{\cdot}5228483$
$3{\cdot}33311$
$3{\cdot}83155$

$$\cdot 49844 = \frac{\gamma_1^2}{\cos^2\beta_1} + \frac{\gamma_2^2}{\cos^2\beta_3} - \frac{2}{\cos^2\beta_2}$$

$\log.\gamma_1 = 9{\cdot}7058945$
$\log. R_3 = 9{\cdot}9940280$
$\cos(\odot_3 - \lambda_1) = 9{\cdot}6371484$
$9{\cdot}3370709$
nearest log. $= 9{\cdot}3370597$
$\cdot 217306$

$\log.\gamma_2 = 0{\cdot}1737912$
$\log. R_1 = 9{\cdot}9948640$
$\log.\cos(\odot_1 - \lambda_3) = 9{\cdot}4449448$
$9{\cdot}6136000$
nearest log. $= 9{\cdot}6135987$
$\cdot 410771$
$\cdot 217306$
$\cdot 628077$

$\log.\gamma_1 = 9{\cdot}7058945$
$\log. R_1 = 9{\cdot}9948640$
$\log.\cos(\odot_1 - \lambda_1) = 9{\cdot}4293286$
$9{\cdot}1300871$
nearest log. $= 9{\cdot}1300763$
$\cdot 134923$

$\log.\gamma_2 = 0{\cdot}1737912$
$\log. R_3 = 9{\cdot}9940280$
$\log.\cos(\odot_3 - \lambda_3) = 9{\cdot}6462604$
$9{\cdot}8140796$
nearest log. $= 9{\cdot}8140744$
$\cdot 651748$
$\cdot 134923$
$\cdot 786671$
$\cdot 628077$
$\cdot 158594$
2
$\cdot 317188$

$\log. R_2 = 9{\cdot}9944260$
$\log.\cos(\odot_2 - \lambda_2) = 9{\cdot}5556761$
$\log. 2 = 0{\cdot}3010300$
$9{\cdot}8511321$
nearest log. $= 9{\cdot}8511299$
$\cdot 709794$
$\cdot 786671$
$\cdot 076877$
2
$\cdot 153754$

$\log. R_1 = 9{\cdot}9948640$
2
$9{\cdot}9897280$
$9{\cdot}9897256$
$R_1^2 = \cdot 976635$

$\log. R_3 = 9{\cdot}9940280$
2
$9{\cdot}9880560$
nearest log. $= 9{\cdot}9880548$
$R_3^2 = \cdot 972873$
$R_1^2 = \cdot 976635$
$R_1^2 + R_3^2 = 1{\cdot}949508$

$$\begin{array}{rl} \log. R_1 = & 9{\cdot}9948640 \\ \log. R_3 = & 9{\cdot}9940280 \\ \log. \cos(\odot_3 - \odot_1) = & 9{\cdot}9932025 \\ \log. 2 = & 0{\cdot}3010300 \\ \hline & 0{\cdot}2831245 \\ \text{nearest log.} = & 0{\cdot}2831202 \\ \hline & 1{\cdot}91922 \\ R_1^2 + R_3^2 = & 1{\cdot}94951 \\ \hline & {\cdot}03029 \end{array}
\qquad
\begin{array}{rl} \log. R_2 = & 9{\cdot}9944260 \\ & 2 \\ \hline & 9{\cdot}9888520 \\ \text{nearest log.} = & 9{\cdot}9888487 \\ \hline R_2^2 = & {\cdot}974657 \\ & 2 \\ \hline & 1{\cdot}949314 \\ R_1^2 + R_3^2 = & 1{\cdot}949500 \\ \hline & {\cdot}000186 \end{array}$$

Supposing $A = 1$, and neglecting the terms multiplied by t^4,

Equation (1) is $\cdot 99657\, \rho_2^2 - \cdot 31719\, \rho_2 + \cdot 03029 = \dfrac{8\,\mu\, t^2}{r_2}$

Equation (2) is $\cdot 49844\, \rho_2^2 - \cdot 15375\, \rho_2 + \cdot 00019 = \dfrac{2\,\mu\, t^2}{r_2}$

Eliminating $\dfrac{1}{r_2}$ between these equations, we get the equation

$$\cdot 99719\, \rho_2^2 - \cdot 29781\, \rho_2 = \cdot 02953$$

This quadratic being solved gives $\rho_2 = \cdot 37715$, and the equation

$r_2^2 = \dfrac{\rho_2^2}{\cos^2 \beta_2} - 2\,\rho_2\, R_2 \cos(\odot_2 - \lambda_2) + R_2^2$ gives $r_2 = \cdot 97160$

This value of r_2, which, however, is not far from the truth, I only consider as a first approximation to be used in finding the values of the factor $1 - \dfrac{\mu\, t^2}{2\, r_2^3}$ and of the quantity $\dfrac{2\,\mu^2\, t^4}{r_2^3}$

$$\begin{array}{rl} \log. t = & 0{\cdot}6989700 \\ & 2 \\ \hline & 1{\cdot}3979400 \\ \log. \mu = & 6{\cdot}4711642 \\ \hline & 7{\cdot}8691042 \\ 3 \log. r_2 = & 9{\cdot}9624661 \\ \hline & 7{\cdot}9066381 \\ \text{nearest log.} = & 7{\cdot}9066367 \\ \hline \dfrac{\mu\, t^2}{r_2^3} = & {\cdot}008066 \end{array}
\qquad
\begin{array}{rl} \log. \dfrac{\mu\, t^2}{r_2^3} = & 7{\cdot}9066381 \\ \log. \mu\, t^2 = & 7{\cdot}8691042 \\ \log. 2 = & 0{\cdot}3010300 \\ \hline & 6{\cdot}0767723 \\ \text{nearest log.} = & 6{\cdot}0767496 \\ \hline \dfrac{2\,\mu^2\, t^4}{r_2^3} = & {\cdot}0001193 \end{array}$$

$$1 - \frac{\mu\, t^2}{2\, r_2^3} = \cdot 995967 = A$$

Although these values of A and $\frac{2\mu^2 t^4}{r_2^3}$ are founded on the value of r_2 obtained above by the very first approximation, the error in what follows, owing to this circumstance, is insensible.

The next step is to divide $\frac{2}{\cos^2 \beta_2}$ by A^2, and $2 R_2 \cos(\odot_2 - \lambda_2)$ by A in equation (2).

$$\begin{array}{rl}
\log. \cdot 995967 = & 9{\cdot}9982450 \\
& 2 \\
\hline
& 9{\cdot}9964900 \\
\log. \frac{2}{\cos^2\beta_2} = & 0{\cdot}5228496 \\
\hline
& 0{\cdot}5263596 \\
\text{nearest log.} = & 0{\cdot}5263522 \\
\hline
& 3{\cdot}36016 \\
& 3{\cdot}83155 \\
\hline
& \cdot 47139
\end{array}
\qquad
\begin{array}{rl}
\log. \cdot 995967 = & 9{\cdot}9982450 \\
\log. 2 R_2 \cos(\odot_2 - \lambda_2) = & 9{\cdot}8511321 \\
\hline
& 9{\cdot}8528871 \\
\text{nearest log.} = & 9{\cdot}8528824 \\
\hline
& \cdot 712667 \\
& \cdot 786671 \\
\hline
& \cdot 074004 \\
& 2 \\
\hline
& \cdot 148008
\end{array}$$

Equation (1) is $\cdot 99657\, A^2 \rho_2^2 - \cdot 31719\, A \rho_2 + \cdot 03029$

$$= \frac{8\mu t^2}{r_2} - \frac{8\mu^2 t^4}{3 r_2^3} + \frac{4\mu s^2 t^4}{r_2^5}$$

Equation (2) is $\cdot 47139\, A^2 \rho_2^2 - \cdot 14801\, A \rho_2 + 00019$

$$= \frac{2\mu t^2}{r_2} - \frac{\mu^2 t^4}{6 r_2^3} + \frac{\mu s^2 t^4}{2 r_2^5}$$

Eliminating $\frac{1}{r_2}$ between these equations, we get

$$\cdot 88899\, A^2 \rho_2^2 - \cdot 27485\, A \rho_2 = \cdot 02953 + \frac{2\mu^2 t^4}{r_2^3} - \frac{2\mu^2 s^2 t^4}{r_2^5}$$

Neglecting $\frac{2\mu s^2 t^4}{r_2^5}$, which is insensible, being equal in this instance to $\cdot 0000042$,

$$\cdot 88899\, A^2 \rho_2^2 - \cdot 27485\, A \rho_2 = \cdot 02953 + \cdot 00012$$

$$= \cdot 02965$$

$$A^2 \rho_2^2 - \cdot 30914\, A \rho_2 = \cdot 033352.$$

whence $A \rho_2 = \cdot 39385 \qquad \rho_2 = \cdot 39544.$

The true value of ρ_2 is $\cdot 39659$; the error therefore of the preceding determination is $-\cdot 00115$.

From $\rho_2 = \cdot 39544$ we find $r_2 = \cdot 97703$; the true value is $\cdot 97739$; the error therefore is $-\cdot 00036$, about $\frac{1}{2713}$ of the quantity to be determined, which is about the proportion of an inch to seventy-five yards.

$$\tan \beta_3 \sin (\lambda_1 - \lambda_2) + \tan \beta_2 \sin (\lambda_3 - \lambda_1) + \tan \beta_1 \sin (\lambda_2 - \lambda_3) = \cdot 0001259$$

$$\tan \beta_1 \sin (\odot_2 - \lambda_3) - \tan \beta_3 \sin (\odot_2 - \lambda_1) = - \cdot 77937$$

These quantities have different signs and $R_2 > r_2$. See p. 15.

It is not sufficient to show that a method gives accurate results in a particular instance, which may easily take place by a compensation of errors; it is necessary to consider attentively the magnitude of the terms which are rejected.

In the expressions given by MM. Legendre and Pontécoulant $4 \mu t^2 \left\{ \frac{2}{R_2} - 1 \right\}$ is substituted for $R_1^2 + R_3^2 - R_1 R_3 \cos (\odot_1 - \odot_3)$. In this example

$$4 \mu t^2 \left\{ \frac{2}{R_2} - 1 \right\} = \cdot 03035$$

$$R_1^2 + R_3^2 - 2 R_1 R_3 \cos (\odot_1 - \odot_3) = \cdot 03029$$

The difference is hardly to be neglected even in this instance, in which the interval t is small; but the error arising from this circumstance is not so great as that which arises from supposing the factor A equal to unity, which evidently amounts in this instance in equation (1) to about ·00159 *. The error which arises from this supposition (namely, that $A = 1$) in equation (2) is much greater. There is no difficulty in retaining the quantity A, in Legendre's equation, because it accompanies ρ_2.

The quantity $\frac{8 \mu^2 t^4}{3 r_2^3}$ is also sensible, being equal to ·000159; the quantity $\frac{4 \mu s^2 t^4}{r_2^5}$, which is less the nearer the comet is to its perihelion at the time of the second observation, is insensible, being equal in this instance to ·0000084.

When the interval between the observations is five days, $\mu t^2 = \cdot 00740$, which introduces into equation (1) the term $\cdot 0296 \times \frac{1}{a}$, and this term even in the case of the comet of Halley (in which a is nearly 20) is not insensible. In such cases as those of the comets of Encke and Biela, the term multiplied by the reciprocal of the semi-axis major would ma-

* The error $= A \rho_2 - \rho_2 = \cdot 39385 - \cdot 39544 = - \cdot 00159$.

terially affect the result. It seems therefore hardly desirable to attempt more than a rough approximation to the remaining elements, when the reciprocal of the semi-axis major is neglected, the period being unknown. Even when the approximate elements are corrected by the equations (14) and (15) of page 11, and the semi-axis major is known approximately, a slight error in that quantity may sensibly affect the remaining elements.

The expressions $\rho_1 = A\,\gamma_1\,\rho_2$, $\rho_3 = A\,\gamma_2\,\rho_2$ are only close approximations. In this example the true value of $\frac{\rho_1}{\rho_2}$ is $\cdot$506114, that of $\frac{\rho_3}{\rho_2}$ is 1$\cdot$48587 $A = \cdot 996038$.

According to the preceding expressions

$$\frac{\rho_1}{\rho_2} = A\,\gamma_1 = \cdot 506023, \text{ error} = -\cdot 00009$$

$$\frac{\rho_3}{\rho_2} = A\,\gamma_2 = 1\cdot 48617, \text{ error} = \cdot 0003$$

More accurate values of the ratios $\rho_1 : \rho_2$, $\rho_3 : \rho_2$ may be obtained analytically. See p. 20.

The errors above, although extremely small, have still a sensible effect on the value of ρ_2 deduced from equations (1) or (2), or the quadratic; fortunately, however, these equations afford values of the unknown quantities sufficiently accurate to be used in obtaining a first approximation to the elements of the orbit.

The equation used in the method of Mr. Ivory is

$$3\cdot 86117\,\rho_1^2 - \cdot 62434\,\rho_1 + \cdot 03029 = \frac{16\,\mu\,t^2}{r_1 + r_3}\left\{1 + \frac{\mu\,t^2}{3(r_1 + r_3)^2}\right\}$$

which is obtained from equation (1), p. 46, substituting $\frac{\rho_1}{A\,\gamma_1}$ for ρ_2, (See p. 26.) and which must be combined with the equations

$$r_1^2 = 3\cdot 0789\,\rho_1^2 - \cdot 53115\,\rho_1 + \cdot 97663$$

$$r_3^2 = 11\cdot 7662\,\rho_1^2 - 2\cdot 5657\,\rho_1 + \cdot 97287$$

In the method whether of Olbers or of Mr. Ivory, A is eliminated; but this is done at the expense of the introduction of a third unknown quantity, it being necessary to determine simultaneously, ρ_1, r_1, and r_3. In the methods of Legendre,

Lagrange and Laplace, ρ_2 and r_2 must be determined simultaneously.

The accurate values of the different quantities obtained from the elements of the orbit, for which see p. 41, are

$$\log.\ \rho_1 = 9{\cdot}3025986 \qquad \log.\ r_1 = 9{\cdot}9986915$$
$$\log.\ \rho_2 = 9{\cdot}5983505 \qquad \log.\ r_2 = 9{\cdot}9900682$$
$$\log.\ \rho_3 = 9{\cdot}7703336 \qquad \log.\ r_3 = 9{\cdot}9846516$$
$$k^2 = {\cdot}99564\, A^2 \rho_2^2 - {\cdot}31756\, A \rho_2 + {\cdot}03029$$
$$\Delta^2 . r^2 = {\cdot}47092\, A^2 \rho_2^2 - {\cdot}14852\, A \rho_2 + {\cdot}00019$$
$$4\,\Delta^2 . r^2 - k^2 = {\cdot}88804\, A^2 \rho_2^2 - {\cdot}27652\, A \rho_2 - {\cdot}02953$$

which may be compared with the equations of p. 46.

$$A = {\cdot}996038 \qquad \log.\ A = 9{\cdot}9982759$$

Calculation of the Elements.

From $\rho_2 = {\cdot}39544 \quad A = {\cdot}995967$ we find

$$\log.\ \rho_1 = 9{\cdot}3012253 \quad r_1^2 = {\cdot}99362 \quad \log.\ r_1 = 9{\cdot}9986101$$
$$\log.\ \rho_3 = 9{\cdot}7691220 \quad r_3^2 = {\cdot}93056 \quad \log.\ r_3 = 9{\cdot}9843722$$

$$\frac{s^2}{\mu} = \frac{(r_3^2 - r_1^2)^2}{16\,\mu\, t^2}\left\{1 + \frac{\mu\, t^2}{3\, r_2^3}\right\} \quad \text{whence } \frac{s^2}{\mu} = {\cdot}033684$$

$$D = r_2 - \frac{s^2}{2\,\mu} = {\cdot}96019; \text{ the accurate value is } {\cdot}9609951.$$

The error of this determination is occasioned by the errors of r_1^2 and r_3^2, and not by the quantities neglected in the equation

$$\frac{s^2}{\mu} = \frac{(r_3^2 - r_1^2)^2}{16\,\mu\, t^2}\left\{1 + \frac{\mu\, t^2}{3\, r_2^3}\right\}$$

$$\cos^2 \frac{v_1}{2} = \frac{D}{r_1} \quad \text{gives } \frac{v_1}{2} = 11^\circ\ 2'\ 58''$$

$$t = \frac{D^{\frac{3}{2}} \sqrt{2}}{\sqrt{\mu}}\left\{\tan \frac{1}{2} v_1 + \frac{1}{3} \tan^3 \frac{1}{2} v_1\right\}$$

gives $\qquad t = 15{\cdot}2968$

According to which, the time of the perihelion passage is November 29·6503.

The equations $\sin \beta' = \frac{\rho \tan \beta}{r}$,

$$\sin C' S T = \frac{\rho \sin C' T S}{r \cos \beta'} = \frac{\rho \sin (\lambda - \odot)}{r \cos \beta'}$$

$$\lambda' = C' S \Upsilon = T S \Upsilon - C' S T = \odot - 180^\circ - C' S T$$

give

$\beta_1' = 16^\circ\ 49'\ 25''$ $\quad C'_1 S T_1 = 11^\circ\ 39'\ 12''$ $\quad \lambda_1' = 41^\circ\ 14'\ 50''$

$\beta_3' = 21\ \ 34\ \ 0$ $\quad C'_3 S T_3 = 35\ \ 57\ \ 58$ $\quad \lambda_3' = 27\ \ 2\ \ 43$

The accurate values calculated from the known elements of the orbit are,

$\beta_1' = 16^\circ\ 52'\ 31''$ $\quad C'_1 S T_1 = 11^\circ\ 41'\ 31''$ $\quad \lambda_1 = 41^\circ\ 12'\ 31''$

$\beta_3' = 21\ \ 36\ \ 55$ $\quad C'_3 S T_3 = 36\ \ 4\ \ 11$ $\quad \lambda_3 = 26\ \ 56\ \ 41$

The equation

$$\tan \nu = \frac{\tan \beta_1' \sin \lambda_3' - \sin \lambda_1' \tan \beta_3'}{\tan \beta_1' \cos \lambda_3' - \cos \lambda_1' \tan \beta_3'} \text{ gives } \nu = 77^\circ\ 14'\ 26''.$$

The equation $\tan \iota = \frac{\tan \beta_1'}{\sin (\lambda_1' - \nu)}$ gives $\iota = 207^\circ\ 13'\ 33''$ or $\iota = 27^\circ\ 13'\ 33''$, *motion retrograde.*

The annexed figure is intended to represent the position of the orbit of the comet in space. $S \Upsilon$ is the line from which longitudes are reckoned, T the earth, S the sun, C the comet, S ☊ the line of nodes, $\Upsilon S ☊ = 77^\circ\ 14'\ 26''$. S D is the line of apsides, $S D = D = \cdot 96019$, $C_1 S D = v_1 = 22^\circ\ 3'\ 56''$.

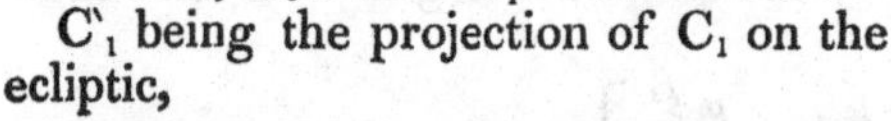

C'_1 being the projection of C_1 on the ecliptic,

$$C'_1 S \Upsilon = \lambda_1' = 41^\circ\ 14'\ 50''$$

$C'_1 S ☊ = 35^\circ\ 59'\ 36''$; by spherical trigonometry

$$\tan C S ☊ = \frac{\tan C' S ☊}{\cos \iota}, \quad C_1 S ☊ = 39^\circ\ 14'\ 39'',$$

$$D S ☊ = 39^\circ\ 14'\ 39'' + 22^\circ\ 3'\ 56'' = 61^\circ\ 18'\ 35''$$

The longitude of the perihelion $= 77^\circ\ 14'\ 26'' - 61^\circ\ 18'\ 35''$
$= 15\ \ 55\ \ 51$

Ex. 2. This example is given by M. Legendre, *Nouvelles*

Méthodes, &c. p. 33, and by Mr. Ivory, *Phil. Tran.* 1814, p. 169. The places refer to the same comet as before, and to the same times, and were observed by Mechain. *Mémoires de l'Acad.* 1780.

1781.			
Nov. 14·3535	$\lambda_1 = 307^\circ\ 14'\ 45''$	$\beta_1 = 55^\circ\ 17'\ 9''$ North.
	$\odot_1 = 232\ 54\ 2$	$\log. R_1 = 9{\cdot}994864$
19·3535	$\lambda_2 = 306\ 51\ 26$	$\beta_2 = 39\ 14\ 48$
	$\odot_2 = 237\ 37\ 4$	$\log. R_2 = 9{\cdot}994426$
24·3535	$\lambda_3 = 306\ 42\ 20$	$\beta_3 = 31\ 4\ 52$
	$\odot_3 = 243\ 0\ 41$	$\log. R_3 = 9{\cdot}994028$

$\log. \gamma_1 = 9{\cdot}7070144$ $\log. \gamma_2 = 0{\cdot}1734360$

Supposing the factor A equal to unity,

Equation (1) is $99015\,\rho_2^2 - {\cdot}31637\,\rho_2 + {\cdot}03029 = \frac{8\,\mu\,t^2}{r_2}$

Equation (2) is ${\cdot}49544\,\rho_2^2 - {\cdot}15376\,\rho_2 + {\cdot}00019 = \frac{2\,\mu\,t^2}{r_2}$

In the notation of M. Legendre ρ is the distance of the comet from the earth (not the curtate distance), if for distinction the curtate distance be called ρ', $\rho' = \rho \cos\beta$ and dividing the coefficients of the former of the two last equations by $8\,\mu\,t^2$, it becomes $10{\cdot}0339\,\rho_2^2 - 4{\cdot}1399\,\rho_2 + {\cdot}51180 = \frac{1}{r_2}$

M. Legendre has

$$10{\cdot}03409\,\rho_2^2 - 4{\cdot}14526\,\rho + {\cdot}512917 = \frac{1}{r}$$

Ex. 3. This example is given in the *Abhandlung uber die leichteste methode*, &c. p. 54, and in the *Berliner Ast. Jahrbuch*, 1833. p. 254.

	h m		
Sept. 4.	14 0	$\lambda_1 = 80^\circ\ 56'\ 11''$	$\beta_1 = 17^\circ\ 51'\ 39''$
		$\odot_1 = 162\ 42\ 5$	$\log. R_1 = {\cdot}003132$
8.	14 0	$\lambda_2 = 101\ 0\ 54$	$\beta_2 = 22\ 5\ 2$
		$\odot_2 = 166\ 35\ 31$	$\log. R_2 = {\cdot}002665$
12.	14 0	$\lambda_3 = 124\ 19\ 22$	$\beta_3 = 22\ 43\ 55$
		$\odot_3 = 170\ 29\ 20$	$\log. R_3 = {\cdot}002184$

$\log. \gamma_1 = {\cdot}0605539$ $\log. \gamma_2 = {\cdot}0013578$

Supposing A equal to unity,

Equation (1) is ${\cdot}65669\,\rho_2^2 - {\cdot}12527\,\rho_2 + {\cdot}01867 = \frac{8\,\mu\,t^2}{r_2}$

Equation (2) is ${\cdot}33038\,\rho_2^2 - {\cdot}06187\,\rho_2 - {\cdot}00007 = \frac{2\,\mu\,t^2}{r_2}$

ρ_2 being the curtate distance of the comet at the second observation. If ρ_1 is the curtate distance of the comet at the first observation,

$$\cdot 49688\,\rho_1^2 - \cdot 10896\,\rho_1 + \cdot 01867 = k^2$$
$$= \frac{16\,\mu\,t^2}{r_1 + r_3}\left\{1 + \frac{\mu\,t^2}{3\,(r_1 + r_3)^2}\right\}$$

which in the method of Olbers (or rather that of Mr. Ivory), is to be combined with the equations

$$\rho_3 = M\,\rho_1 \qquad M = \frac{\gamma_2}{\gamma_1} \qquad \log. M = 9{\cdot}9408039$$

$$r_1^2 = \frac{\rho_1^2}{\cos^2\beta_1} - 2\,R_1\,\rho_1\cos(\odot_1 - \lambda_1) + R_1^2$$

$$r_3^2 = \frac{\rho_3^2}{\cos^2\beta_3} - 2\,R_3\,\rho_3\cos(\odot_3 - \lambda_3) + R_3^2$$

In the treatise of Olbers, p. 56, the equation given is

$$49702\,\rho_1^2 - \cdot 10954\,\rho_1 + \cdot 01868 = k^2$$

$\log. M = 9{\cdot}940836$, p. 55.

Ex. 4. The following example is worked at length in the Memoirs of the Astronomical Society, vol. iv. part i.

	h			
May 1.	22	$\lambda_1 = 172°\ 22'\ 16''$		$\beta_1 = 31°\ 22'\ 34''$
		$\odot_1 = 41\ \ 9\ \ 7$	log.	$R_1 = \cdot 0038021$
18.	10	$\lambda_2 = 157\ 52\ 30$		$\beta_2 = 15\ 54\ 44$
		$\odot_2 = 57\ \ 4\ \ 0$	log.	$R_2 = \cdot 0053859$
June 3.	22	$\lambda_3 = 157\ 36\ 22$		$\beta_3 = 13\ \ 5\ 21$
		$\odot_3 = 72\ 53\ 29$	log.	$R_3 = \cdot 0064878$

$a = 18{\cdot}018467$

$\log. \gamma_1 = 9{\cdot}4258236 \qquad \log. \gamma_2 = 0{\cdot}2438063$

Supposing A equal to unity,

Equation (1) is

$$2{\cdot}3006\,\rho_2^2 - 1{\cdot}6382\,\rho_2 + \cdot 30631 = 4\,\mu\,t^2\left\{\frac{2}{r_2} - \frac{1}{a}\right\}$$

Equation (2) is

$$1{\cdot}1744\,\rho_2^2 - \cdot 7328\,\rho_2 - \cdot 00223 = 2\,\mu\,t^2\left\{\frac{1}{r_2} - \frac{1}{a}\right\}$$

When the semi-axis major is known, the trouble of the numerical calculation is not at all increased by taking it into account.

THE END.

PRINTED BY RICHARD TAYLOR,
RED LION COURT, FLEET STREET.

ON

THE THEORY OF THE MOON,

AND ON

THE PERTURBATIONS OF THE PLANETS.

BY J. W. LUBBOCK, ESQ. F.R.S.

LONDON:

PRINTED FOR C. KNIGHT, PALL MALL EAST; AND
J. AND J. J. DEIGHTON, CAMBRIDGE.

1833.

PRINTED BY RICHARD TAYLOR,
RED LION COURT, FLEET STREET.

PREFACE.

THE researches of mathematicians who, since the time of the immortal Newton, have contributed to improve the science of physical astronomy, are described in the fifth volume of the *Mécanique céleste*, in the introduction to the *Théorie analytique du système du monde*, and in the *Essai historique sur le problème des trois corps.* I shall therefore briefly state in this introduction, wherein the methods of this treatise differ from those previously employed.

The analytical investigation of the theory of the moon was undertaken nearly at the same time by Clairaut, D'Alembert, and Euler, towards the middle of the last century. Laplace adopted the method of Clairaut, and carried the approximation much further; so that his coefficients agree far better with observation. The approximation has since been pushed to an almost incredible extent by M. Damoiseau, without any alteration however in the method employed.

MM. Carlini and Plana presented to the Bureau des Longitudes of France at the same time with M. Damoiseau, a memoir upon the Theory of the Moon; and they have long announced their intention of publishing an elaborate work upon this subject, the very great importance of which may be inferred from various notices which are given in the *Correspondance Astronomique* of Baron de Zach. The memoir of MM. Carlini and Plana has not yet, I believe, been published; but it appears from what is stated in *Zach's Corr.* vol. iv. p. 4, that they also integrate by successive approximations the differential equations, in which the true longitude is the independent variable, but that they do not employ the method of indeterminate coefficients.

Laplace insists particularly upon the necessity of having recourse to the equations in which the true longitude is the independent variable: but notwithstanding the respect which is due to so great authority, I am convinced, after much reflection, that the method which I have submitted to the Royal Society, and which forms the groundwork of this Essay, is on many accounts to be preferred. According to this method, the expressions for the parallax, and the true longitude obtained by the reversion of series in the method of Clairaut, are arrived at directly. The differential equations which I employ, and in which the mean longitude is the independent variable, are far more simple than those in which the true longitude is the independent variable; and I avoid the substitution of the true longitude of the sun in terms of the true longitude of the moon, which is necessary in the other method. The developments required in order to take into account the terms of the order of the square of the disturbing force are tedious, but they may all be formed with great facility, and with scarcely any risk of error, by means of the Table which I have given for the purpose. This Table may be used with equal advantage in either method.

In order to give numerical examples, I have shown how, when the most sensible terms are retained, some of the coefficients may be found; but even with the advantages which this method possesses, to determine the coefficients of the inequalities with that degree of accuracy which the present state of astronomy demands, will require great labour, and the sacrifice of more time than I can devote to this subject; nor must it be forgotten, that if the method originally proposed by Clairaut has at last been brought to perfection, this result has been obtained through the efforts of the greatest mathematicians, continued for more than half a century. The accuracy however of the Lunar tables is a matter of such extreme importance, and the calculations required in order to deduce from theory the numerical values of the coefficients of the inequalities are of such prodigious complexity, that no method of verification should be neglected by which the exactness of the coefficients now in use can be ascertained, or any new inequalities can be detected; and although, indeed, the identity of results obtained by different methods can be anticipated *à priori*, yet it is always interesting to trace their coincidences.

It seems, moreover, desirable to introduce into the science of physical astronomy a uniform system, by employing, if possible, methods which embrace the motions of all the heavenly bodies. With this view I have shown how the perturbations of the planets and of the satellites of Jupiter may be obtained from the equations employed in my theory of the moon; and how, when the square of the disturbing function is neglected, the table before referred to may be used in the planetary theory, both in the integrations required, and also in the development of the disturbing function.

The resulting expressions for the planetary perturbations differ in form from those of the *Mécanique céleste.* I obtain the inequalities of the reciprocal of the radius vector, by the method of indeterminate coefficients, from the equation

$$\frac{d^2 r^2}{2\,d t^2} - \frac{\mu}{r} + \frac{\mu}{a} + 2\int d R + r\left(\frac{d R}{d r}\right) = 0$$

without the intervention of an auxiliary variable, as in the *Mécanique céleste*, vol. i.; and I obtain the inequalities of longitude from the equation

$$d\lambda = \frac{h}{r^2}\left\{1 - \frac{1}{h}\int \frac{d R}{d \lambda} d t\right\} d t$$

instead of the equation

$$\delta\lambda = \frac{\dfrac{2 r d.\delta r + d r \delta r}{a^2 n\, d t} + \dfrac{a n}{\mu}\left\{3\iint d t\, d R + 2\int r\left(\dfrac{d R}{d r}\right) d t\right\}}{\sqrt{1-e^2}}$$

employed in the *Mécanique céleste,* and in other works where the same methods are adopted.

I have shown how the development of the disturbing function may be obtained in a more simple form than that given in the *Mé-*

canique céleste, through the binomial theorem. The expression which at first results is very complicated; but admits of simplification, in consequence of equations of condition which obtain between the quantities of which the general symbol is b. This method is not readily applicable to the terms which are multiplied by the higher powers of the eccentricities; but I have given another method, according to which, the terms multiplied by any given powers of the eccentricities are obtained in terms of those of the inferior orders, and which is peculiarly advantageous in the lunar theory.

The stability of a system of bodies in motion subject to the law of mutual gravitation which obtains in nature, is a question of great interest. It appears to me that it may be established through considerations different from those hitherto employed, namely by observing that the quantities which have been called c and g in the Lunar Theory are *rational*, so that no imaginary angles or exponentials are introduced in the expressions for the coordinates of the body in motion, which are thus functions of periodic quantities. In the theory of the moon I apprehend this theorem is certainly independent of the direction of the moon's motion.

Various other questions are treated incidentally in the following pages; my object being to give a connected, but brief sketch of some researches printed in the *Philosophical Transactions*, and to improve them, as far as I am able.

23, *St. James's Place, Feb.* 1833.

CONTENTS.

ON

THE THEORY OF THE MOON,

AND ON

THE PERTURBATIONS OF THE PLANETS.

1. Let x, y, z denote the rectangular co-ordinates of a planet.

r	....	distance from the sun
r'	....	distance from the sun projected upon the plane $x\ y$
* λ	....	longitude reckoned upon the plane of its orbit
λ'	....	longitude reckoned upon the plane $x\ y$
s	....	tangent of the latitude
υ	......	a variable, which in the elliptic theory is the eccentric anomaly
m	......	the mass
a	....	semiaxis major
e	....	eccentricity
ϖ	....	longitude of the perihelion
ε	....	longitude of the epoch
ν	....	longitude of the ascending node
ι	....	inclination of the orbit to the plane $x\ y$
α	......	a constant quantity which accompanies υ

(the above all "of a planet")

M ...... the mass of the sun.

* Laplace uses the letter v to denote longitude, u the eccentric anomaly, and φ the inclination of the orbit to a fixed plane; but as v is very frequently used to

$$M + m = \mu, \quad \sqrt{\frac{\mu}{a^3}} = n.$$

$x = \grave{r} \cos \grave{\lambda}$, $y = \grave{r} \sin \grave{\lambda}$, $z = \grave{r}\, s$, and in the elliptic motion $\grave{r}^2 \mathrm{d} \grave{\lambda} = h \,\mathrm{d} t.$

$$R = m_{,} \left\{ \frac{x x_{,} + y y_{,} + z z_{,}}{\{x_{,}^2 + y_{,}^2 + z_{,}^2\}^{\frac{3}{2}}} - \frac{1}{\{(x - x_{,})^2 + (y - y_{,})^2 + (z - z_{,})^2\}^{\frac{1}{2}}} \right\}$$

$$= m_{,} \left\{ \frac{\grave{r} \{\cos(\grave{\lambda} - \grave{\lambda}_{,}) + s s_{,}\}}{\grave{r}_{,}^2 (1 + s_{,}^2)^{\frac{3}{2}}} - \frac{1}{\left\{\grave{r}^2 (1 + s^2) - 2 \grave{r} \grave{r}_{,} \{\cos(\grave{\lambda} - \grave{\lambda}_{,}) + s s_{,}\} + \grave{r}_{,}^2 (1 + s_{,}^2)\right\}^{\frac{1}{2}}} \right\}$$

$$= - m_{,} \left\{ \begin{aligned} & \frac{1}{\grave{r}_{,} \sqrt{1 + s_{,}^2}} + \frac{\grave{r}^2}{4 \grave{r}_{,}^3} \{1 + 3 \cos(2\grave{\lambda} - 2\grave{\lambda}_{,}) + 12\, s s_{,} \cos(\grave{\lambda} - \grave{\lambda}_{,}) - 2 s^2\} \\ & + \frac{\grave{r}^3}{8 \grave{r}_{,}^4} \{3 (1 - 4 s^2) \cos(\grave{\lambda} - \grave{\lambda}_{,}) + 5 \cos(3\grave{\lambda} - 3\grave{\lambda}_{,})\} \\ & + \frac{\grave{r}^4}{64 \grave{r}_{,}^5} \{9 + 20 \cos(2\grave{\lambda} - 2\grave{\lambda}_{,}) + 35 \cos(4\grave{\lambda} - 4\grave{\lambda}_{,})\} \end{aligned} \right\}$$

$$\left(\frac{\mathrm{d} R}{\mathrm{d} \grave{r}}\right) = m_{,} \left\{ \frac{\cos(\grave{\lambda} - \grave{\lambda}_{,}) + s s_{,}}{\grave{r}_{,}^2 (1 + s_{,}^2)^{\frac{3}{2}}} + \frac{\grave{r} (1 + s^2) - \grave{r}_{,} \{\cos(\grave{\lambda} - \grave{\lambda}_{,}) + s s_{,}\}}{\left\{\grave{r}^2 (1 + s^2) - 2 \grave{r} \grave{r}_{,} \{\cos(\grave{\lambda} - \grave{\lambda}_{,}) + s s_{,}\} + \grave{r}_{,}^2 (1 + s_{,}^2)\right\}^{\frac{3}{2}}} \right\}$$

$$\left(\frac{\mathrm{d} R}{\mathrm{d} \grave{\lambda}}\right) = - m_{,} \left\{ \frac{\grave{r} \sin(\grave{\lambda} - \grave{\lambda}_{,})}{\grave{r}_{,}^2 (1 + s'^2)^{\frac{3}{2}}} - \frac{\grave{r} \grave{r}_{,} \sin(\grave{\lambda} - \grave{\lambda}_{,})}{\left\{\grave{r}^2 (1 + s^2) - 2 \grave{r} \grave{r}_{,} \{\cos(\grave{\lambda} - \grave{\lambda}_{,}) + s s_{,}\} + \grave{r}_{,}^2 (1 + s_{,}^2)\right\}^{\frac{3}{2}}} \right\}$$

$$\left(\frac{\mathrm{d} R}{\mathrm{d} s}\right) = m_{,} \left\{ \frac{\grave{r}\, s_{,}}{\grave{r}_{,}^2 (1 + s_{,}^2)^{\frac{3}{2}}} + \frac{\grave{r}^2 s - \grave{r} \grave{r}_{,} s_{,}}{\left\{\grave{r}^2 (1 + s^2) - 2 \grave{r} \grave{r}_{,} \{\cos \grave{\lambda} - \grave{\lambda}_{,}) + s s_{,}\} + \grave{r}_{,}^2 (1 + s_{,}^2)\right\}^{\frac{3}{2}}} \right\}$$

---

signify velocity, and φ geographical latitude, and as the letters of the Greek alphabet are generally used for angles, I have taken the letters λ, υ, and ι for these quantities. The letter R is used throughout this treatise in the same acceptation as in the *Mécanique céleste*. I have used the grave accent to denote projection.

Let P be the place of the planet m, S the place of the sun, S N the intersection of the orbits of m and $m_{\prime}$, S L the line from which longitudes are reckoned, $P_{\prime}'$ the projection of $P_{\prime}$ upon the plane of the orbit of P; then if the plane $x\,y$ coincide with the orbit of m, $SP = r$, $PSL = \lambda$, $NSL = \nu_{\prime}$, $P_{\prime}SN + NSL = \lambda_{\prime}$, $P_{\prime}'SL = \lambda_{\prime}'$, $SP_{\prime}' = r_{\prime}'$, $SP_{\prime} = r_{\prime} = r_{\prime}'\,(1 + s_{\prime}^2)^{\frac{1}{2}}$

$$R = m_{\prime}\left\{\frac{SP \times SP_{\prime} \cos PSP_{\prime}}{SP_{\prime}^3} - \frac{1}{\{SP^2 - 2\,SP \times SP_{\prime} \cos PSP_{\prime} + SP_{\prime}^2\}^{\frac{1}{2}}}\right\}$$

$$\left(\frac{dR}{dr}\right) = m_{\prime}\left\{\frac{SP_{\prime}' \cos PSP_{\prime}'}{SP_{\prime}^3} + \frac{SP - SP_{\prime}' \cos PSP_{\prime}'}{\{SP^2 - 2\,SP \times SP_{\prime}' \cos PSP_{\prime}' + SP_{\prime}^2\}^{\frac{3}{2}}}\right\}$$

$$r\left(\frac{dR}{dr}\right) = m_{\prime}\left\{\frac{SP \times SP_{\prime} \cos PSP_{\prime}}{SP_{\prime}^3} + \frac{SP^2 - SP \times SP_{\prime} \cos PSP_{\prime}}{\{SP^2 - 2\,SP \times SP_{\prime}' \cos PSP_{\prime}' + SP_{\prime}^2\}^{\frac{3}{2}}}\right\}$$

$$\left(\frac{dR}{d\lambda}\right) = -\,m_{\prime}\left\{\frac{SP \times SP_{\prime}' \sin PSP_{\prime}'}{SP_{\prime}^3} - \frac{SP \times SP_{\prime}' \sin PSP_{\prime}'}{\{SP^2 - 2\,SP \times SP_{\prime}' \cos PSP_{\prime}' + SP_{\prime}^2\}^{\frac{3}{2}}}\right\}$$

$$\left(\frac{dR}{ds}\right) = m_{\prime}\left\{\frac{SP \times P_{\prime}P_{\prime}'}{SP_{\prime}^3} - \frac{SP \times P_{\prime}P_{\prime}'}{\{SP^2 - 2\,SP \times SP_{\prime}' \cos PSP_{\prime}' + SP_{\prime}^2\}^{\frac{3}{2}}}\right\}$$

$$\cos PSP_{\prime} = \cos(\lambda - \nu_{\prime}) \cos(\lambda_{\prime} - \nu_{\prime}) + \cos\iota_{\prime} \sin(\lambda - \nu_{\prime}) \sin(\lambda_{\prime} - \nu_{\prime})$$

$$= \cos^2\frac{\iota_{\prime}}{2} \cos(\lambda - \lambda_{\prime}) + \sin^2\frac{\iota_{\prime}}{2} \cos(\lambda + \lambda_{\prime} - 2\,\nu_{\prime})$$

$$\tan(\lambda_{\prime}' - \nu_{\prime}) = \cos\iota_{\prime} \tan(\lambda_{\prime} - \nu_{\prime})$$

$$\sin(\lambda_{\prime}' - \nu_{\prime}) = \frac{\cos\iota_{\prime} \tan(\lambda_{\prime} - \nu_{\prime})}{(1 + \cos^2\iota_{\prime} \tan^2(\lambda_{\prime} - \nu_{\prime}))^{\frac{1}{2}}}$$

$$\cos(\lambda_{\prime}' - \nu_{\prime}) = \frac{1}{(1 + \cos^2\iota_{\prime} \tan^2(\lambda_{\prime} - \nu_{\prime}))^{\frac{1}{2}}}$$

$$rr_{\prime}' \sin(\lambda - \lambda_{\prime}') = rr_{\prime}' \{\sin\lambda \cos\lambda_{\prime}' - \cos\lambda \sin\lambda_{\prime}'\}$$

$$\cos\lambda_{\prime}' = \cos(\lambda_{\prime}' - \nu_{\prime}) \cos\nu_{\prime} - \sin(\lambda_{\prime}' - \nu_{\prime}) \sin\nu_{\prime}$$

$$= \frac{\cos(\lambda_{,}-\nu_{,})\cos\nu_{,}-\sin(\lambda_{,}-\nu_{,})\sin\nu_{,}\left(1-2\sin^2\frac{\iota_{,}}{2}\right)}{(1-\sin^2\iota_{,}\sin^2(\lambda_{,}-\nu_{,}))}$$

$$= \frac{\cos\lambda_{,}+2\sin^2\frac{\iota_{,}}{2}\sin(\lambda_{,}-\nu_{,})\sin\nu_{,}}{(1-\sin^2\iota_{,}\sin^2(\lambda_{,}-\nu_{,}))^{\frac{1}{2}}}$$

$$\sin\lambda_{,}' = \sin(\lambda_{,}'-\nu_{,})\cos\nu_{,}+\cos(\lambda_{,}'-\nu_{,})\sin\nu$$

$$= \frac{\sin(\lambda_{,}-\nu_{,})\cos\nu_{,}\left(1-2\sin^2\frac{\iota_{,}}{2}\right)+\cos(\lambda_{,}-\nu_{,})\sin\nu_{,}}{(1-\sin^2\iota_{,}\sin^2(\lambda_{,}-\nu_{,}))^{\frac{1}{2}}}$$

$$\sin P_{,}SP_{,}' = \sin\iota_{,}\sin(\lambda_{,}-\nu_{,}), \qquad r_{,}' = r_{,}\cos P_{,}SP_{,}'$$

$$= r_{,}\left(1-\sin^2\iota_{,}\sin^2(\lambda_1-\nu_1)\right)^{\frac{1}{2}}$$

Therefore,

$$rr_{,}'\sin(\lambda-\lambda_{,}') = rr_{,}\left\{\sin(\lambda-\lambda_{,})+2\sin^2\frac{\iota_{,}}{2}\sin(\lambda_{,}-\nu_{,})\cos(\lambda-\nu_{,})\right\}$$

$$= rr_{,}\left\{\cos^2\frac{\iota_{,}}{2}\sin(\lambda-\lambda_{,})+\sin^2\frac{\iota_{,}}{2}\sin(\lambda+\lambda_{,}-2\nu_{,})\right\}$$

Similarly it may be shown that

$$rr_{,}'\cos(\lambda-\lambda_{,}') = rr_{,}\left\{\cos^2\frac{\iota_{,}}{2}\cos(\lambda-\lambda_{,})+\sin^2\frac{\iota_{,}}{2}\cos(\lambda+\lambda_{,}-2\nu_{,})\right\}$$

$$SP_{,} = \frac{a_{,}(1-e_{,}^2)}{1+e_{,}\cos(P_{,}SN+NSL-\varpi_{,})} = \frac{a_{,}(1-e_{,}^2)}{1+e_{,}\cos(\lambda_{,}-\varpi_{,})}$$

$$\frac{d^2x}{dt^2}+\frac{\mu x}{r^3}+\left(\frac{dR}{dx}\right)=0$$

$$\frac{d^2y}{dt^2}+\frac{\mu y}{r^3}+\left(\frac{dR}{dy}\right)=0$$

$$\frac{d^2z}{dt^2}+\frac{\mu z}{r^3}+\left(\frac{dR}{dz}\right)=0$$

$$\frac{r'^2\,d\lambda'^2+dr'^2(1+s^2)+2r'\,s\,dr'\,ds+r'^2\,ds^2}{dt^2}$$

$$-\frac{2\mu}{r'(1+s^2)^{\frac{1}{2}}}+\frac{\mu}{a}+2\int dR=0$$

dR being the differential of R with regard only to the co-ordinates of the planet m.

$$\frac{(1+s^2)\,\mathrm{d}^2 r' - r'\,\mathrm{d}\,\lambda'^2 + 2\,s\,\mathrm{d}\,r'\,\mathrm{d}\,s + r'\,s\,\mathrm{d}^2 s}{\mathrm{d}\,t^2} + \frac{\mu}{r'(1+s^2)^{\frac{1}{2}}} + \left(\frac{\mathrm{d}\,R}{\mathrm{d}\,r'}\right) = 0$$

$$\frac{\mathrm{d}.\frac{r'^2\,\mathrm{d}\,\lambda'}{\mathrm{d}\,t}}{\mathrm{d}\,t} + \left(\frac{\mathrm{d}\,R}{\mathrm{d}\,\lambda'}\right) = 0$$

$$\frac{r'\,s\,\mathrm{d}^2 r' + 2\,r'\,\mathrm{d}\,r'\,\mathrm{d}\,s + r'^2\,\mathrm{d}^2 s}{\mathrm{d}\,t^2} + \frac{\mu\,s}{r'(1+s^2)^{\frac{3}{2}}} + \left(\frac{\mathrm{d}\,R}{\mathrm{d}\,s}\right) = 0$$

$$\frac{r'\,\mathrm{d}^2 r' - r'^2\,\mathrm{d}\,\lambda'^2}{\mathrm{d}\,t^2} + \frac{\mu}{r'(1+s^2)^{\frac{1}{2}}} + r'\left(\frac{\mathrm{d}\,R}{\mathrm{d}\,r'}\right) - s\left(\frac{\mathrm{d}\,R}{\mathrm{d}\,s}\right) = 0$$

$$\frac{r'^2\,\mathrm{d}^2 s + 2\,r'\,\mathrm{d}\,r'\,\mathrm{d}\,s - r'^2\,s\,\mathrm{d}\,\lambda'^2}{\mathrm{d}\,t^2} + (1+s^2)\left(\frac{\mathrm{d}\,R}{\mathrm{d}\,s}\right) - r'\,s\left(\frac{\mathrm{d}\,R}{\mathrm{d}\,r'}\right) = 0$$

$$\frac{\mathrm{d}^2 . r'^2(1+s^2)}{2\,\mathrm{d}\,t^2} - \frac{\mu}{r'(1+s^2)^{\frac{1}{2}}} + \frac{\mu}{a} + 2\int \mathrm{d}\,R + r'\left(\frac{\mathrm{d}\,R}{\mathrm{d}\,r'}\right) = 0$$

Making λ the independent variable instead of t,

$$r'\frac{\mathrm{d}^2 r'}{\mathrm{d}\,t^2} - r'\frac{\mathrm{d}\,r'\,\mathrm{d}^2 t}{\mathrm{d}\,t^3} - r'^2\frac{\mathrm{d}\,\lambda'^2}{\mathrm{d}\,t^2} + \frac{\mu}{r'(1+s^2)^{\frac{1}{2}}} + r'\left(\frac{\mathrm{d}\,R}{\mathrm{d}\,r'}\right) - s\left(\frac{\mathrm{d}\,R}{\mathrm{d}\,s}\right) = 0$$

$$r'^4\frac{\mathrm{d}\,\lambda'^2}{\mathrm{d}\,t^2} = h^2 - 2\int r'^2\left(\frac{\mathrm{d}\,R}{\mathrm{d}\,\lambda'}\right)\mathrm{d}\,\lambda',$$ h being a constant,

$$4\,r'^3 d\,r'\frac{\mathrm{d}\,\lambda'^2}{\mathrm{d}\,t^2} - 2\,r'^4\frac{\mathrm{d}\,\lambda'^2\,\mathrm{d}^2 t}{\mathrm{d}\,t^3} = -\,2\,r'^2\left(\frac{\mathrm{d}\,R}{\mathrm{d}\,\lambda'}\right)\mathrm{d}\,\lambda'$$

$$\left\{\frac{\mathrm{d}^2 . \frac{1}{r'}}{\mathrm{d}\,\lambda'^2} + \frac{1}{r'}\right\}\left\{1 - \frac{2}{h^2}\int r'^2\left(\frac{\mathrm{d}\,R}{\mathrm{d}\,\lambda'}\right)\mathrm{d}\,\lambda'\right\} - \frac{\mu}{h^2(1+s^2)^{\frac{1}{2}}}$$
$$-\,\frac{r'}{h^2}\left\{r'\left(\frac{\mathrm{d}\,R}{\mathrm{d}\,r'}\right) - s\left(\frac{\mathrm{d}\,R}{\mathrm{d}\,s}\right) - \frac{1}{r'}\left(\frac{\mathrm{d}\,R}{\mathrm{d}\,\lambda'}\right)\frac{\mathrm{d}\,r'}{\mathrm{d}\,\lambda'}\right\} = 0$$

$$\left\{\frac{\mathrm{d}^2 s}{\mathrm{d}\,\lambda'^2} + s\right\}\left\{1 - \frac{2}{h^2}\int r'^2\left(\frac{\mathrm{d}\,R}{\mathrm{d}\,\lambda'}\right)\mathrm{d}\,\lambda'\right\}$$
$$+\,\frac{r'^2}{h^2}\left\{(1+s^2)\left(\frac{\mathrm{d}\,R}{\mathrm{d}\,s}\right) - r'\,s\left(\frac{\mathrm{d}\,R}{\mathrm{d}\,r'}\right) - \left(\frac{\mathrm{d}\,R}{\mathrm{d}\,\lambda'}\right)\frac{\mathrm{d}\,s}{\mathrm{d}\,\lambda'}\right\} = 0$$

These equations are identical with the equations (L) of the *Mécanique céleste*, vol. i. p. 151, and those of the *Mechanism of the heavens*, p. 425, observing that

$$\frac{\mathrm{d}\,R}{\mathrm{d}\,r} = \frac{\mathrm{d}\,R}{\mathrm{d}\,u}\,\frac{\mathrm{d}\,u}{\mathrm{d}\,r'} = -\,\frac{\mathrm{d}\,R}{r'^2\,\mathrm{d}\,u}, \qquad \frac{\mathrm{d}\,r}{\mathrm{d}\,\lambda'} = \frac{\mathrm{d}\,u\,\mathrm{d}\,r}{\mathrm{d}\lambda'\,\mathrm{d}u} = -\,\frac{r^2\mathrm{d}\,u}{\mathrm{d}\,\lambda'}$$

In the latter work the letter R represents the same quantity as in the *Mécanique céleste*, but with a contrary sign.

When the disturbing force is neglected,

$$\frac{d.r'^2 d\lambda'}{dt'} = 0, \quad \frac{d^2.\frac{1}{r'}}{d\lambda'^2} + \frac{1}{r'} - \frac{\mu}{h^2(1+s^2)^{\frac{3}{2}}} = 0, \quad \frac{d^2 s}{d\lambda'^2} + s = 0$$

of which equations the integrals are,

$$r'^2 d\lambda' = h\,dt, \quad \frac{1}{r'} = \frac{\mu \cos \iota^2}{h^2}\left\{(1+s^2)^{\frac{1}{2}} + e\cos(\lambda' - \varpi)\right\}$$

$$s = \tan \iota \sin(\lambda' - \nu).$$

If $dt = \sqrt{\frac{a}{\mu}}\, r'\,dv$, and v be taken for the independent variable,

$$r'\frac{d^2 r'}{dt^2} = r' d r' \frac{d^2 t}{dt^3} - r'^2 \frac{d\lambda'^2}{dt^2} + \frac{\mu}{r(1+s^2)^{\frac{1}{2}}} + r'\left(\frac{dR}{dr'}\right) - s\left(\frac{dR}{ds}\right) = 0$$

$$d^2 t = \left(\frac{a}{\mu}\right)^{\frac{1}{2}} d r' dv$$

$$\frac{d^2 r'}{dv^2} + \left(\frac{d.r's}{dv}\right)^2 - \frac{a}{(1+s^2)^{\frac{1}{2}}} + r' + \frac{ar'}{\mu}\left\{2\int dR + r'\left(\frac{dR}{dr'}\right)\right.$$

$$\left. - s\left(\frac{dR}{ds}\right)\right\} = 0$$

If the orbit of the planet m coincide with the plane $x\,y$, s is of the order of the disturbing force, of which therefore neglecting the square, $r' = r$, $\lambda' = \lambda$.

$$\frac{d^2 r}{dv^2} - a + r + \frac{ar}{\mu}\left\{2\int dR + r\left(\frac{dR}{dr}\right)\right\} = 0.$$

When the disturbing force is neglected, the integral of this equation is $r = a\{1 - e\cos(v - \alpha)\}$, v being the eccentric anomaly.

$$nt + \varepsilon - \varpi = v - \alpha - e\sin(v - \alpha)$$

$$\tan\frac{\lambda - \varpi}{2} = \left\{\frac{1+e}{1-e}\right\}^{\frac{1}{2}} \tan\frac{v-\alpha}{2}.$$

If Q be put for the quantity $\frac{ar}{\mu}\left\{2\int dR + r\left(\frac{dR}{dr}\right)\right\}$ and the constant α, which may afterwards be replaced, be omitted for the present,

$$r = a\{1 - e\cos v\} - \sin v\int Q\cos v\,dv + \cos v\int Q\sin v\,dv$$

$$dt = \sqrt{\frac{a}{\mu}}\left\{a\{1 - e\cos \upsilon\} - \sin \upsilon \int Q \cos \upsilon \, d\upsilon\right.$$
$$\left. + \cos \upsilon \int Q \sin \upsilon \, d\upsilon\right\} d\upsilon$$

$$nt + \varepsilon - \varpi = \upsilon - e \sin \upsilon - \frac{1}{a}\left\{\int Q \, d\upsilon - \cos \upsilon \int Q \cos \upsilon \, d\upsilon\right.$$
$$\left. - \sin \upsilon \int Q \sin \upsilon \, d\upsilon\right\}$$

If $\upsilon = \mathrm{f}\,(n\,t + \varepsilon - \varpi)$ in the elliptic theory, then neglecting the square of the disturbing force,

$$\upsilon = \mathrm{f}\,(n\,t + \varepsilon - \varpi) + \frac{d.\mathrm{f}\,(n\,t + \varepsilon - \varpi)}{d\,t}\,\frac{1}{a}\left\{\int Q\, d\upsilon\right.$$
$$\left. - \cos \upsilon \int Q \cos \upsilon \, d\upsilon - \sin \upsilon \int Q \sin \upsilon \, d\upsilon\right\}$$

If $\delta\,\upsilon$, $\delta\,r$ denote the values of these parts of r and υ which are due to the first power of the disturbing force,

$$\delta\,\upsilon = \frac{d\upsilon}{an\,dt}\left\{\int Q\, d\upsilon - \cos \upsilon \int Q \cos \upsilon \, d\upsilon - \sin \upsilon \int Q \sin \upsilon \, d\upsilon\right\}$$

$$\delta\,r = a\,e \sin \upsilon\, d\upsilon - \sin \upsilon \int Q \cos \upsilon \, d\upsilon + \cos \upsilon \int Q \sin \upsilon \, d\upsilon$$
$$= \frac{\sin \upsilon}{1 - e\cos \upsilon}\int Q\,\{e - \cos \upsilon\}\, d\upsilon - \frac{e - \cos \upsilon}{1 - e\cos \upsilon}\int Q \sin \upsilon \, d\upsilon.$$

In the elliptic theory,

$$d\upsilon = \frac{n\,dt}{1 - e\cos \upsilon}, \qquad \frac{\cos \upsilon - e}{1 - e\cos \upsilon} = \cos \lambda, \qquad \frac{\sin \upsilon\,(1 - e^2)^{\frac{1}{2}}}{1 - e\cos \upsilon} = \sin \lambda$$

Therefore,

$$\delta r = \frac{an\cos\lambda \int r\sin\lambda\left\{2\int dR + r\left(\frac{dR}{dr}\right)\right\}dt - an\sin\lambda\int r\cos\lambda\left\{2\int dR + r\left(\frac{dR}{dr}\right)\right\}dt}{\mu\,(1 - e^2)^{\frac{1}{2}}}$$

which is the equation X. of the *Mécanique céleste*, vol. i. p. 258.

Multiplying the equation of p. 5, l. 3, by $\frac{2\,r'^2\,d\,\lambda'}{d\,t}$, and integrating, $\frac{r'^4\,d\,\lambda'^2}{d\,t^2} + 2\int r'^2\left(\frac{dR}{d\,\lambda'}\right)d\,\lambda' = h_0^2$.

If $h^2 = h_0^2 - 2\int r'^2\left(\frac{dR}{d\,\lambda'}\right)d\,\lambda'$, h being variable, and h_0 the value of h at a given epoch, $h\,d\,h = -\,r'^2\left(\frac{d\,R}{d\,\lambda'}\right)d\,\lambda'$, $r^2\,d\,\lambda' = h\,d\,t$, and making λ' the independent variable instead of t,

$$2\,r'\,d\,r'\,d\,\lambda' = d\,h\,d\,t + h\,d^2\,t$$

$$r'\frac{d^2 r}{d\,t^2} - \frac{r'\,d\,r\,d^2 t}{d\,t^3} - \frac{r'^2\,d\,\lambda'^2}{d\,t^2} + \frac{\mu}{r'(1+s^2)^{\frac{1}{2}}} + r'\left(\frac{d\,R}{d\,r'}\right) - s\left(\frac{d\,R}{d\,s}\right) = 0$$

$$r'\left(\frac{d^2 r'}{d\,t^2}\right) + \frac{r'\,d\,r'\,d\,h}{h\,d\,t^2} - \frac{2\,r'^2\,d\,r'^2\,d\,\lambda}{h\,d\,t^3} - \frac{r'^2\,d\,\lambda'^2}{d\,t^2} + \frac{\mu}{r'(1+s^2)^{\frac{1}{2}}}$$

$$+\,r'\left(\frac{d R}{d\,r'}\right) - s\left(\frac{d\,R}{d\,s}\right) = 0$$

$$\frac{d^2.\frac{1}{r'}}{d\,\lambda'^2} + \frac{1}{r'} - \frac{\mu}{h^2(1+s^2)^{\frac{3}{2}}} - \frac{r'}{h^2}\left\{r'\left(\frac{d R}{d\,r'}\right) - s\left(\frac{d\,R}{d\,s}\right)\right.$$

$$\left. - \frac{1}{r'}\left(\frac{d\,R}{d\,\lambda'}\right)\frac{d\,r'}{d\,\lambda'}\right\} = 0$$

$$\frac{d^2\,s}{d\,\lambda'^2} + s + \frac{r'^2}{h^2}\left\{(1+s^2)\left(\frac{d R}{d\,s}\right) - r's\left(\frac{d\,R}{d\,r'}\right) - \left(\frac{d R}{d\,\lambda}\right)\frac{d\,s}{d\,\lambda'}\right\} = 0$$

If all the constants in the elliptic integrals are supposed to vary, subject to the condition that they still satisfy these differential equations, and that the form of the first differential coefficients $\frac{d\,r}{d\,\lambda'}$, $\frac{d\,s}{d\,\lambda'}$, remains unaltered,

$$\frac{d.\frac{1}{r'}}{d\,\lambda'} = \frac{\mu\cos\iota^2}{h^2}\left\{\frac{s}{(1+s^2)^{\frac{1}{2}}}\frac{d\,s}{d\,\lambda'} - e\sin(\lambda' - \varpi)\right\},$$

$$\frac{d\,s}{d\,\lambda'} = \tan\iota\cos(\lambda' - \nu)$$

$$-\,2\,\{(1+s^2)^{\frac{1}{2}} + e\cos(\lambda' - \varpi)\}\,\{h^2\sin\iota\,d\,\iota + \cos\iota\,h\,d\,h\}$$
$$+\,h^2\cos\iota\cos(\lambda' - \varpi)\,d\,e + h^2\cos\iota\,e\sin(\lambda' - \varpi)\,d\,\varpi = 0$$

$$-\,2\left\{\frac{s\tan\iota\cos(\lambda' - \nu)}{(1+s^2)^{\frac{1}{2}}} - e\sin(\lambda' - \varpi)\right\}\;\{h^2\sin\iota\,d\,\iota + \cos\iota\,h\,d\,h\}$$
$$-\,h^2\cos\iota\sin(\lambda' - \varpi)\,d\,e + h^2\cos\iota\,e\cos(\lambda' - \varpi)\,d\,\varpi$$
$$-\,\frac{\cos\iota\,s\,r'^2}{(1+s^2)^{\frac{1}{2}}}\left\{(1+s^2)\left(\frac{d\,R}{d\,s}\right) - r'\left(\frac{d\,R}{d\,r'}\right) - \left(\frac{d\,R}{d\,\lambda'}\right)\left(\frac{d\,s}{d\,\lambda'}\right)\right\}d\,\lambda'$$
$$-\,\frac{h^2\,r}{\mu\cos\iota}\left\{r'\left(\frac{d\,R}{d\,r'}\right) - s\left(\frac{d\,R}{d\,s}\right) - \frac{1}{r'}\left(\frac{d\,R}{d\,\lambda'}\right)\left(\frac{d\,r'}{d\,\lambda'}\right)\right\}d\,\lambda' = 0$$

$$\sin(\lambda' - \nu)\frac{d\,\iota}{\cos\iota^2} - \tan\iota\cos(\lambda' - \nu)\,d\,\nu = 0$$

$$\cos(\lambda' - \nu)\frac{d\,\iota}{\cos\iota^2} + \tan\iota\sin(\lambda' - \nu)\,d\,\nu + \frac{r'^2}{h^2}\left\{(1+s^2)\left(\frac{d\,R}{d\,s}\right) - r' s\left(\frac{d\,R}{d\,r'}\right) - \left(\frac{d\,R}{d\,\lambda'}\right)\left(\frac{d\,s}{d\,\lambda'}\right)\right\} d\,\lambda' = 0$$

$$h\,d\,h = -r'^2\left(\frac{d\,R}{d\,\lambda'}\right) d\,\lambda'$$

Whence by elimination,

$$h^2\cos\iota\,d\,e + 2\left\{(1+s^2)^{\frac{1}{2}}\cos(\lambda' - \varpi) + e - \frac{\tan\iota\, s\cos(\lambda' - \nu)\sin(\lambda' - \varpi)}{(1+s^2)^{\frac{1}{2}}}\right\}$$

$$\left\{\sin\iota\cos^2\iota\, r'^2\cos(\lambda' - \nu)\left\{(1+s^2)\left(\frac{d\,R}{d\,s}\right) - r' s\left(\frac{d\,R}{d\,r'}\right) - \left(\frac{d\,R}{d\,\lambda'}\right)\left(\frac{d\,s}{d\,\lambda'}\right)\right\} + \cos\iota\, r'^2\left(\frac{d\,R}{d\,\lambda'}\right)\right\} d\,\lambda'$$

$$+ \frac{s\cos\iota\, r'^2\sin(\lambda' - \varpi)}{(1+s^2)^{\frac{1}{2}}}\left\{(1+s^2)\left(\frac{d\,R}{d\,s}\right) - r' s\left(\frac{d\,R}{d\,r'}\right) - \left(\frac{d\,R}{d\,\lambda'}\right)\left(\frac{d\,s}{d\,\lambda'}\right)\right\} d\,\lambda$$

$$+ \frac{h^2 r'}{\mu\cos\iota}\sin(\lambda' - \varpi)\left\{r\left(\frac{d\,R}{d\,r'}\right) - s\left(\frac{d\,R}{d\,s}\right) - \frac{1}{r'}\left(\frac{d\,R}{d\,\lambda'}\right)\left(\frac{d\,r}{d\,\lambda'}\right)\right\} d\,\lambda' = 0$$

$$h^2\cos\iota\, e\,d\,\varpi + 2\left\{(1+s^2)^{\frac{1}{2}}\sin(\lambda' - \varpi) + \frac{s\tan\iota\cos(\lambda' - \nu)\cos(\lambda' - \varpi)}{(1+s^2)^{\frac{1}{2}}}\right\}$$

$$\left\{\sin\iota\cos^2\iota\, r'^2\cos(\lambda' - \nu)\left\{(1+s^2)\left(\frac{d\,R}{d\,s}\right) - s\left(\frac{d\,R}{d\,r'}\right) - \left(\frac{d\,R}{d\,\lambda'}\right)\left(\frac{d\,s}{d\,\lambda'}\right)\right\} + \cos\iota\, r'^2\left(\frac{d\,R}{d\,\lambda'}\right)\right\} d\,\lambda'$$

$$- \frac{s\cos\iota\, r'^2\cos(\lambda - \varpi)}{(1+s^2)^{\frac{1}{2}}}\left\{(1+s^2)\left(\frac{d\,R}{d\,s}\right) - r'\left(\frac{d\,R}{d\,r'}\right) - \left(\frac{d\,R}{d\,\lambda'}\right)\left(\frac{d\,s}{d\,\lambda'}\right)\right\} d\,\lambda'$$

$$- \frac{h^2 r'}{\mu\cos\iota}\cos(\lambda' - \varpi)\left\{r'\left(\frac{d\,R}{d\,r'}\right) - s\left(\frac{d\,R}{d\,s}\right)\right.$$

$$-\frac{1}{r'}\left(\frac{dR}{d\lambda'}\right)\left(\frac{dr'}{d\lambda'}\right)\Big\}d\lambda=0$$

$$dv+\frac{r'^2\sin(\lambda'-v)}{h^2\tan\iota}\Big\{(1+s^2)\left(\frac{dR}{ds}\right)-r's\left(\frac{dR}{dr'}\right)$$

$$-\left(\frac{dR}{d\lambda'}\right)\left(\frac{ds}{d\lambda'}\right)\Big\}d\lambda'=0$$

$$d\iota+\frac{r^2\cos\iota^2\cos(\lambda'-v)}{h^2}\Big\{(1+s^2)\left(\frac{dR}{ds}\right)-r's\left(\frac{dR}{dr'}\right)$$

$$-\left(\frac{dR}{d\lambda'}\right)\left(\frac{ds}{d\lambda'}\right)\Big\}d\lambda'=0$$

If $x=\frac{r\cos\lambda'}{(1+s^2)^{\frac{1}{2}}}$, $\quad y=\frac{r\sin\lambda'}{(1+s^2)^{\frac{1}{2}}}$, $\quad z=\frac{rs}{(1+s^2)^{\frac{1}{2}}}$,

$$\frac{dr^2+\frac{r^2}{1+s^2}\left(\frac{ds^2}{1+s^2}+d\lambda'^2\right)}{dt^2}-\frac{2\mu}{r}+\frac{\mu}{a}+2\int dR=0$$

$$\frac{d^2r}{dt^2}-\frac{r}{1+s^2}\,\frac{\left(\frac{ds^2}{1+s^2}+d\lambda'^2\right)}{dt^2}-\frac{\mu}{r^2}+\left(\frac{dR}{dr}\right)=0$$

$$\frac{d.\frac{r^2}{1+s^2}\cdot\frac{d\lambda'}{dt}}{dt}+\left(\frac{dR}{d\lambda'}\right)=0$$

$$\frac{r^2d^2s+2r\,dr\,ds-\frac{2r^2s\,ds^2}{(1+s^2)}+r^2s\,d\lambda^2}{dt^2}+(1+s^2)^2\left(\frac{dR}{ds}\right)=0$$

Of which equations the integrals are, when the disturbing function is neglected;

$$\frac{r^2}{1+s^2}.d\lambda'=h\,dt,\qquad \frac{1}{r}=\frac{\mu\cos\iota^2}{h^2(1+s^2)^{\frac{1}{2}}}\Big\{(1+s^2)^{\frac{1}{2}}+e\cos(\lambda'-\varpi)\Big\}$$

$$s=\tan\iota\sin(\lambda'-v)$$

If $dt=\sqrt{\frac{a}{\mu}}\,r\,dv$, and v be taken for the independent variable,

$$\frac{d^2r}{dv^2}-a+r+\frac{ar}{\mu}\Big\{2\int dR+r\left(\frac{dR}{dr}\right)\Big\}=0$$

This equation is of remarkable simplicity: in the elliptic motion $r=a\{1-e'\cos(v-\alpha)\}$.

e' is accented for the present, in order to distinguish it from e.

If the constants in the elliptic integrals are supposed to vary, subject to the condition that they still satisfy these differential equations, and that the form of the first differential coefficient $\frac{d\,r}{d\,v}$ remains unaltered,

$$d^2 t = \frac{r\,d\,v\,d\,a}{2\sqrt{a\,\mu}} + \sqrt{\frac{a}{\mu}}\,d\,r\,d\,v, \quad \frac{r\,d^2 r}{d\,t^2} = \frac{r\,d\,r^2}{d\,t^2} - \frac{r\,d\,r\,d^2 t}{d\,t^3}$$

$$\frac{d^2 r}{d\,v^2} - a + r + \frac{a\,r^2}{\mu}\left(\frac{d\,R}{d\,r}\right) - \frac{d\,r\,d\,a}{2\,a\,d\,v^2} = 0$$

$$(1 - e' \cos(v - \alpha))\,d\,a - a\cos(v - \alpha)\,d\,e' - a\,e' \sin(v - \alpha)\,d\,\alpha = 0$$

$$e' \sin(v - \alpha)\,d\,a + a\sin(v - \alpha)\,d\,e' - a\,e' \cos(v - \alpha)\,d\,\alpha$$

$$+ \frac{a\,r^2}{\mu}\left(\frac{d\,R}{d\,r}\right) d\,v + \frac{a^2\,e}{\mu} \sin(v - \alpha)\,d\,R = 0$$

$$d\,a = -\frac{2\,a^2}{\mu}\,d\,R$$

$$d\,e' = \frac{a}{\mu}\{2\,e' - 2\cos(v - \alpha) - e' \sin(v - \alpha)^2\}\,d\,R$$

$$- \frac{r^2}{\mu} \sin(v - \alpha)\left(\frac{d\,R}{d\,r}\right) d\,v$$

$$e'\,d\,\alpha = \frac{a}{\mu} \sin(v - \alpha)\{e' \cos(v - \alpha) - 2\}\,d\,R$$

$$+ \frac{r^2}{\mu} \cos(v - \alpha)\left(\frac{d\,R}{d\,r}\right) d\,v$$

$$\int n\,d\,t + \varepsilon - \varpi = v - \alpha - e' \sin(v - \alpha)$$

$$d\,\varepsilon - d\,\varpi = -\,d\,\alpha - \sin(v - \alpha)\,d\,e' + e' \cos(v - \alpha)\,d\,\alpha.$$

The equation of condition which obtains between the constants a, e, ϖ, ν, ι and h, may be found from the equation

$$\frac{d\,r^2 + \frac{r^2}{1 + s^2}\left(\frac{d\,s^2}{1 + s^2} + d\,\lambda'^2\right)}{d\,t^2} - \frac{2\,\mu}{r} + \frac{\mu}{a} = 0$$

$$\frac{d\,r^2}{d\,t^2} + \frac{h^2}{r^2 \cos \iota^2} - \frac{2\,\mu}{r} + \frac{\mu}{a} = 0$$

which gives

$$\frac{\mu \cos \iota^2}{h^2} \left\{ e^2 \cos^2 \iota \sin^2 (\nu - \varpi) - \{1 + e \cos (\nu - \varpi)\} \right.$$

$$\left. \{1 - e \cos (\nu - \varpi)\} \right\} + \frac{1}{a} = 0$$

Equating the values of r which have been found,

$$\frac{h^2 \sqrt{1+s^2}}{\mu \cos \iota^2 \{\sqrt{1+s^2} + e \cos (\lambda' - \varpi)\}} = a \{1 - e' \cos (\upsilon - \alpha)\}$$

Fig. 1.

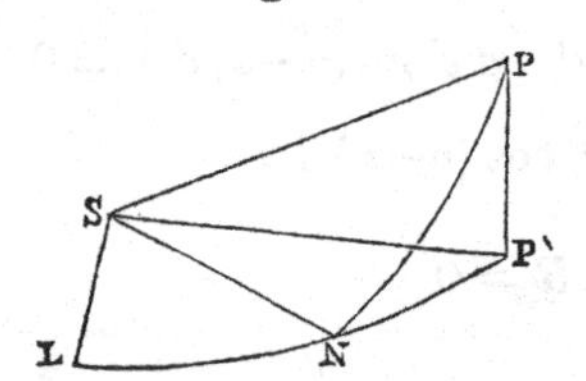

Fig. 2.

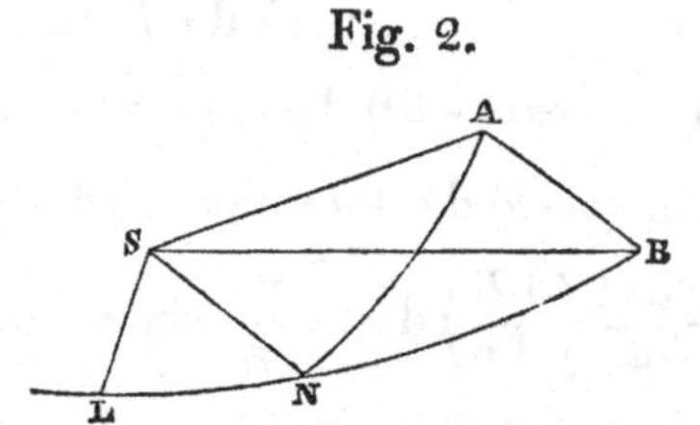

Let P be the place of the planet, P' its projection on the fixed plane L N P' (fig. 1 and 2), S N the line of nodes, S L the line from which longitudes are reckoned. The angle L S P' $= \lambda'$. Let S A be the line of apsides (fig. 2), and let S A B be a plane cutting the plane of the orbit at right angles, so that the angle

$$\text{S A B} = 90^\circ, \text{ A N B} = \iota, \text{ B S N} = \varpi - \nu.$$

$$\frac{d r^2}{d t^2} + \frac{h^2}{r^2 \cos \iota^2} - \frac{2 \mu}{r} + \frac{\mu}{a} = 0$$

$$r = a \{1 - e' \cos (\upsilon - \alpha)\}$$

When r is a maximum or minimum, $\dfrac{d r}{d t} = 0$,

$$\frac{a h^2}{\mu \cos \iota^2} - 2 a r + r^2 = 0, \text{ whence } r = a \pm \sqrt{a - \frac{h^2}{\mu \cos \iota^2}}$$

$$r = a (1 \pm e') \qquad \frac{h^2}{\mu \cos \iota^2} = a (1 - e'^2)$$

By the equation of p. 12, line 1,

$$\frac{h^2}{\mu \cos \iota^2} = a \{1 - e^2 + e^2 \sin^2 \iota \sin^2 (\nu - \varpi)\}$$

$$e'^2 = e^2 \left\{ 1 - \sin \iota^2 \sin^2 (\nu - \varpi) \right\} = e^2 \cos^2 \text{A S B}$$

The equations of p. 9. are susceptible of simplification when the square and higher powers of the disturbing function are neglected. In this case, if the orbit be supposed to coincide with the plane $x\,y$, $\tan\iota = 0$,

$$\frac{1}{r} = \frac{\mu}{h^2}\left\{1 + e\cos(\lambda - \varpi)\right\}$$

$$\frac{\mathrm{d}^2.\frac{1}{r}}{\mathrm{d}\lambda^2} + \frac{1}{r} - \frac{\mu}{h^2} - \frac{r}{h^2}\left\{r\left(\frac{\mathrm{d}R}{\mathrm{d}r}\right) - \frac{1}{r}\left(\frac{\mathrm{d}R}{\mathrm{d}\lambda}\right)\frac{\mathrm{d}r}{\mathrm{d}\lambda}\right\} = 0$$

$$-2\left\{1 + e\cos(\lambda - \varpi)\right\}h\,\mathrm{d}h + h^2\cos(\lambda - \varpi)\,\mathrm{d}e$$
$$+ h^2 e\sin(\lambda - \varpi)\,\mathrm{d}\varpi = 0$$

$$2e\sin(\lambda - \varpi)\,h\,\mathrm{d}h - h^2\sin(\lambda - \varpi)\,\mathrm{d}e + h^2 e\cos(\lambda - \varpi)\,\mathrm{d}\varpi$$
$$- \frac{r}{h^2}\left\{r\left(\frac{\mathrm{d}R}{\mathrm{d}r}\right) - \frac{1}{r}\left(\frac{\mathrm{d}R}{\mathrm{d}r}\right)\frac{\mathrm{d}r}{\mathrm{d}\lambda}\right\}\mathrm{d}\lambda = 0$$

$$h^2\,\mathrm{d}e + \left\{2\cos(\lambda - \varpi) + e + e\cos^2(\lambda - \varpi)\right\}r^2\left(\frac{\mathrm{d}R}{\mathrm{d}\lambda}\right)\mathrm{d}\lambda$$
$$+ h^2 r^2\,\frac{\sin(\lambda - \varpi)}{\mu}\left(\frac{\mathrm{d}R}{\mathrm{d}r}\right)\mathrm{d}\lambda = 0$$

$$h^2 e\,\mathrm{d}\varpi + \left\{2 + e\cos(\lambda - \varpi)\right\}\sin(\lambda - \varpi)\,r^2\left(\frac{\mathrm{d}R}{\mathrm{d}\lambda}\right)\mathrm{d}\lambda$$
$$- h^2 r^2\cos(\lambda - \varpi)\left(\frac{\mathrm{d}R}{\mathrm{d}r}\right)\mathrm{d}\lambda = 0$$

$$\mathrm{d}\iota + r^2\,\frac{\cos(\lambda - \nu)}{\mu}\left(\frac{\mathrm{d}R}{\mathrm{d}s}\right)\mathrm{d}\lambda = 0$$

$$\mathrm{d}\nu + r^2\,\frac{\sin(\lambda - \nu)}{\mu}\left(\frac{\mathrm{d}R}{\mathrm{d}s}\right)\mathrm{d}\lambda = 0$$

In order to give an application of these formulæ, suppose

$$R = \frac{m_{/}}{r}, \quad \text{then } \frac{\mathrm{d}R}{\mathrm{d}r} = -\frac{m_{/}}{r^2}, \quad \frac{\mathrm{d}R}{\mathrm{d}\lambda} = 0, \quad \mathrm{d}h = 0$$

$$\mathrm{d}e - \frac{m_{/}}{\mu}\sin(\lambda - \varpi)\,\mathrm{d}\lambda = 0$$

$$e\,\mathrm{d}\varpi + \frac{m_{/}}{\mu}\cos(\lambda - \varpi)\,\mathrm{d}\lambda = 0$$

$$\sin\varpi\,\mathrm{d}e + e\cos\varpi\,\mathrm{d}\varpi + \frac{m_{/}}{\mu}\cos\lambda\,\mathrm{d}\lambda = 0$$

$$\cos\varpi\, d e - e \sin\varpi\, d\varpi - \frac{m_{\prime}}{\mu} \sin\lambda\, d\lambda = 0$$

Integrating

$$e \sin\varpi = -\frac{m_{\prime}}{\mu} \sin\lambda + \text{constant}$$

$$e \cos\varpi = -\frac{m_{\prime}}{\mu} \cos\lambda + \text{constant}$$

$$\frac{1}{r} = \frac{\mu}{h^2}\left\{1 + e\cos(\lambda - \varpi)\right\}$$

$$= \frac{\mu}{h^2}\left\{1 + e(\cos\lambda\cos\varpi + \sin\lambda\sin\varpi)\right\}$$

if $\varpi = 0$ when $\lambda = 0$,

$$e \sin\varpi = -\frac{m_{\prime}}{\mu}\sin\lambda, \quad e\cos\varpi = -\frac{m}{\mu}\cos\lambda + (e) + \frac{m_{\prime}}{\mu}$$

(e) being a constant.

$$\frac{1}{r} = \frac{\mu}{h^2}\left\{1 - \frac{m_{\prime}}{\mu} + \left\{(e) + \frac{m_{\prime}}{\mu}\right\}\cos\lambda\right\}$$

which agrees with the known form of the result.

When $\iota = 0$, as before, by the equation, p. 12, line 14, $\mu a(1 - e^2) = h^2$; and by the equation, line 15, $e = e'$.

$$d e = \frac{a}{\mu}\left\{2e - 2\cos(v - \alpha) - e\sin(v - \alpha)^2\right\} d R$$

$$- \frac{r^2}{\mu}\sin(v - \alpha)\left(\frac{d R}{d r}\right) d v$$

$$e\, d\alpha = \frac{a}{\mu}\left\{e\cos(v - \alpha) - 2\right\}\sin(v - \alpha)\, d R$$

$$+ \frac{r^2}{\mu}\cos(v - \alpha)\left(\frac{d R}{d r}\right) d v$$

$$\int n\, d t + \epsilon - \varpi = v - \alpha - e\sin(v - \alpha)$$

$$d\epsilon - d \quad = -\{1 - e\cos(v - \alpha)\}\, d\alpha - \sin(v - \alpha)\, d e$$

$$d\epsilon - d\varpi = -\frac{r^2}{a^2\sqrt{1 - e^2}} d\varpi - \left\{\frac{r}{a(1 - e^2)} + 1\right\}\sin(v - \alpha)\, d e$$

$$\frac{\frac{h^2}{\mu}}{1 + e\cos(\lambda - \varpi)} = a\{1 - e\cos(v - \alpha)\}$$

$$\mathrm{d}\,e = -\frac{a\,n\,\mathrm{d}t}{\mu\,\sqrt{1-e^2}}\left\{2\cos(\lambda-\varpi) + e + e\cos(\lambda-\varpi)^2\right\}\left(\frac{\mathrm{d}R}{\mathrm{d}\lambda}\right)$$

$$-\frac{a^2\,n\,\mathrm{d}t\,\sqrt{1-e^2}}{\mu}\sin(\lambda-\varpi)\left(\frac{\mathrm{d}R}{\mathrm{d}r}\right)$$

$$e\,\mathrm{d}\varpi = -\frac{a\,n\,\mathrm{d}t}{\mu\,\sqrt{1-e^2}}\left\{2 + e\cos(\lambda-\varpi)\right\}\sin(\lambda-\varpi)\left(\frac{\mathrm{d}R}{\mathrm{d}\lambda}\right)$$

$$+\frac{a^2\,n\,\mathrm{d}t\,\sqrt{1-e^2}}{\mu}\cos(\lambda-\varpi)\left(\frac{\mathrm{d}R}{\mathrm{d}r}\right)$$

The two last equations, obtained indirectly, are given in the *Mécanique céleste*, vol. i. p. 347, and in the *Mechanism of the Heavens*, p. 226.

$$\mathrm{d}\,\epsilon - \mathrm{d}\varpi = -\sqrt{1-e^2}\,\mathrm{d}\varpi + \frac{2\,a^2\,n\,(1-e^2)\,\mathrm{d}t}{\mu\,\{1+e\cos(\lambda-\varpi)\}}\left(\frac{\mathrm{d}R}{\mathrm{d}r}\right)$$

The last equations serve to determine the perturbations of a comet.

Let $(\Delta\,e)_n$ be the variation of any element e during the variation $\Delta\,v$ of v at any given epoch n, then neglecting the square and higher powers of $\Delta\,v$,

$$(\Delta e)_n = \left(\frac{\mathrm{d}e}{\mathrm{d}v}\right)_n \Delta\,v$$

If the values of $(\Delta\,e)_n$ be calculated for the epochs $0, 1, 2 \ldots m$ corresponding to the values v, $v + i\,\Delta\,v$, $v + 2\,i\,\Delta\,v$, &c., differing from each other by $i\,\Delta\,v$, then the whole variation $(\delta\,e)$ of e corresponding to the variation $i\,\Delta\,v$ of v

$$= i\left\{(\Delta\,e)_0 + (\Delta\,e)_1 \ldots\ldots + (\Delta\,e)_{m-1}\right\}$$

$$+\frac{i-1}{2}\left\{(\Delta e)_m - (\Delta\,e)_0\right\}$$

$$-\frac{(i-1)(i+1)}{12\,i}\left\{(\Delta^2 e)_m - (\Delta^2\,e)_0\right\} + \&\text{c.}$$

$$= i\left\{\left(\frac{\mathrm{d}e}{\mathrm{d}v}\right)_0 + \left(\frac{\mathrm{d}e}{\mathrm{d}v}\right)_1 \ldots\ldots + \left(\frac{\mathrm{d}e}{\mathrm{d}v}\right)_{m-1}\right\}\Delta\,v$$

$$+\frac{i-1}{2}\left\{\left(\frac{\mathrm{d}e}{\mathrm{d}v}\right)_m - \left(\frac{\mathrm{d}e}{\mathrm{d}v}\right)_0\right\}\Delta\,v$$

$$-\frac{(i-1)(i+1)}{12\,i}\left\{\Delta\left(\frac{\mathrm{d}e}{\mathrm{d}v}\right)_m - \Delta\left(\frac{\mathrm{d}e}{\mathrm{d}v}\right)_0\right\}\Delta\,v.$$

When the interval $\Delta\,\upsilon$ is indefinitely diminished, $i\,\Delta\,\upsilon$ is still equal to the variation of υ between the epochs for which the quantities $\left(\frac{d\,e}{d\,\upsilon}\right)_0$, $\left(\frac{d\,e}{d\,\upsilon}\right)_1$ &c. are calculated, and

$$\delta\,e = i\,\Delta\,\upsilon \left\{ \left(\frac{d\,e}{d\,\upsilon}\right)_0 + \left(\frac{d\,e}{d\,\upsilon}\right)_1 \cdots\cdots + \left(\frac{d\,e}{d\,\upsilon}\right)_{m-1} \right.$$
$$\left. + \frac{1}{2}\left\{ \left(\frac{d\,e}{d\,\upsilon}\right)_m - \left(\frac{d\,e}{d\,\upsilon}\right)_0 \right\} - \frac{1}{12}\left\{ \Delta\left(\frac{d\,e}{d\,\upsilon}\right)_m - \Delta\left(\frac{d\,e}{d\,\upsilon}\right)_0 \right\} + \&c. \right\}$$

If the radius be taken for unity, and $i\,\Delta\,\upsilon$ is the mth part of the circumference; $i\,\Delta\,\upsilon = \frac{2 \times 3{\cdot}14159}{m}$, or, in other words, the resulting values of $\delta\,e$ and $\delta\,a$ in the equations given above must be multiplied by $2 \times 3{\cdot}14159$, and divided by $360°$ expressed in the same unit as $i\,\Delta\,\upsilon$.

n is equal to the angular circumference divided by the periodic time expressed in the same unit as t; so that if a degree be taken as the unity of angular circumference, $n = 360°$ divided by the periodic time expressed in the same unit as t.

The preceding series, which is to be found in the *Mécanique céleste*, vol. iv. p. 206, is substantially identical with that given in the *Traité elémentaire de calcul différentiel et de calcul intégral* of Lacroix, p. 578,

$$\int y\,d\,x = h\,\Sigma y + \frac{h}{2}\,y - \frac{h^2}{12}\frac{d\,y}{d\,x} + \&c.$$

In order to take a very simple application of this series, let $y = a\,x$, $P_1\,P_2 = h = 1$, and suppose it were required to determine *by quadratures* the area of the triangle $O\,P_m\,M_m = \frac{a\,x^2}{2}$.

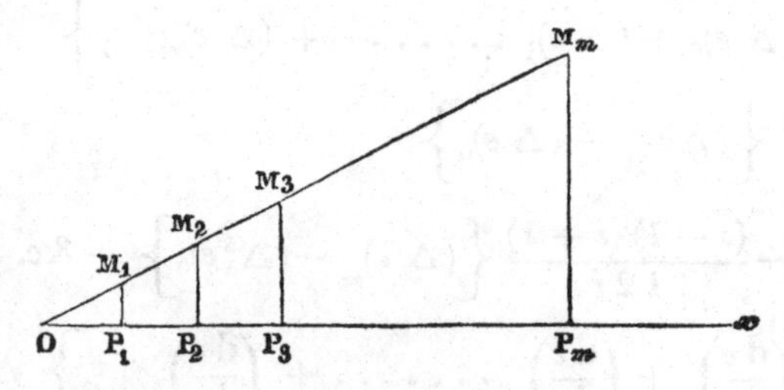

According to the preceding expression

$$\int y\,d\,x = P_1\,M_1 + P_2\,M_2 \cdots\cdots\cdots P_{m-1}\,M_{m-1}$$
$$+ \frac{P_m\,M_m}{2} - \frac{a}{12} + \text{constant}$$

constant $= \frac{a}{12}$ and

$$\int y \, d x = P_1 M_1 + P_2 M_2 \ldots\ldots\ldots + P_{m-1} M_{m-1} + \frac{P_m M_m}{2}$$

$$= a \left\{ x_1 + x_2 \ldots\ldots\ldots + x_{m-1} + \frac{x_m}{2} \right.$$

Suppose O $Pm = x_m = 4$, then

$$\int y \, d x = a \left\{ 1 + 2 + 3 + \frac{4}{2} \right\}$$

$$= 8 \, a = \frac{a \, x^2}{2}$$

In the same manner if $y = a \, x^2$, after determining the constant,

$$\int y \, d x = \Sigma y + \frac{y}{2} - \frac{a \, x}{6}$$

and between the same limits as before

$$\int y \, d x = a \left\{ 1 + 2^2 + 3^2 + \frac{4^2}{2} - \frac{4}{6} \right\}$$

$$= \frac{64 \, a}{3} = \frac{a \, x^3}{3}$$

The expression

$$h \, \Sigma y = \int y \, d x + h \, A y + h^2 B \frac{d y}{d x} + h^3 C \frac{d^2 y}{d x^2} + \&c.$$

is of very extensive application. The quantities A, B, C, &c. may be determined directly from the equation itself, by taking $y = x^n$ and making $x = 1 = h$ after integration, because then the term $\Sigma \, y$ vanishes.

By making $y = x$, $y = x^2$, $y = x^3$, &c. successively, we get

$$\frac{1}{2} + A = 0$$

$$\frac{1}{3} + A + 2 \, B = 0$$

$$\frac{1}{4} + A + 3 \, B + 3 \, . \, 2 \, C = 0$$

$$\frac{1}{5} + A + 4B + 4.3C + 4.3.2D = 0$$

whence $A = -\frac{1}{2}$, $B = \frac{1}{12}$, $C = 0$, $D = -\frac{1}{720}$

The preceding methods are used to determine by quadratures the perturbations of a comet. When, however, the perturbations are required through only a short interval, as during the time of its *apparition*, then the equations which I have given (*On the determination of the distance of a comet from the earth*, p. 40,) present advantages. By means of these equations the perturbations of the rectangular coordinates δx, δy, δz, and $\delta \frac{dx}{dt}$, $\delta \frac{dy}{dt}$, and $\delta \frac{dz}{dt}$ are obtained; and from these the changes in the elements of the osculating ellipse may be easily found, if required.

In the elliptic movement or first approximation

$\lambda = nt + \epsilon +$ a series of sines of arcs multiples of nt &c.

$\lambda_{,} = n_{,}t + \epsilon_{,} +$ a series of sines of arcs multiples of $n_{,}t$ &c.

$\frac{a}{r} =$ constant + a series of cosines.

$s =$ a series of sines.

$s_{,} =$ a series of sines.

These values being substituted in the equations of p. 9 give $\frac{de}{d\lambda}$, $\frac{dh}{d\lambda}$ and $\frac{d\iota}{d\lambda}$ each equal to a series of sines, and $\frac{d\varpi}{d\lambda}$, $\frac{d\nu}{d\lambda}$ and $\frac{d\epsilon}{d\lambda}$ each equal to a series of cosines + a constant quantity.

These equations, however, cannot be integrated directly, and the conclusions which I formerly attempted to deduce from them with respect to the stability of the system are liable to objection; because admitting that the form of the values of the quantities a, e, ϖ &c. continues the same in the successive approximations, it remains to show that none of the angles which occur under the sign *sine* or *cosine* become imaginary and give rise to exponentials.

If the sun or primary be a spheroid, ω the angle which the plane of the sun's equator makes with the plane of the orbit of the planet; and if the longitude be reckoned from the line of intersection of the sun's equator with the orbit of the planet; R is increased by the quantity $c\left\{\frac{3\sin^2\omega\sin^2\lambda - 1}{r^3}\right\}$, c being a constant dependent upon the figure of the sun; but the partial differential coefficients of this quantity, which are introduced into the values of $\mathrm{d}\, e$, $\mathrm{d}\, \varpi$, &c. do not change the form of the expressions for those quantities.

If the planet move in a medium which resists according to any power n of the velocity, if c be a constant and v the velocity, the term $2c\int v^{n+1}\,\mathrm{d}t$ must be added to $2\int \mathrm{d}R$,

$$c v^{n-1}\left\{(1+s^2)\frac{\mathrm{d}r'}{\mathrm{d}t}+r's\frac{\mathrm{d}s}{\mathrm{d}t}\right\} \text{ to } \left(\frac{\mathrm{d}R}{\mathrm{d}r'}\right),$$

$$c v^{n-1} r'^2\frac{\mathrm{d}\lambda'}{\mathrm{d}t} \text{ to } \left(\frac{\mathrm{d}R}{\mathrm{d}\lambda'}\right), \text{ and } c v^{n-1} r'\frac{\mathrm{d}.r's}{\mathrm{d}t} \text{ to } \left(\frac{\mathrm{d}R}{\mathrm{d}s}\right)$$

in the equations of p. 9.

If the orbit of the planet be supposed to coincide with the plane $x\,y$, so that $s = 0$, then by the equations of p. 9 after reductions

$$\mathrm{d}a = -2ca\left(\frac{\mu}{a}\right)^{\frac{n-1}{2}}\left\{\frac{1+e\cos v}{1-e\cos v}\right\}^{\frac{n+1}{2}}\frac{(1-e\cos v)}{n}\,\mathrm{d}v$$

$$\mathrm{d}e = -2c\left(\frac{\mu}{a}\right)^{\frac{n-1}{2}}\left\{\frac{1+e\cos v}{1-e\cos v}\right\}^{\frac{n-1}{2}}\left(\frac{1-e^2}{n}\right)\cos v\,\mathrm{d}v$$

$$e\,\mathrm{d}\varpi = -2c\left(\frac{\mu}{a}\right)^{\frac{n-1}{2}}\left\{\frac{1+e\cos v}{1-e\cos v}\right\}^{\frac{n-1}{2}}\sqrt{\frac{1-e^2}{n}}\sin v\,\mathrm{d}v$$

$$\mathrm{d}\epsilon-\mathrm{d}\varpi=2c\left(\frac{\mu}{a}\right)^{\frac{n-1}{2}}\left\{\frac{1+e\cos v}{1-e\cos v}\right\}^{\frac{n-1}{2}}\frac{\sin v}{n}\left\{\frac{1-e^3\cos v}{e}\right\}\mathrm{d}v$$

The form of these equations differs from that which obtained before, now the variations of ϖ and ϵ are periodical, while those of a and e have terms which vary with the time. The periods of the periodic inequalities of all the elliptic constants due to the action of the resisting medium are fractional parts of the periodic time of the planet.

2. If the origin of t coincides with the instant of the perihelion passage, by Lagrange's theorem, since $nt = v - e\sin v$,

$$\cos v = \cos nt - e\sin^2 nt - \frac{e^2}{2}\frac{\mathrm{d}.\sin^2 nt}{\mathrm{d}.nt} + \&\text{c.}$$

$$\sin v = \sin nt + e\sin nt\cos nt + \frac{e^2}{2}\frac{\mathrm{d}.\sin^2 nt\cos nt}{\mathrm{d}.nt}$$

If the origin of t does not coincide with the perihelion passage, $nt+\epsilon-\varpi$ must be substituted for nt; but as ϵ always accompanies nt, it may be suppressed at present for convenience, and afterwards replaced.

$$\cos v = \left\{1 - \frac{3}{8} e^2\right\} \cos(nt - \varpi) - \frac{e}{2}$$

$$+ \frac{e}{2} \cos(2nt - 2\varpi) + \frac{3}{8} e^2 \cos(3nt - 3\varpi)$$

$$\sin v = \left\{1 - \frac{e^2}{8}\right\} \sin(nt - \varpi)$$

$$+ \frac{e}{2} \sin(2nt - 2\varpi) + \frac{3}{8} e^2 \sin(3nt - 3\varpi)$$

$$\cos(v + \varpi) = \cos v \cos \varpi - \sin v \sin \varpi$$

$$= \left\{1 - \frac{e^2}{4}\right\} \cos nt - \frac{e}{2} \cos \varpi$$

$$+ \frac{e}{2} \cos(2nt - \varpi) + \frac{3}{8} e^2 \cos(3nt - 2\varpi)$$

$$\sin(v + \varpi) = \sin v \cos \varpi + \cos v \sin \varpi$$

$$= \left\{1 - \frac{e^2}{4}\right\} \sin nt - \frac{e}{2} \sin \varpi$$

$$+ \frac{e}{2} \sin(2nt - \varpi) + \frac{3}{8} e^2 \sin(3nt - 2\varpi)$$

$$r \cos \lambda = r \cos(\lambda - \varpi) \cos \varpi - r \sin(\lambda - \varpi) \sin \varpi$$

$$= a \left\{(\cos v - e) \cos \varpi - (1 - e^2)^{\frac{1}{2}} \sin v \sin \varpi\right\}$$

$$r \sin \lambda = r \sin(\lambda - \varpi) \cos \varpi + r \cos(\lambda - \varpi) \sin \varpi$$

$$= a \left\{(1 - e^2)^{\frac{1}{2}} \sin v \cos \varpi + (\cos v - e) \sin \varpi\right\}$$

$$\begin{matrix} r \cos \lambda \\ r \sin \lambda \end{matrix} = a \left\{\left(1 - \frac{e^2}{4}\right) \begin{matrix} \cos \\ \sin \end{matrix} (v + \varpi) - e \begin{matrix} \cos \\ \sin \end{matrix} \varpi \pm \frac{e^2}{4} \begin{matrix} \cos \\ \sin \end{matrix} (v - \varpi)\right\}$$

Let $nt - n_{,}t$ be called τ *

$cnt - \varpi$ .... ξ

$c_{,}n_{,}t - \varpi_{,}$ .... $\xi_{,}$

$gnt - \nu$ .... η

---

* In the Philosophical Transactions I have called these quantities t, x, z and y, in order to agree with the notation of M. Damoiseau. So much inconvenience, however, arises from using the letter t with two different significations, that I have reluctantly here adopted other letters.

The quantities c and g may be considered as equal to unity in the planetary theory: the difference is of the order of the disturbing force, as will be shown hereafter.

$$r\,r_{,}\,{}^{\cos}_{\sin}\,(\lambda-\lambda_{,})=a\,a_{,}\Big\{\Big(1-\frac{e^2}{2}-\frac{e_{,}^2}{2}\Big){}^{\cos}_{\sin}\,\tau$$

$$-\frac{3}{2}\,e\,{}^{\cos}_{\sin}\,(\tau-\xi)+\frac{e}{2}\,{}^{\cos}_{\sin}\,(\tau+\xi)$$

$$+\frac{e_{,}}{2}\,{}^{\cos}_{\sin}\,(\tau-\xi_{,})-\frac{3}{2}\,e_{,}\,{}^{\cos}_{\sin}\,(\tau+\xi_{,})-\frac{e^2}{8}\,{}^{\cos}_{\sin}\,(\tau-2\,\xi)$$

$$+\frac{3}{8}\,e^2\,{}^{\cos}_{\sin}\,(\tau+2\xi)-\frac{3}{4}\,e\,e_{,}\,{}^{\cos}_{\sin}\,(\tau-\xi-\xi_{,})-\frac{3}{4}\,e\,e_{,}\,{}^{\cos}_{\sin}\,(\tau+\xi+\xi_{,})$$

$$+\frac{9}{4}\,e\,e_{,}\,{}^{\cos}_{\sin}\,(\tau-\xi+\xi_{,})+\frac{e\,e_{,}}{4}\,{}^{\cos}_{\sin}\,(\tau+\xi-\xi_{,})+\frac{3}{8}\,e_{,}^2\,{}^{\cos}_{\sin}\,(\tau-2\,\xi_{,})$$

$$+\frac{e_{,}^2}{8}\,{}^{\cos}_{\sin}\,(\tau+2\,\xi_{,})\Big\}$$

$$\frac{r}{a}=1+\frac{e^2}{2}-e\left(1-\frac{3\,e^2}{8}\right)\cos\xi-\frac{e^2}{2}\left(1-\frac{2\,e^2}{3}\right)\cos 2\,\xi$$

$$-\frac{3\,e^3}{8}\cos 3\,\xi-\frac{e^4}{3}\cos 4\,\xi.$$

This equation is given in the *Mécanique céleste*, vol. i. p. 179. The following may be obtained by involution:

$$a^5\,r^{-5}=1+5\,e^2\left(1+\frac{21}{8}\,e^2\right)+5\,e\left(1+\frac{27}{8}\,e^2\right)\cos\xi$$

$$+\;10\,e\left(1+\frac{31}{12}\,e^2\right)\cos 2\,\xi$$

$$+\;\frac{145}{8}\,e^3\cos 3\,\xi+\frac{745}{48}\,e^4\cos 4\,\xi$$

$$a^4\,r^{-4}=1+3\,e^2+4\,e\cos\xi+7\,e^2\cos 2\,\xi$$

$$a^3\,r^{-3}=1+\frac{3}{2}\,e^2\left(1+\frac{5}{4}\,e^2\right)+3\,e\left(1+\frac{9}{8}\,e^2\right)\cos\xi$$

$$+\;\frac{9}{2}\,e^2\left(1+\frac{7}{9}\,e^2\right)\cos 2\,\xi+\frac{53}{8}\,e^3\cos 3\,\xi+\frac{77}{8}\,e^4\cos 4\,\xi$$

$$a^2\,r^{-2}=1+\frac{e^2}{2}\left(1+\frac{3}{4}\,e^2\right)+2\,e\left(1+\frac{3}{8}\,e^2\right)\cos\xi$$

$$+\frac{5}{2}e^2\left(1+\frac{2}{15}e^2\right)\cos 2\xi+\frac{13}{4}e^3\cos 3\xi+\frac{103}{24}e^4\cos 4\xi$$

$$a\,r^{-1}=1+e\left(1-\frac{e^2}{8}\right)\cos\xi+e^2\left(1-\frac{e^2}{3}\right)\cos 2\xi$$

$$+\frac{9}{8}e^3\cos 3\xi+\frac{4}{3}e^4\cos 4\xi$$

$$\frac{r^2}{a^2}=1+\frac{3e^2}{2}-2e\left(1-\frac{e^2}{8}\right)\cos\xi-\frac{e^2}{2}\left(1\quad\frac{e^2}{3}\right)\cos 2\xi$$

$$-\frac{e^3}{4}\cos 3\xi-\frac{e^4}{6}\cos 4\xi$$

$$\frac{r^3}{a^3}=1+3e^2\left(1+\frac{e^2}{8}\right)-3e\left(1+\frac{3}{8}e^2\right)\cos\xi$$

$$+\frac{5}{8}e^4\cos 2\xi+\frac{e^3}{8}\cos 3\xi+\frac{e^4}{8}\cos 4\xi$$

$$\frac{r^4}{a^4}=1+5e^2-4e\cos\xi+e^2\cos 2\xi$$

Let

$$\left\{1-\frac{2a}{a_{\prime}}\cos\theta+\frac{a^2}{a_{\prime}^2}\right\}^{-\frac{n}{2}}=\frac{1}{2}b_{n,0}+b_{n,1}\cos\theta+b_{n,2}\cos 2\theta$$

$$+\ \&c.$$

Then

$$\frac{1}{2}b_{1,0}=1+\left(\frac{1}{2}\right)^2\frac{a^2}{a_{\prime}^2}+\left(\frac{3}{2.4}\right)^2\frac{a^4}{a_{\prime}^4}+\left(\frac{3.5}{2.4.6}\right)\frac{a^6}{\epsilon_{\prime}^6}+\ \&c.$$

$$b_{1,1}=\frac{a}{a_{\prime}}+\frac{1.3}{2.4}\frac{a^3}{a_{\prime}^3}+\frac{1.3.3.5}{2.4.4.6}\frac{a^5}{a_{\prime}^5}+\ \&c.$$

$$b_{1,2}=\frac{3}{4}\frac{a^2}{a_{\prime}^2}+\frac{1.3.5}{2.4.6}\frac{a^4}{a_{\prime}^4}+\frac{1.3.3.5.7}{2.4.4.6.8}\frac{a^6}{a_{\prime}^6}+\ \&c.$$

$$b_{1,3}=\frac{3.5}{4.6}\frac{a^3}{a_{\prime}^3}+\frac{1.3.5.7}{2.4.6.8}\frac{a^5}{a_{\prime}^5}+\frac{1.3.3.5.7.9}{2.4.4.6.8.10}\frac{a^7}{a_{\prime}^7}+\ \&c.$$

$$\frac{1}{2}b_{3,0}=1+\left(\frac{3}{2}\right)^2\frac{a^2}{a_{\prime}^2}+\left(\frac{3.5}{2.4}\right)^2\frac{a^4}{a_{\prime}^4}+\left(\frac{3.5.7}{2.4.6}\right)^2\frac{a^6}{a_{\prime}^6}+\ \&c.$$

$$b_{3,1}=\frac{3a}{a_{\prime}}+\frac{3.3.5}{2.4}\frac{a^3}{a_{\prime}^3}+\frac{3.5.3.5.7}{2.4.4.6}\frac{a^5}{a_{\prime}^5}+\ \&c.$$

$$b_{3,2} = \frac{3.5}{4}\frac{a^2}{a_i^2} + \frac{3.3.5.7}{2.4.6}\frac{a^4}{a_i^4} + \frac{3.5.3.5.7}{2.4.4.6.8}\frac{a^6}{a_i^6} + \&c.$$

$$b_{3,3} = \frac{3.5.7}{4.6}\frac{a^3}{a_i^3} + \frac{3.3.5.7.9}{2.4.6.8}\frac{a^5}{a_i^5} + \frac{3.5.3.5.7.9.11}{2.4.4.6.8.10}\frac{a^7}{a_i^7} + \&c.$$

$$\frac{1}{2} b_{5,0} = 1 + \left(\frac{5}{2}\right)^2 \frac{a^2}{a_i^2} + \left(\frac{5.7}{2.4}\right)^2 \frac{a^4}{a_i^4} + \left(\frac{5.7.9}{2.4.6}\right)^2 \frac{a^6}{a_i^6} + \&c.$$

$$b_{5,1} = 5\,\frac{a}{a_i} + \frac{5.5.7}{2.4}\frac{a^3}{a_i^3} + \frac{5.7.5.7.9}{2.4.4.6}\frac{a^5}{a_i^5} + \&c.$$

$$b_{5,2} = \frac{5.7}{4}\frac{a^2}{a_i^2} + \frac{5.5.7.9}{2.4.6}\frac{a^4}{a_i^4} + \frac{5.7.5.7.9.11}{2.4.4.6.8}\frac{a^6}{a_i^6} + \&c.$$

$$b_{5,3} = \frac{5.7.9}{4.6}\frac{a^3}{a_i^3} + \frac{5.5.7.9.11}{2.4.6.8}\frac{a^5}{a_i^5} + \frac{5.7.5.7.9.11.13}{2.4.4.6.8.10}\frac{a^7}{a_i^7}$$

$$R = m_i \left\{ \frac{r r_i' \cos(\lambda - \lambda_i')}{r_i^3} - \frac{1}{\{r^2 - 2 r r_i' \cos(\lambda - \lambda_i') + r_i^2\}^{\frac{1}{2}}} \right\}$$

and since

$$r r_i' \cos(\lambda - \lambda_i') = r r_i \left\{ \cos^2\frac{\iota_i}{2} \cos(\lambda - \lambda_i) + \sin^2\frac{\iota_i}{2} \cos(\lambda + \lambda_i - 2\nu_i) \right\},$$

See p. 4;

R = terms independent of b

$$- \frac{m_i}{a_i}\left\{ \frac{1}{2} b_{1,0} + b_{1,1}\cos\tau + b_{1,2}\cos 2\tau + \&c. \right\}$$

$$+ \frac{m_i}{2 a_i^3}\left\{ \frac{1}{2} b_{3,0} + b_{3,1}\cos\tau + b_{3,2}\cos 2\tau + \&c. \right\}$$

$$\left\{ a^2 \left\{ \frac{3}{2} e^2 - 2e\cos\xi - \frac{e^2}{2}\cos 2\xi \right\}^2 - 2aa_i \left\{ \left(-\sin^2\frac{\iota_i}{2} - \frac{e^2 + e_i^2}{2} \right) \cos\tau \right. \right.$$

$$- \frac{3}{2} e\cos(\tau - \xi) + \frac{e}{2}\cos(\tau + \xi) + \frac{e}{2}\cos(\tau + \xi_i) - \frac{3}{2} e_i \cos(\tau - \xi_i)$$

$$+ \frac{e^2}{8}\cos(\tau - 2\xi) + \frac{3}{8} e^2 \cos(\tau + 2\xi) - \frac{3}{4} ee_i \cos(\tau - \xi - \xi_i)$$

$$- \frac{3}{4} ee_i \cos(\tau + \xi + \xi_i) + \frac{9}{4} ee_i \cos(\tau - \xi + \xi_i) + \frac{ee_i}{4}\cos(\tau + \xi - \xi_i)$$

$$+ \frac{3}{8} e_i^2 \cos(\tau - 2\xi_i) + \frac{e_i^2}{8}\cos(\tau + 2\xi_i) + \sin^2\frac{\iota_i}{2}\cos(2\tau - \eta_i) \bigg\}$$

$$+a_{,}^{2}\left\{\frac{3}{2}e_{,}^{2}-2e_{,}\cos\xi_{,}-\frac{e_{,}^{2}}{2}\cos 2\xi_{,}\right\}+\&c.\Bigg\}$$

$$-\frac{3}{2.4}\frac{m_{,}}{a_{,}^{5}}\left\{\frac{1}{2}b_{5,0}+b_{5,1}\cos\tau+b_{5,2}\cos 2\tau+\&c.\right\}$$

$$\Big\{2a^{2}e\cos\xi-3aa_{,}e\cos(\tau-\xi)+aa_{,}e\cos(\tau+\xi)$$

$$+aa_{,}e_{,}\cos(\tau-\xi_{,})-3aa_{,}e_{,}\cos(\tau+\xi_{,})+2a_{,}^{2}e_{,}\cos\xi_{,}\Big\}^{2}+\&c.$$

The following equations of condition subsist between the quantities b,

$$\frac{(2i-n)}{2}b_{n,i}=(i-1)\frac{(a^{2}+a_{,}^{2})}{a_{,}^{2}}b_{n,i-1}-\frac{(2i-4+n)}{2}\frac{a}{a_{,}}b_{n,i-2}$$

$$b_{n,i}=\frac{(a^{2}+a_{,}^{2})}{a_{,}^{2}}b_{n+2,i}-\frac{a}{a_{,}}b_{n+2,i-1}-\frac{a}{a_{,}}b_{n+2,i+1}$$

$$\left(1-\frac{a^{2}}{a_{,}^{2}}\right)^{2}b_{n,i}=\frac{(2i-2+n)}{(n-2)}\left(1+\frac{a^{2}}{a_{,}^{2}}\right)b_{n-2,i}$$

$$-\frac{2(2i-n+4)}{n-2}\frac{a}{a_{,}}b_{n-2,i+1}$$

$$i\,b_{n,i}=\frac{n\,a}{2\,a_{,}}\left\{b_{n+2,i-1}-b_{n+2,i+1}\right\}$$

Of these equations the three former are given in the *Mécanique céleste*, vol. i. p. 268. See also *The mechanism of the heavens*, p. 243.

We have besides,

$$\frac{a\,d\,b_{n,i}}{d^{2}}=-\frac{n\,a}{a_{,}}\left\{\frac{a}{a_{,}}b_{n+2,i}-\frac{1}{2}b_{n+2,i+1}-\frac{1}{2}b_{n+2,i+1}\right\}$$

In the notation of the *Mécanique céleste*,

$$b_{n,i}=b_{\frac{s}{2}}^{(i)}$$

The quantities $b_{1,0}$, $b_{1,1}$, from which all the other quantities b_{3}, b_{5}, &c. depend, may be obtained at once from Table IX. in the *Exercices de calcul intégral*, by M. Legendre, vol. iii. See also vol. i. p. 171 of the same work.

$$\left(1+\frac{a}{a_{,}}\right)b_{1,0}=\frac{4}{\pi}\int\frac{d\,\phi}{\Delta}\qquad\left(1+\frac{a}{a_{,}}\right)b_{1,1}=\frac{2}{\pi}\int-\frac{d\,\phi\cos 2\,\phi}{\Delta}$$

the integrals being taken from $\phi = 0$ to $\phi = \frac{1}{2}\pi$.

$\Delta = \sqrt{(1 - c^2 \sin^2 \phi)}$

$c^2 = \frac{4\,a\,a_{,}}{(a+a_{,})^2} = \frac{4\,\alpha}{(1+\alpha)^2}$, α* being $= \frac{a}{a_{,}}$ as in the notation of the *Mécanique céleste.*

$$b_{1,0} = \frac{4}{\pi\,(1+\alpha)}\,\mathrm{F}^{1} \qquad b_{1,1} = \frac{2}{\pi\,(1+\alpha)}\left\{\frac{2}{c^2}\,(\mathrm{F}^{1} - \mathrm{E}^{1}) - \mathrm{F}^{1}\right\}$$

In the theory of Jupiter disturbed by Saturn, $\alpha = {\cdot}54531725$; and hence in this instance if $c = \sin\theta$, $\theta = 72^\circ\ 53'\ 17''$.

By interpolation, I find from Table IX. p. 424,

$\mathrm{F}\,(72^\circ\ 53'\ 18'') = 2{\cdot}6460986$

and $b_{1,0} = 2{\cdot}180214$, which differs but slightly from the value of $b_{1,0}$ given by Laplace, viz. $2{\cdot}1802348$.

By developing the value of R, given p. 23, an expression is obtained which is complicated, but admits of many reductions in consequence of the equations of condition between the quantities b.

The first term of R for instance

$$= m_{,}\left\{-\frac{b_{1,0}}{2a_{,}} + \frac{3\,(a^2 e^2 + a_{,}^2 e_{,}^2)}{2\,.\,4\,a_{,}^3}\,b_{3,0} + \frac{a\,a_{,}}{2\,a_{,}^3}\left(\sin^2\frac{\iota_{,}}{2} + \frac{(e^2 + e_{,}^2)}{2}\right) b_{3,1}\right.$$

$$-\frac{3}{4\,.\,4\,a_{,}^5}\left(2\,a^4 e^2 + 5\,a^2 a_{,}^2\,(e^2 + e_{,}^2) + 2\,a_{,}^4 e_{,}^2\right) b_{5,0}$$

$$\left. + \frac{3\,.\,2}{2\,.\,4\,a_{,}^5}\,(a^2 e^2 + a_{,}^2 e_{,}^2)\,a\,a_{,}\,b_{5,1} + \frac{1\,.\,3\,.\,3}{2\,.\,4\,.\,2}\,\frac{a^2 a_{,}^2}{a_{,}^5}\,(e^2 + e_{,}^2)\,b_{5,2}\right\}$$

$$= m_{,}\left\{-\frac{b_{1,0}}{2\,a_{,}} + \frac{3\,(a^2 e^2 + a_{,}^2 e_{,}^2)}{2\,.\,4\,a_{,}^3}\,b_{3,0} + \frac{a\,a_{,}}{2\,a_{,}^3}\left(\sin^2\frac{\iota_{,}}{2} + \left(\frac{e^2 + e_{,}^2}{2}\right)\right) b_{3,1}\right.$$

$$-\frac{3\,.\,2}{2\,.\,4}\,\frac{(a^2 + a_{,}^2)}{a_{,}^2}\,\frac{(a^2 e^2 + a_{,}^2 e_{,}^2)}{2\,a_{,}^3}\,b_{5,0} - \frac{3\,.\,3}{4\,.\,4}\,\frac{a^2 a_{,}^2}{a_{,}^5}\,(e^2 + e_{,}^2)\,b_{5,0}$$

$$\left. + \frac{3\,.\,2}{2\,.\,4}\,\frac{(a^2 e^2 + a_{,}^2 e_{,}^2)}{a_{,}^3}\,\frac{a}{a_{,}}\,b_{5,1} + \frac{3\,.\,3}{2\,.\,4\,.\,2}\,\frac{a^2 a_{,}^2}{a_{,}^5}\,(e^2 + e_{,}^2)\,b_{5,2}\right\}$$

and since

$$b_{3,0} = \frac{(a^2 + a_{,}^2)}{a_{,}^2}\,b_{5,0} - 2\,\frac{a}{a_{,}}\,b_{5,1} \quad \text{See p. 24.}$$

---

* ρ in the notation of Woodhouse's *Astronomy*, vol. ii. p. 287.

$$b_{3,1} = \frac{3\,a}{2a_{\prime}}\left\{b_{5,0}-b_{5,2}\right\}$$

this term reduces itself to

$$m_{\prime}\left\{-\frac{b_{1,0}}{2\,a_{\prime}}+\frac{a\,a_{\prime}}{2\,a_{\prime}^{3}}\left(\sin^2\frac{\iota_{\prime}}{2}+\frac{e^2+e_{\prime}^2}{2}\right)b_{3,1}-\frac{3}{2\,.\,4}\frac{a\,a_{\prime}}{a_{\prime}^{3}}(e^2+e_{\prime}^2)\,b_{3,1}\right\}$$

$$= m_{\prime}\left\{-\frac{b_{1,0}}{2\,a_{\prime}}+\frac{a}{2\,a_{\prime}^{2}}\left(\sin^2\frac{\iota_{\prime}}{2}-\frac{e^2+e_{\prime}^2}{4}\right)b_{3,1}\right\}$$

Again, the coefficient of $\cos(\varpi - \varpi_{\prime})$

$$= m_{\prime}\left\{-\frac{9}{8}\frac{a}{a_{\prime}^{3}}b_{3,0}-\frac{a}{8\,a_{\prime}^{2}}b_{3,2}+\frac{3\,.\,6}{2\,.\,4}\frac{(a^2+a_{\prime}^2)}{2\,a_{\prime}^{5}}a\,a_{\prime}\,b_{5,0}\right.$$

$$\left.-\frac{3\,.\,7}{2.4.2}\frac{a^2\,a_{\prime}^2}{a_{\prime}^5}b_{5,1}-\frac{3}{2\,.\,4}\frac{(a^2+a_{\prime}^2)}{a_{\prime}^5}a\,a_{\prime}\,b_{5,2}-\frac{3}{2.4.2}\frac{a^2\,a_{\prime}^2}{a_{\prime}^5}b_{5,3}\right\}ee_{\prime}$$

$$= m_{\prime}\left[-\frac{9}{8}\frac{a}{a_{\prime}^{2}}b_{3,0}-\frac{a}{8\,a_{\prime}^{2}}b_{3,2}+\frac{3\,.\,6}{2\,.\,4}\frac{a}{a_{\prime}^{2}}\left\{\frac{(a^2+a_{\prime}^2)}{2\,a_{\prime}^{2}}b_{5,0}-\frac{a}{a_{\prime}}b_{5,1}\right\}\right.$$

$$-\frac{3}{8}\frac{a}{a_{\prime}^{2}}\left\{\frac{(a^2+a_{\prime}^2)}{a_{\prime}^{2}}b_{5,2}-\frac{a}{a_{\prime}}b_{5,1}-\frac{a}{a_{\prime}}b_{5,3}\right\}$$

$$\left.+\frac{9}{16}\frac{a}{a_{\prime}^{3}}\left\{b_{5,1}-b_{5,3}\right\}\right]$$

$$= m_{\prime}\left\{-\frac{9}{8}\frac{a}{a_{\prime}^{2}}b_{3,0}-\frac{a}{8\,a_{\prime}^{2}}b_{3,2}+\frac{9}{8}\frac{a}{a_{\prime}^{2}}b_{3,0}-\frac{3}{8}\frac{a}{a_{\prime}^{2}}b_{3,2}\right.$$

$$\left.+\frac{2\,.\,3}{8}\frac{a}{a_{\prime}^{2}}b_{3,2}\right\}ee_{\prime}$$

$$= m_{\prime}\frac{a}{4\,a_{\prime}^{2}}b_{3,2}$$

So that the part of R which is independent of $n\,t$, $n_{\prime}t$

$$= m_{\prime}\left\{-\frac{b_{1,0}}{2a_{\prime}}+\frac{a}{2\,a_{\prime}^{2}}\left(\sin^2\frac{\iota_{\prime}}{2}-\frac{e^2+e_{\prime}^2}{4}\right)b_{3,1}+\frac{a}{4\,a_{\prime}^{2}}b_{3,2}ee_{\prime}\cos(\varpi-\varpi_{\prime})\right\}$$

$$= m_{\prime}\left[-\frac{b_{1,0}}{2a_{\prime}}+\frac{a}{2\,a_{\prime}^{2}}\left(\sin^2\frac{\iota_{\prime}}{2}-\frac{e^2+e_{\prime}^2}{4}\right)b_{3,1}\right.$$

$$\left.-\left\{\frac{3\,a}{4\,a_{\prime}^{2}}b_{3,0}-\frac{a^2+a_{\prime}^2}{2\,a_{\prime}^{3}}b_{3,1}\right\}e\,e_{\prime}\cos(\varpi-\varpi_{\prime})\right]$$

In the general case, when ι_1, ι_2 are the inclinations of the orbits of the planets P and $P_{\prime}$ to any plane, the direction of which is arbitrary,

$$\cos \iota_{\prime} = \cos \iota_1 \cos \iota_2 + \sin \iota_1 \sin \iota_2 \cos (\nu_1 - \nu_2)$$

$$\sin^2 \frac{\iota_{\prime}}{2} = \frac{1 - \cos \iota_1 \cos \iota_2 - \sin \iota_1 \sin \iota_2 \cos (\nu_1 - \nu_2)}{2}$$

The part of R which is independent of $n t$, $n_{\prime} t$

$$= m_{\prime} \left\{ -\frac{b_{1,0}}{2a_{\prime}} + \frac{a}{4a_{\prime}^2} \left\{ 1 - \cos \iota_1 \cos \iota_2 - \sin \iota_1 \sin \iota_2 \cos (\nu_1 - \nu_2) \right. \right.$$

$$\left. \left. - \frac{e^2 + e_{\prime}^2}{2} \right\} b_{3,1} - \left\{ \frac{3a}{4a_{\prime}^2} b_{3,0} - \frac{(a^2 + a_{\prime}^2)}{2a_{\prime}^3} b_{3,1} \right\} e e_{\prime} \cos (\varpi - \varpi_{\prime}) \right\}$$

$$\cos \iota = \frac{1}{\sqrt{1 + \tan^2 \iota}} = 1 - \tfrac{1}{2} \tan^2 \iota, \quad \sin \iota = \frac{\tan \iota}{\sqrt{1 + \tan^2 \iota}} = \tan \iota \text{ nearly.}$$

$$= m_{\prime} \left\{ -\frac{b_{1,0}}{2a_{\prime}} + \frac{a}{8a_{\prime}^2} \left\{ (\tan^2 \iota_1 + \tan^2 \iota_2 - 2 \tan \iota_1 \tan \iota_2 \cos (\nu_1 - \nu_2) \right. \right.$$

$$\left. \left. - e^2 - e_{\prime}^2 \right\} b_{3,1} - \left\{ \frac{3a}{4a_{\prime}^2} b_{3,0} - \frac{(a^2 + a_{\prime}^2)}{2a_{\prime}^3} b_{3,1} \right\} e e_{\prime} \cos (\varpi - \varpi_{\prime}) \right\}$$

$$= m_{\prime} \left\{ -\frac{b_{1,0}}{2a_{\prime}} + \frac{a}{8a_{\prime}^2} \left\{ (\tan \iota_1 \cos \nu_1 - \tan \iota_2 \cos \nu_2)^2 \right. \right.$$

$$\left. + (\tan \iota_1 \sin \nu_1 - \tan \iota_2 \sin \iota_2)^2 - e^2 - e_{\prime}^2 \right\} b_{3,1}$$

$$\left. - \left\{ \frac{3a}{4a_{\prime}^2} b_{3,0} - \frac{(a^2 + a_{\prime}^2)}{2a_{\prime}^3} b_{3,1} \right\} e e_{\prime} \cos (\varpi - \varpi_{\prime}) \right\}.$$

and if $\tan \iota \sin \nu = p$, $\tan \iota \cos \nu = q$, this quantity

$$= m_{\prime} \left\{ -\frac{b_{1,0}}{2a_{\prime}} + \frac{a}{8a_{\prime}^2} \left\{ (p_1 - p_2)^2 + (q_1 - q_2)^2 - e^2 - e_{\prime}^2 \right\} b_{3,1} \right.$$

$$\left. - \left\{ \frac{3a}{4a_{\prime}^2} b_{3,0} - \frac{a^2 + a_{\prime}^2}{2a_{\prime}^3} b_{3,1} \right\} e e_{\prime} \cos (\varpi - \varpi_{\prime}) \right\}$$

which evidently agrees with the result given by M. de Pontécoulant, *Théor. anal.* vol. i. p. 363. All the other coefficients of terms multiplied by the squares and products of the eccentricities are susceptible of reductions similar to those in the two preceding pages, and finally;

$$R = m_{\prime}\left\{\frac{a}{a_{\prime}^2}\left(\cos^2\frac{\iota_{\prime}}{2} - \frac{e^2+e_{\prime}^2}{2}\right\}\cos(nt - n_{\prime}t)\right.$$

$$-\frac{3\,m_{\prime}}{2}\frac{a}{a_{\prime}^2}e\cos(n_{\prime}t - \varpi) + \frac{m_{\prime}a}{2a_{\prime}^2}e\cos(2\,nt - n_{\prime}t - \varpi)$$

$$+\frac{2\,m_{\prime}a}{a_{\prime}^2}e_{\prime}\cos(nt - 2\,n_{\prime}t + \varpi_{\prime})$$

$$+\frac{m_{\prime}a}{8\,a_{\prime}^2}e^2\cos(nt+n_{\prime}t-2\,\varpi) + \frac{3\,m_{\prime}a}{8\,a_{\prime}^2}e^2\cos(3\,nt-n_{\prime}t-2\,\varpi)$$

$$-\frac{3\,m_{\prime}a}{a_{\prime}^2}e\,e_{\prime}\cos(2\,n_{\prime}t - \varpi - \varpi_{\prime})$$

$$+\frac{m_{\prime}a}{a_{\prime}^2}e\,e_{\prime}\cos(2\,nt - 2\,n_{\prime}t - \varpi + \varpi_{\prime})$$

$$+\frac{27}{8}\frac{m_{\prime}a}{a_{\prime}^2}e_{\prime}^2\cos(nt - 3\,n_{\prime}t + 2\,\varpi_{\prime})$$

$$+\frac{m_{\prime}a}{8a_{\prime}^2}e_{\prime}^2\cos(nt+n_{\prime}t-2\,\varpi_{\prime}) + \frac{m_{\prime}a}{a_{\prime}^2}\sin^2\frac{\iota_{\prime}}{2}\cos(nt+n_{\prime}t-2\,\nu_{\prime})$$

$$+m_{\prime}\Sigma\left\{-\frac{b_{1,i}}{2\,a_{\prime}} + \frac{a}{4\,a_{\prime}^2}\sin^2\frac{\iota_{\prime}}{2}\left(b_{3,i-1} + b_{3,i+1}\right)\right.$$

$$+\frac{a\,(e^2+e_{\prime}^2)}{16\,a_{\prime}^2}\Big((3\,i-1)\,b_{3,i-1}$$

$$\left.-(3\,i+1)\,b_{3,i+1}\Big)\right\}\cos i\,(nt - n_{\prime}t)$$

$$+m_{\prime}\Sigma\left\{-\frac{a}{4\,a_{\prime}^2}b_{3,i-1} - \frac{a^2}{2\,a_{\prime}^3}b_{3,i}\right.$$

$$\left.+\frac{3\,a}{4\,a_{\prime}^2}b_{3,i+1}\right\}e\cos\Big(i\,(nt - n_{\prime}t) + nt - \varpi\Big)$$

$$+m_{\prime}\Sigma\left\{\frac{3}{4}\frac{a}{a_{\prime}^2}b_{3,i-1} - \frac{1}{2\,a_{\prime}}b_{3,i}\right.$$

$$\left.-\frac{a}{4\,a_{\prime}^2}b_{3,i+1}\right\}e_{\prime}\cos\Big(i\,(nt - n_{\prime}t) + n_{\prime}t - \varpi_{\prime}\Big)$$

$$+m_{\prime}\Sigma\left\{-\frac{(2+i)}{16}\frac{a}{a_{\prime}^2}b_{3,i-1} - \frac{(1+i)}{2}\frac{a^2}{a_{\prime}^3}b_{3,i}\right.$$

$$+\left.\frac{(8+9i)}{16}\frac{a}{a_i^2}b_{3,i+1}\right\}e^2\cos\left(i(nt-n_it)+2nt-2\varpi\right)$$

$$+m_i\Sigma\left\{\frac{(3+9i)}{8}\frac{a}{a_i^2}b_{3,i-1}-\frac{i}{a_i}b_{3,i}\right.$$

$$-\left.\frac{(1+i)}{8}\frac{a}{a_i^2}b_{3,i+1}\right\}ee_i\cos\left(i(nt-n_it)+nt+n_it-\varpi-\varpi_i\right)$$

$$+m_i\Sigma\left\{-\frac{(1+3i)}{8}\frac{a}{a_i^2}b_{3,i-1}\right.$$

$$+\left.\frac{3(1+i)}{8}\frac{a}{a_i^2}b_{3,i+1}\right\}ee_i\cos\left(i(nt-n_it)+nt-n_it-\varpi+\varpi_i\right)$$

$$+m_i\Sigma\left\{\frac{(8-9i)}{16}\frac{a}{a_i^2}b_{3,i-1}+\frac{(1-i)}{2a_i}b_{3,i}\right.$$

$$-\left.\frac{(2-i)}{16}\frac{a}{a_i^2}b_{3,1+i}\right\}e_i^2\cos\left(i(nt-n_it)+2n_it-2\varpi\right)$$

$$-m_i\Sigma\frac{a}{2a_i^2}b_{3,i-1}\sin^2\frac{\iota_i}{2}\cos\left(i(nt-n_it)+2n_it-2\nu_i\right)$$

i being every whole number, positive and negative and zero, and observing that $b_{m,n}=b_{m,-n}$. The preceding expression differs in form from that given in the *Mécanique céleste*, and in the *Théorie analytique du système du monde*, vol. i. p. 463, and is more simple. M. de Pontécoulant has given two examples of reductions applied to the terms upon which the *secular inequalities* depend; but these reductions may be extended to all the terms, and then the expression is obtained which I have given above.

The terms in $r\frac{dR}{dr}$ multiplied by e and e_i admit of similar reductions. Considering only these terms,

$$r\left(\frac{dR}{dr}\right)=-\frac{3m_i}{2}\frac{a}{a_i^2}e\cos(n_it-\varpi)+\frac{m_ia}{2a_i^2}e\cos(2nt-n_it-\varpi)$$

$$+\frac{2m_ia}{a_i^2}e_i\cos(nt-2n_it+\varpi_i)$$

$$+m_i\Sigma\left\{-\frac{i}{4}\frac{a}{a_i^2}b_{3,i-1}+\frac{(1+2i)}{2}\frac{a^2}{a_i^3}b_{3,i}\right.$$

$$-\left.\frac{3i}{4}\frac{a}{a_i^2}b_{3,i+1}\right\}e\cos\left(i(nt-n_it)+nt-\varpi\right)$$

$$+ m_{,} \Sigma \left\{ - \frac{3(1+i)}{4} \frac{a}{a_{,}^{2}} b_{3,i-1} + \frac{i\,a}{a_{,}} b_{3,i} + \frac{(1-i)}{4} b_{3,i+1} \right\} e_{,} \cos \left(i\,(n\,t - n_{,}\,t) + n_{,}\,t - \varpi_{,} \right)$$

If $a > a_{,}$, and

$$\left\{ 1 - \frac{a_{,}}{a} \cos\theta + \frac{a_{,}^{2}}{a^{2}} \right\}^{-\frac{1}{2}} = \frac{1}{2} b_{1,0} + b_{1,1} \cos\theta + b_{1,2} \cos 2\theta + \&c.$$

$$\left\{ 1 - \frac{a_{,}}{a} \cos\theta + \frac{a_{,}^{2}}{a^{2}} \right\}^{-\frac{3}{2}} = \frac{1}{2} b_{3,0} + b_{3,1} \cos\theta + b_{3,2} \cos 2\theta + \&c.$$

the value of R may easily be inferred from the value which it has in the former case. Considering only the terms multiplied by the eccentricities,

$$\left(\frac{dR}{dr}\right) = - \frac{3\,m_{,}}{2} \frac{a}{a_{,}^{2}} e \cos (n_{,}\,t - \varpi) + \frac{m_{,}}{2} \frac{a}{a_{,}^{2}} e \cos (2\,n\,t - n_{,}\,t - \varpi)$$

$$+ \frac{2\,m_{,}\,a}{a_{,}^{2}} e_{,} \cos (n\,t - 2\,n_{,}\,t + \varpi_{,})$$

$$+ m_{,} \Sigma \left\{ - \frac{i}{4} \frac{a_{,}}{a^{2}} b_{3,i-1} + \frac{(1+2\,i)}{2\,a} b_{3,i} - \frac{3\,i}{4} \frac{a_{,}}{a^{2}} b_{3,i+1} \right\} e \cos \left(i\,(n\,t - n_{,}\,t) + n\,t - \varpi \right)$$

$$+ m_{,} \Sigma \left\{ - \frac{3\,(1+i)}{4} \frac{a_{,}}{a^{2}} b_{3,i-1} + \frac{i\,a_{,}^{2}}{a^{3}} b_{3,i} + \frac{(1-i)}{4} \frac{a_{,}}{a^{2}} b_{3,i+1} \right\} e_{,} \cos \left(i\,(n\,t - n_{,}\,t) + n_{,}\,t - \varpi_{,} \right)$$

All these expressions are to a certain extent arbitrary, on account of the equation which connects $b_{3,i-1}$, $b_{3,i}$, and $b_{3,i+1}$

$$\frac{(2\,i+1)}{2} \frac{a}{a_{,}} b_{3,i+1} = \frac{i\,(a^{2} + a_{,}^{2})}{a_{,}^{2}} b_{3,i} - \frac{(2\,i-1)}{2} \frac{a}{a_{,}} b_{3,i-1}$$

3. The preceding equations, which obtain in the planetary theory, are also applicable to the theory of the moon, provided

The mass of the sun	$= m_{,}$
The mass of the earth + the mass of the moon .	$= \mu$
The distance of the moon from the earth	$= r$
The distance of the sun from the earth	$= r_{,}$
The longitude of the moon	$= \lambda'$

The longitude of the sun . . . . $= \lambda_{,}$
The tangent of the moon's latitude . . . . . . $= s$

I shall now proceed to consider particularly the theory of the moon.

In future I shall distinguish the arguments by the numerical indices which are contained in the first column of the following Table. The numbers in the second column are the indices of the arguments in M. Damoiseau's *Mémoire sur la théorie de la lune.*

0	...	0	42	73	$2\tau-3\xi-\xi_{,}$	84	......	$\xi+\xi_{,}+2\eta$
1	30	2τ *	43	...	$2\tau+3\xi+\xi_{,}$	85	......	$2\tau-\xi-\xi_{,}-2\eta$
2	1	ξ	44	26	$3\xi-\xi_{,}$	86	......	$2\tau-\xi-\xi_{,}+2\eta$
3	31	$2\tau-\xi$ †	45	...	$2\tau-3\xi+\xi_{,}$	87	......	$2\tau+\xi+\xi_{,}-2\eta$
4	32	$2\tau+\xi$	46	...	$2\tau+3\xi-\xi_{,}$	88	......	$2\tau+\xi+\xi_{,}+2\eta$
5	16	$\xi_{,}$ ‡	47	...	$2\xi+2\xi_{,}$	89	......	$\xi-\xi_{,}-2\eta$
6	33	$2\tau-\xi_{,}$	48	75	$2\tau-2\xi-2\xi_{,}$	90	......	$\xi-\xi_{,}+2\eta$
7	34	$2\tau+\xi_{,}$	49	...	$2\tau+2\xi+2\xi_{,}$	91	......	$2\tau-\xi+\xi_{,}-2\eta$
8	2	2ξ	50	...	$2\xi-2\xi_{,}$	92	......	$2\tau-\xi+\xi_{,}+2\eta$
9	35	$2\tau-2\xi$	51	...	$2\tau-2\xi+2\xi_{,}$	93	......	$2\tau+\xi-\xi_{,}-2\eta$
10	36	$2\tau+2\xi$	52	...	$2\tau+2\xi-2\xi_{,}$	94	......	$2\tau+\xi-\xi_{,}+2\eta$
11	19	$\xi+\xi_{,}$	53	...	$\xi+3\xi_{,}$	95	......	$2\xi_{,}-2\eta$
12	41	$2\tau-\xi-\xi_{,}$	54	...	$2\tau-\xi-3\xi_{,}$	96	......	$2\xi_{,}+2\eta$
13	42	$2\tau+\xi+\xi_{,}$	55	...	$2\tau+\xi+3\xi_{,}$	97	......	$2\tau-2\xi_{,}-2\eta$
14	18	$\xi-\xi_{,}$	56	...	$\xi-3\xi_{,}$	98	......	$2\tau-2\xi_{,}+2\eta$
15	39	$2\tau-\xi+\xi_{,}$	57	...	$2\tau-\xi+3\xi_{,}$	99	......	$2\tau+2\xi_{,}+2\eta$
16	40	$2\tau+\xi-\xi_{,}$	58	...	$2\tau+\xi-3\xi_{,}$	100	......	$2\tau+2\xi_{,}+2\eta$
17	17	$2\xi_{,}$	59	...	$4\xi_{,}$	101	80	τ §
18	43	$2\tau-2\xi_{,}$	60	...	$2\tau-4\xi_{,}$	102	81	$\tau-\xi$
19	44	$2\tau+2\xi_{,}$	61	...	$2\tau+4\xi_{,}$	103	82	$\tau+\xi$
20	4	3ξ	62	3	2η	104	83	$\tau-\xi_{,}$
21	45	$2\tau-3\xi$	63	37	$2\tau-2\eta$	105	84	$\tau+\xi_{,}$
22	46	$2\tau+3\xi$	64	38	$2\tau+2\eta$	106	85	$\tau-2\xi$
23	21	$2\xi+\xi_{,}$	65	5	$\xi-2\eta$	107	86	$\tau+2\xi$
24	53	$2\tau-2\xi-\xi_{,}$	66	6	$\xi+2\eta$	108	91	$\tau-\xi-\xi_{,}$
25	54	$2\tau+2\xi+\xi_{,}$	67	49	$2\tau-\xi-2\eta$	109	92	$\tau+\xi+\xi_{,}$
26	20	$2\xi-\xi_{,}$	68	47	$2\tau-\xi+2\eta$	110	89	$\tau-\xi+\xi_{,}$
27	51	$2\tau-2\xi+\xi_{,}$	69	48	$2\tau+\xi-2\eta$	111	......	$\tau+\xi-\xi_{,}$
28	52	$2\tau+2\xi-\xi_{,}$	70	50	$2\tau+\xi+2\eta$	112	......	$\tau-2\xi_{,}$
29	23	$\xi+2\xi_{,}$	71	24	$\xi_{,}-2\eta$	113	......	$\tau+2\xi_{,}$
30	59	$2\tau-\xi-2\xi_{,}$	72	25	$\xi_{,}+2\eta$	114	......	$\tau-2\eta$
31	...	$2\tau+\xi+2\xi_{,}$	73	57	$2\tau-\xi_{,}-2\eta$	115	......	$\tau+2\eta$
32	22	$\xi-2\xi_{,}$	74	56	$2\tau-\xi_{,}+2\eta$	116	100	3τ
33	61	$2\tau-\xi+2\xi_{,}$	75	55	$2\tau+\xi_{,}-2\eta$	117	101	$3\tau-\xi$
34	60	$2\tau+\xi-2\xi_{,}$	76	58	$2\tau+\xi_{,}+2\eta$	118	102	$3\tau+\xi$
35	...	$3\xi_{,}$	77	7	$2\xi-2\eta$	119	103	$3\tau-\xi_{,}$
36	...	$2\tau-3\xi_{,}$	78	8	$2\xi+2\eta$	120	104	$3\tau+\xi_{,}$
37	...	$2\tau+3\xi_{,}$	79	65	$2\tau-2\xi-2\eta$	121	......	$3\tau-2\xi$
38	9	4ξ	80	63	$2\tau-2\xi+2\eta$	122	......	$3\tau+2\xi$
39	67	$2\tau-4\xi$	81	64	$2\tau+2\xi-2\eta$	123	......	$3\tau-\xi-\xi_{,}$
40	...	$2\tau+4\xi$	82	...	$2\tau+2\xi+2\eta$	124	......	$3\tau+\xi+\xi_{,}$
41	27	$3\xi+\xi_{,}$	83	...	$\xi+\xi_{,}-2\eta$	125	......	$3\tau-\xi+\xi_{,}$

* Variation. † Evection. ‡ Annual Equation. § Parallactic inequality.

126	......	$3\tau+\xi-\xi_{\prime}$	147	...	$2\tau-\eta$	167	...	$\xi+\xi_{\prime}-\eta$
127	......	$3\tau-2\xi_{\prime}$	148	...	$2\tau+\eta$	168	...	$\xi+\xi_{\prime}+\eta$
128	......	$3\tau+2\xi_{\prime}$	149	...	$\xi-\eta$	169	...	$2\tau-\xi-\xi_{\prime}-\eta$
129	......	$3\tau-2\eta$	150	...	$\xi+\eta$	170	...	$2\tau-\xi-\xi_{\prime}+\eta$
130	......	$3\tau+2\eta$	151	...	$2\tau-\xi-\eta$	171	...	$2\tau+\xi+\xi_{\prime}-\eta$
131	120	4τ	152	...	$2\tau-\xi+\eta$	172	...	$2\tau+\xi+\xi_{\prime}+\eta$
132	121	$4\tau-\xi$	153	...	$2\tau+\xi-\eta$	173	...	$\xi-\xi_{\prime}-\eta$
133	122	$4\tau+\xi$	154	...	$2\tau+\xi+\eta$	174	...	$\xi-\xi_{\prime}+\eta$
134	123	$4\tau-\xi_{\prime}$	155	...	$\xi_{\prime}-\eta$	175	...	$2\tau-\xi+\xi_{\prime}-\eta$
135	124	$4\tau+\xi_{\prime}$	156	...	$\xi_{\prime}+\eta$	176	...	$2\tau-\xi+\xi_{\prime}+\eta$
136	125	$4\tau-2\xi$	157	...	$2\tau-\xi_{\prime}-\eta$	177	...	$2\tau+\xi-\xi_{\prime}-\eta$
137	126	$4\tau+2\xi$	158	...	$2\tau-\xi_{\prime}+\eta$	178	...	$2\tau+\xi-\xi_{\prime}+\eta$
138	131	$4\tau-\xi-\xi_{\prime}$	159	...	$2\tau+\xi_{\prime}-\eta$	179	...	$2\xi_{\prime}-\eta$
139	......	$4\tau+\xi+\xi_{\prime}$	160	...	$2\tau+\xi_{\prime}+\eta$	180	...	$2\xi_{\prime}+\eta$
140	129	$4\tau-\xi+\xi_{\prime}$	161	...	$2\xi-\eta$	181	...	$2\tau-2\xi_{\prime}-\eta$
141	......	$4\tau+\xi-\xi_{\prime}$	162	...	$2\xi+\eta$	182	...	$2\tau-2\xi_{\prime}+\eta$
142	......	$4\tau-2\xi_{\prime}$	163	...	$2\tau-2\xi-\eta$	183	...	$2\tau+2\xi_{\prime}-\eta$
143	......	$4\tau+2\xi_{\prime}$	164	...	$2\tau-2\xi+\eta$	184	...	$2\tau+2\xi_{\prime}+\eta$
144	127	$4\tau-2\eta$	165	...	$2\tau+2\xi-\eta$	185	...	$\tau-\eta$
145	......	$4\tau+2\eta$	166	...	$2\tau+2\xi+\eta$	186	...	$\tau+\eta$
146	......	η						

$$\cos 2\tau \cos 2\tau = \frac{1}{2}\cos 4\tau + \frac{1}{2}$$
$$[131] \qquad [0]$$

$$\cos 2\tau \cos \xi = \frac{1}{2}\cos(2\tau+\xi) + \frac{1}{2}\cos(-2\tau+\xi)$$
$$[4] \qquad [-3]$$

Hence the argument 131 is produced *by addition* through the square of the term corresponding to argument 1. Similarly the argument 4 is produced *by addition* through the multiplication of argument 1 by argument 2, and the argument -3 *by subtraction.*

These results, extended through all the arguments, are exhibited in the Table at the end of this work, by means of which any multiplication of one series by another may be effected, and any coefficient obtained at pleasure, when the approximation is limited to those terms which the table reaches, as will be shown by repeated examples. By means of this Table all the developments which are required in the theory of the moon, and which would otherwise be extremely laborious, not to say impracticable, may be obtained with very great facility, and with little, if any, risk of error.

This Table may also be used in forming the developments required in the method employed by MM. Laplace and Damoiseau. For this purpose it is merely necessary to make

$$\begin{aligned} \tau &= \lambda' - \lambda_{\prime} \text{ instead of } nt - n_{\prime}t \\ \xi &= c\lambda' - \varpi \; . \; . \; . \; . \; cnt - \varpi \\ \xi_{\prime} &= c\lambda_{\prime} - \varpi_{\prime} . \; . \; . \; . \; c_{\prime}n_{\prime}t - \varpi_{\prime} \\ \text{and } \eta &= g\lambda' - \nu \; . \; . \; . \; . \; gnt - \nu \end{aligned}$$

the indices remaining the same.

$$R = m_{,} \left\{ \frac{r^{`} r_{,} \cos(\lambda - \lambda_{,})}{r_{,}^{3}} - \frac{1}{\{r^{2} - 2 r^{`} r_{,} \cos(\lambda^{`} - \lambda_{,}) + r_{,}^{2}\}^{\frac{1}{2}}} \right\}$$

$$= m_{,} \left\{ -\frac{1}{r_{,}} + \frac{r^{2}}{2 r_{,}^{3}} - \frac{3}{8} \frac{\{2 r^{`} r_{,} \cos(\lambda^{`} - \lambda_{,}) - r^{2}\}^{2}}{r_{,}^{5}} \right.$$

$$\left. - \frac{15}{18} \frac{\{2 r^{`} r_{,} \cos(\lambda^{`} - \lambda_{,}) - r^{2}\}^{3}}{r_{,}^{7}} \right\}$$

$$= m_{,} \left\{ -\frac{1}{r_{,}} + \frac{r^{2}}{2 r_{,}^{3}} - \frac{3}{2} \frac{r^{`2} r_{,}^{2}}{r_{,}^{5}} \cos(\lambda^{`} - \lambda_{,})^{2} \right.$$

$$\left. + \frac{3}{2} \frac{r^{2} r^{`} r_{,}}{r_{,}^{5}} \cos(\lambda - \lambda_{,}) - \frac{5}{2} \frac{r^{`3} r_{,}^{3}}{r_{,}^{7}} \cos(\lambda^{`} - \lambda_{,})^{3} \right\}$$

$$= m_{,} \left\{ -\frac{1}{r_{,}} - \frac{r^{`2}}{4 r_{,}^{3}} \left\{ 1 + 3 \cos(2 \lambda^{`} - 2 \lambda_{,}) - 2 s^{2} \right\} \right.$$

$$\left. - \frac{r^{`3}}{8 r_{,}^{4}} \left\{ 3 (1 - 4 s^{2}) \cos(\lambda^{`} - \lambda_{,}) + 5 \cos(3 \lambda - 3 \lambda_{,}) \right\} \right\}$$

Let $R = m_{,} \left\{ R_0 + R_1 \cos \tau + e R_2 \cos \xi + e R_3 \cos(2 \tau - \xi) + \&c. \right\}$

[1] [2] [3]

$$r^{-1} = 1 + e\left(1 - \frac{e^{2}}{8}\right) \cos \xi + e^{2} \left(1 - \frac{e^{2}}{3}\right) \cos 2 \xi$$

[2] [8]

$$+ \frac{9}{8} e^{3} \cos 3 \xi + \frac{4}{3} e^{4} \cos 4 \xi$$

[20] [35]

$$\frac{d r}{d e} = e - \left(1 - \frac{9}{8} e^{2}\right) \cos \xi - e \left(1 - \frac{4 e^{2}}{3}\right) \cos 2 \xi$$

[2] [8]

$$- \frac{9}{8} e^{2} \cos 3 \xi - \frac{4 e^{3}}{3} \cos 4 \xi$$

[20] [35]

$$\frac{d r}{r d e} = \frac{e}{2} \left(1 + \frac{e^{2}}{4}\right) - \left(1 - \frac{9}{8} e^{2}\right) \cos \xi$$

[0] [2]

$$- \frac{3}{2} e \left(1 - \frac{11}{9} e^{2}\right) \cos 2 \xi$$

[8]

$$- \frac{17}{8} e^{2} \cos 3 \xi - \frac{71}{24} e^{3} \cos 4 \xi$$

[20] [35]

$$\frac{d\lambda}{de} = 2\left(1 - \frac{3e^2}{8}\right)\sin\xi + \frac{5}{2}e\left(1 - \frac{28}{15}e^2\right)\sin 2\xi$$
[2] [8]

$$+ \frac{13}{4}e^2 \sin 3\xi + \frac{103}{24}e^2 \sin 4\xi$$
[20] [35]

$$\frac{dR}{de} = \frac{dR\,dr}{dr\,de} + \frac{dR\,d\lambda}{d\lambda\,de}$$

$$= \frac{r\,dR}{dr}\frac{dr}{r\,de} + \frac{dR}{d\lambda}\frac{d\lambda}{de}$$

$$\frac{r\,dR}{dr} = \frac{a\,dR}{da} \qquad \frac{dR}{d\lambda} = \frac{dR}{d\tau}.$$ See *Méc. cél.* vol. i. p. 267.

Considering the terms in R multiplied by $\frac{a^2}{a_,^3}$, $\frac{r\,dR}{dr} = 2R$

$$R_0 = -\frac{a^2}{4a_,^3} + \&c. \quad R_1 = -\frac{3a^2}{4a^3} + \&c.$$

$$R_{101} = -\frac{3a^3}{8a_,^4} + \&c. \qquad R_{116} = -\frac{5a^3}{8a_,^4} + \&c.$$

Multiplying by means of the Table, neglecting the squares of the eccentricities, we find

$$R_2 = -\frac{a\,dR_0}{da} = \frac{a^2}{2a_,^3}$$

$$R_3 = -\frac{a\,dR_1}{2da} - 2R_1 = \frac{9a^2}{4a_,^3}$$

$$R_4 = -\frac{a\,dR_1}{2da} + 2R_1 = -\frac{3a^2}{4a_,^3}$$

Neglecting the cubes of the eccentricities, we have

$$2R_8 = -\frac{a\,dR_2}{2da} - \frac{3a\,dR_0}{2da} = \frac{a^2}{4a_,^3}$$

$$2R_9 = -\frac{a\,dR_3}{2da} - 2R_3 - \frac{3a\,dR_1}{4da} - \frac{5.2R_1}{4} = -\frac{15a^2}{4a_,^3}$$

$$2R_{10} = -\frac{a\,dR_4}{2da} + 2R_4 - \frac{3a\,dR_1}{4da} + \frac{5.2R_1}{4} = -\frac{3a^2}{2a_,^2}$$

These equations may be formed at once from the Table by inspection, taking care to write R with the sign + in the term multiplied by $\frac{dR}{d\tau}$ when the index is found in the upper line in the Table, as in the case of the argument (10); and with the sign — when in the lower, as in the case of the argument (9). The term multiplied by $\frac{a\,\mathrm{d}R}{\mathrm{d}a}$ always takes its sign from the factor arising from $\frac{\mathrm{d}r}{r\,\mathrm{d}e}$.

By means of the Table, any term in R depending on the eccentricities may be found at pleasure, and the development, *Phil. Trans.* 1831, p. 263, may be verified with great facility; thus,

$$4\,R_{38} = -\frac{a\,\mathrm{d}R_{20}}{2\,\mathrm{d}a} - \frac{3\,a\,\mathrm{d}R_8}{4\,\mathrm{d}a} - \frac{17\,a\,\mathrm{d}R_2}{16\,\mathrm{d}a} - \frac{71\,a\,\mathrm{d}R_0}{24\,\mathrm{d}a}$$

I find, on reference to the development in question,

$$R_{20} = \frac{a^2}{16\,a_i^3} \quad R_8 = \frac{a^2}{8\,a_i^3} \quad R_2 = \frac{a^2}{2\,a_i^3} \quad R_0 = -\frac{a^2}{4\,a_i^3}$$

whence

$$a\frac{\mathrm{d}R_{20}}{\mathrm{d}a} = \frac{a^2}{8\,a_i^3} \quad \frac{a\,\mathrm{d}R_8}{\mathrm{d}a} = \frac{a^2}{4\,a_i^3} \quad \frac{a\,\mathrm{d}R_2}{\mathrm{d}a} = \frac{a^2}{a_i^3} \quad \frac{a\,\mathrm{d}R_0}{\mathrm{d}a} = -\frac{a^2}{2\,a_i^3}$$

$$4\,R_{38} = \left\{-\frac{1}{2.8} - \frac{3}{4.4} - \frac{17}{16} + \frac{71}{24.2}\right\}\frac{a^2}{a_i^3} = \frac{4\,a^2}{24a_i^3} \qquad R_{38} = \frac{a^2}{24\,a_i^3}$$

The terms depending upon the inclination of the moon's orbit may be obtained in a similar manner.

If R be considered as a function of r', λ' and s, we have

$$\frac{\mathrm{d}R}{\mathrm{d}\gamma} = \frac{r'\mathrm{d}R}{\mathrm{d}r'}\,\frac{\mathrm{d}r'}{r'\mathrm{d}\gamma} + \frac{\mathrm{d}R}{\mathrm{d}\lambda'}\,\frac{\mathrm{d}\lambda'}{\mathrm{d}\gamma} + \frac{\mathrm{d}R}{\mathrm{d}s}\,\frac{\mathrm{d}s}{\mathrm{d}\gamma}$$

$$\frac{\mathrm{d}r'}{r'\mathrm{d}\gamma} = \frac{\gamma}{2} - \frac{\gamma}{2}(1 - 4\,e^2)\cos 2\eta + \gamma\,e\cos(\xi - 2\eta)$$

[62] [65]

$$-\gamma\,e\cos(\xi + 2\eta) - \frac{3}{8}\gamma\,e^2\cos(2\xi - 2\eta)$$

[66] [77]

$$+\frac{13}{8}\gamma\,e^2\cos(2\xi + 2\eta)$$

[78]

$$\frac{\mathrm{d}\lambda'}{\mathrm{d}\gamma} = -\gamma\,(1 - 4\,e^2)\sin 2\eta + \gamma\,e\sin(\xi - 2\eta) - 3\,\gamma\,e\sin(\xi + 2\eta)$$

[62] [65] [66]

$$-\frac{13}{2}\gamma\,e^2\sin(2\xi + 2\eta)\Big\}$$

[78]

$$\frac{d\,s}{d\,\gamma} = (1 - e^2)\sin\eta + e\sin(\xi - \eta) + e\sin(\xi + \eta)$$

[146] [149] [150]

$$+ \frac{e^2}{8}\sin(2\xi - \eta) + \frac{9}{8}e^2\sin(2\xi + \eta)$$

[161] [162]

If R be considered as a function of r, λ' and s,

$$\frac{d\,R}{d\,\gamma} = \frac{d\,R}{d\,\lambda'}\frac{d\,\lambda'}{d\,\gamma} + \frac{d\,R}{d\,s}\frac{d\,s}{d\,\gamma}$$

In this case

$$\frac{d\,R}{d\,s} = m_{\prime}\left\{\frac{r^2}{r_{\prime}^3} + \frac{r^2}{2r_{\prime}^3}\{1+\cos(2\lambda-2\lambda_{\prime})\}\right\} s \text{ nearly}$$

$$\frac{r^2}{2r_{\prime}^3} + \frac{r^2}{4r_{\prime}^3}\{1+3\cos(2\lambda-2\lambda_{\prime})\}$$

$$= \frac{a^2}{a_{\prime}^3}\left\{\frac{3}{4}\left\{1 + \frac{3}{2}e^2 + \frac{3}{2}e_{\prime}^2\right\}\right.$$

$$+ \frac{3}{4}\left\{1 - \frac{5}{2}e^2 - \frac{5}{2}e_{\prime}^2\right\}\cos 2\tau - \frac{3}{2}e\cos\xi$$

$$- \frac{9}{4}e\cos(2\tau-\xi) + \frac{3}{4}e\cos(2\tau+\xi) + \frac{9}{4}e_{\prime}\cos\xi_{\prime}$$

$$+ \frac{21}{8}e_{\prime}\cos(2\tau-\xi_{\prime}) - \frac{3}{8}e_{\prime}\cos(2\tau+\xi_{\prime}) - \frac{3}{8}e^2\cos 2\xi$$

$$+ \frac{15}{8}e^2\cos(2\tau-2\xi) + \frac{3}{4}e^2\cos(2\tau+2\xi) - \frac{9}{4}ee_{\prime}\cos(\xi+\xi_{\prime})$$

$$- \frac{63}{8}ee_{\prime}\cos(2\tau-\xi-\xi_{\prime}) - \frac{3}{8}ee_{\prime}\cos(2\tau+\xi+\xi_{\prime})$$

$$- \frac{9}{4}ee_{\prime}\cos(\xi-\xi_{\prime}) + \frac{9}{8}ee_{\prime}\cos(2\tau-\xi+\xi_{\prime})$$

$$\left. + \frac{21}{8}ee_{\prime}\cos(2\tau+\xi-\xi_{\prime}) + \frac{27}{8}e_{\prime}^2\cos 2\xi_{\prime} + \frac{51}{8}e_{\prime}^2\cos(2\tau-2\xi_{\prime})\right\}$$

The expression for $\frac{d\,R}{d\,s}$ is given in the *Philosophical Transactions*, 1832, p. 6.

By proceeding in the way explained above, all the coefficients in the development of R are made to depend ultimately upon R_0, R_1,

R_{101}, &c. This method is far more simple than that which I employed in the *Philosophical Transactions*, 1832, for the development of R in the Lunar Theory. A verification may be obtained, as far as the terms depending upon the squares of the eccentricities, by substituting for the quantities b in the expression p. 28 their values in series.

Finally,

$$R = m_{,} \Big\{ - \frac{1}{r_{,}} - \frac{1}{4} \Big\{ 1 + \frac{3}{2} e^2 + \frac{3}{2} e_{,}^2 + \frac{9}{4} e^2 e_{,}^2 + \frac{15}{8} e_{,}^4 - \frac{3}{2} \gamma^2$$

[0]

$$- \frac{9}{4} \gamma^2 e^2 - \frac{9}{4} \gamma^2 e_{,}^2 + \frac{39}{8} \gamma^4 \Big\} \frac{a^2}{a_{,}^3}$$

$$- \frac{3}{4} \Big\{ 1 - \frac{5}{2} e^2 - \frac{5}{2} e_{,}^2 + \frac{23}{16} e^4 + \frac{25}{4} e^2 e_{,}^2$$

$$+ \frac{13}{16} e_{,}^4 \Big\} \cos^4 \frac{\iota}{2} \frac{a^2}{a_{,}^3} \cos 2\tau$$

[1]

$$+ \frac{1}{2} \Big\{ 1 - \frac{e^2}{8} - \frac{3}{2} e_{,}^2 - \frac{3}{2} \gamma^2 \Big\} \frac{a^2}{a_{,}^3} e \cos \xi$$

[2]

$$+ \frac{9}{4} \Big\{ 1 - \frac{13}{24} e^2 - \frac{5}{2} e_{,}^2 \Big\} \cos^4 \frac{\iota}{2} \frac{a^2}{a_{,}^3} e \cos (2\tau - \xi)$$

[3]

$$- \frac{3}{4} \Big\{ 1 - \frac{19}{8} e^2 - \frac{5}{2} e_{,}^2 \Big\} \cos^4 \frac{\iota}{2} \frac{a^2}{a_{,}^3} e \cos (2\tau + \xi)$$

[4]

$$- \frac{3}{4} \Big\{ 1 + \frac{3}{2} e^2 + \frac{9}{8} e_{,}^2 - \frac{3}{2} \gamma^2 \Big\} \frac{a^2}{a_{,}^3} e_{,} \cos \xi_{,}$$

[5]

$$- \frac{21}{8} \Big\{ 1 - \frac{5}{2} e^2 - \frac{123}{56} e_{,}^2 \Big\} \cos^4 \frac{\iota}{2} \frac{a^2}{a_{,}^3} e_{,} \cos (2\tau - \xi_{,})$$

[6]

$$+ \frac{3}{8} \Big\{ 1 - \frac{5}{2} e^2 - 4 e_{,}^2 \Big\} \cos^4 \frac{\iota}{2} \frac{a^2}{a_{,}^3} e_{,} \cos (2\tau + \xi_{,})$$

[7]

$$+ \frac{1}{8} \Big\{ 1 - \frac{e^2}{3} + \frac{3}{2} e_{,}^2 - \frac{3}{2} \gamma^2 \Big\} \frac{a^2}{a_{,}^3} e^2 \cos 2\xi$$

[8]

$$-\frac{15}{8}\left\{1-\frac{5}{2}e_{,}^{2}\right\}\cos^4\frac{\iota}{2}\frac{a^2}{a_{,}^3}e^2\cos(2\tau-2\xi)$$
[9]

$$-\frac{3}{4}\left\{1-\frac{5}{2}e^2-\frac{5}{2}e_{,}^{2}\right\}\cos^4\frac{\iota}{2}\frac{a^2}{a_{,}^3}e^2\cos(2\tau+2\xi)$$
[10]

$$+\frac{3}{4}\left\{1-\frac{e^2}{8}+\frac{9}{8}e_{,}^{2}-\frac{3}{2}\gamma^2\right\}\frac{a^2}{a_{,}^3}ee_{,}\cos(\xi+\xi_{,})$$
[11]

$$+\frac{63}{8}\left\{1-\frac{91}{168}e^2-\frac{123}{56}e_{,}^{2}\right\}\cos^4\frac{\iota}{2}\frac{a^2}{a_{,}^3}ee_{,}\cos(2\tau-\xi-\xi_{,})$$
[12]

$$+\frac{3}{8}\left\{1-\frac{19}{8}e^2-\frac{e_{,}^{2}}{8}\right\}\cos^4\frac{\iota}{2}\frac{a^2}{a_{,}^3}ee_{,}\cos(2\tau+\xi+\xi_{,})$$
[13]

$$+\frac{3}{4}\left\{1-\frac{e^2}{8}+\frac{9}{8}e_{,}^{2}-\frac{3}{2}\gamma^2\right\}\frac{a^2}{a_{,}^2}ee_{,}\cos(\xi-\xi_{,})$$
[14]

$$-\frac{9}{8}\left\{1-\frac{13}{24}e^2-\frac{e_{,}^{2}}{8}\right\}\cos^4\frac{\iota}{2}\frac{a^2}{a_{,}^3}ee_{,}\cos(2\tau-\xi+\xi_{,})$$
[15]

$$-\frac{21}{8}\left\{1-\frac{19}{8}e^2-\frac{123}{56}e_{,}^{2}\right\}\cos^4\frac{\iota}{2}\frac{a^2}{a_{,}^3}ee_{,}\cos(2\tau+\xi-\xi_{,})$$
[16]

$$-\frac{9}{8}\left\{1+\frac{3}{2}e^2+\frac{7}{9}e_{,}^{2}-\frac{3}{2}\gamma^3\right\}\frac{a^2}{a_{,}^3}e_{,}^{2}\cos 2\xi_{,}$$
[17]

$$-\frac{51}{8}\left\{1-\frac{5}{2}e^2-\frac{115}{51}e_{,}^{2}\right\}\cos^4\frac{\iota}{2}\frac{a^2}{a_{,}^3}e_{,}^{2}\cos(2\tau-2\xi_{,})\Bigg\}$$
[18]

$$=\frac{m_{,}a^2}{a_{,}^3}\Bigg\{-\frac{34}{137}-\frac{20}{27}\cos 2\tau+\frac{38}{77}e\cos\xi+\frac{38}{17}e\cos(2\tau-\xi)$$
[0] [1] [2] [3]

$$-\frac{20}{27}e\cos(2\tau+\xi)-\frac{32}{43}e_{,}\cos\xi_{,}-\frac{70}{27}e_{,}\cos(2\tau-\xi_{,})$$
[4] [5] [6]

$$+\frac{10}{27}e_{,}\cos(2\tau+\xi_{,})+\frac{10}{81}e^2\cos 2\xi-\frac{28}{15}e^2\cos(2\tau-2\xi)$$
[7] [8] [9]

$$- \frac{20}{27} e^2 \cos(2\tau + 2\xi) + \frac{20}{27} e e_, \cos(\xi + \xi_,)$$
[10] [11]

$$+ \frac{180}{23} e e_, \cos(2\tau - \xi - \xi_,) + \frac{10}{27} e e_, \cos(2\tau + \xi + \xi_,)$$
[12] [13]

$$+ \frac{20}{27} e e_, \cos(\xi + \xi_,) - \frac{66}{59} e e_, \cos(2\tau - \xi + \xi_,)$$
[14] [15]

$$- \frac{83}{32} e e_, \cos(2\tau + \xi - \xi_,) - \frac{67}{60} e_,^2 \cos 2\xi_,$$
[16] [17]

$$- \frac{233}{37} e_,^2 \cos(2\tau - 2\xi_,) - \frac{16}{43} \gamma^2 \cos 2\eta$$
[18] [62]

$$- \frac{26}{69} \gamma^2 \cos(2\tau - 2\eta) \Big\} \text{ nearly.}$$
[63]

The expression for R may be given in the following compendious form, by writing the indices of the arguments between brackets, instead of the cosines of the corresponding arguments. The coefficients are equal to zero of those terms which are wanting.

$$R = \frac{m_, a^2}{a_,^3} \Big\{ - \frac{34}{137} [0] - \frac{20}{27} [1] + \frac{38}{77} e [2] + \frac{38}{17} e [3] - \frac{20}{27} e [4]$$

$$- \frac{32}{43} e_, [5] - \frac{70}{27} e_, [6] + \frac{10}{27} e_, [7] + \frac{10}{81} e^2 [8] - \frac{28}{15} e^2 [9]$$

$$- \frac{20}{27} e^2 [10] + \frac{20}{27} e e_, [11] + \frac{180}{23} e e_, [12] + \frac{10}{27} e e_, [13]$$

$$+ \frac{20}{27} e e_, [14] - \frac{66}{59} e e_, [15] - \frac{83}{32} e e_, [16] - \frac{67}{70} e_,^2 [17]$$

$$- \frac{233}{17} e_,^2 [18] + \frac{1}{16} e^3 [20] + \frac{7}{32} e^3 [21]$$

$$- \frac{25}{32} e^3 [22] + \frac{3}{16} e^2 e_, [23] - \frac{105}{16} e^2 e_, [24] + \frac{3}{8} e^2 e_, [25]$$

$$+ \frac{3}{16} e^2 e_, [26] + \frac{15}{16} e^2 e_, [27] - \frac{21}{8} e^2 e_, [28] + \frac{9}{8} e e_,^2 [29]$$

$$+\frac{153}{8}ee_{,}^{2}[30]+\frac{9}{8}ee_{,}^{2}[32]-\frac{51}{8}ee_{,}^{2}[34]-\frac{53}{32}e_{,}^{3}[35]$$

$$-\frac{845}{64}e_{,}^{3}[36]-\frac{1}{64}e_{,}^{3}[37]+\frac{1}{24}e^{4}[38]+\frac{3}{64}e^{4}[39]$$

$$-\frac{27}{32}e^{4}[40]+\frac{3}{32}e^{3}e_{,}[41]+\frac{49}{64}e^{3}e_{,}[42]+\frac{25}{64}e^{3}e_{,}[43]$$

$$+\frac{3}{32}e^{3}e_{,}[44]-\frac{7}{64}e^{3}e_{,}[45]-\frac{175}{64}e^{3}e_{,}[46]+\frac{9}{32}e^{2}e_{,}^{2}[47]$$

$$-\frac{255}{16}e^{2}e_{,}^{2}[48]+\frac{9}{32}e^{2}e_{,}^{2}[50]-\frac{51}{8}e^{2}e_{,}^{2}[52]+\frac{53}{32}[53]$$

$$+\frac{2535}{64}ee_{,}^{3}[54]-\frac{1}{64}ee_{,}^{3}[55]+\frac{53}{32}ee_{,}^{3}[56]+\frac{3}{64}ee_{,}^{3}[57]$$

$$+\frac{45}{64}ee_{,}^{3}[58]+\frac{591}{64}e_{,}^{4}[59]-\frac{2453}{128}e_{,}^{4}[60]+\frac{741}{128}e_{,}^{4}[61]$$

$$-\frac{16}{43}\gamma^{2}[62]-\frac{26}{69}\gamma^{2}[63]+\frac{9}{8}\gamma^{2}e[65]-\frac{3}{8}\gamma^{2}e[66]$$

$$+\frac{3}{8}\gamma^{2}e[67]+\frac{3}{8}\gamma^{2}e[69]-\frac{9}{16}\gamma^{2}e_{,}[71]-\frac{9}{16}\gamma^{2}e_{,}[72]$$

$$-\frac{21}{16}\gamma^{2}e_{,}[73]+\frac{3}{16}\gamma^{2}e_{,}[75]-\frac{15}{16}\gamma^{2}e^{2}[77]-\frac{3}{8}\gamma^{2}e^{2}[78]$$

$$+\frac{3}{32}\gamma^{2}e^{2}[79]+\frac{3}{32}\gamma^{2}e^{2}[81]+\frac{27}{16}\gamma^{2}ee_{,}[83]$$

$$-\frac{9}{16}\gamma^{2}ee_{,}[84]+\frac{21}{16}\gamma^{2}ee_{,}[85]-\frac{3}{16}\gamma^{2}ee_{,}[87]$$

$$+\frac{27}{16}\gamma^{2}ee_{,}[89]-\frac{9}{16}\gamma^{2}ee_{,}[90]-\frac{3}{16}\gamma^{2}ee_{,}[91]$$

$$+\frac{21}{16}\gamma^{2}ee_{,}[93]-\frac{51}{64}\gamma^{2}e_{,}^{2}[95]-\frac{45}{64}\gamma^{2}e_{,}^{2}[96]$$

$$-\frac{195}{64}\gamma^{2}e_{,}^{2}[97]+\frac{3}{64}\gamma^{2}e_{,}^{2}[99]\Big\}$$

$$+\frac{m_{,}a^{3}}{a_{,}^{4}}\Big\{-\frac{3}{8}[101]+\frac{15}{16}e[102]+\frac{3}{16}e[103]-\frac{9}{8}e_{,}[104]$$

$$-\frac{3}{8}e\,[105]-\frac{33}{64}e^2\,[106]+\frac{9}{64}e^2\,[107]+\frac{45}{16}[108]$$

$$+\frac{3}{16}ee_,[109]+\frac{15}{16}ee_,[110]+\frac{9}{16}ee_,[111]-\frac{159}{64}e_,^2[112]$$

$$-\frac{33}{64}e_,^2[113]-\frac{9}{16}\gamma^2[114]-\frac{15}{32}\gamma^2[115]-\frac{5}{8}[116]$$

$$+\frac{45}{16}e[117]-\frac{15}{16}e[118]-\frac{25}{8}e_,[119]+\frac{5}{8}e_,[120]$$

$$-\frac{285}{64}e^2\,[121]-\frac{75}{64}e^2\,[122]-\frac{225}{16}ee_,\,[123]$$

$$+\frac{15}{16}ee_,\,[124]-\frac{45}{16}ee_,[125]-\frac{75}{16}ee_,\,[126]$$

$$-\frac{635}{64}e_,^2\,[127]-\frac{5}{64}e_,^2\,[128]-\frac{15}{32}\gamma^2[129]\Big\}$$

Similarly

$$\frac{dR}{ds}=\frac{m_,a^2}{a_,^3}\gamma\Big\{\frac{204}{137}[146]-\frac{20}{27}[147]+\frac{20}{27}[148]+\frac{3}{2}e[149]$$

$$-\frac{3}{2}e[150]+\frac{9}{4}e[151]-\frac{9}{4}e[152]-\frac{3}{4}e[153]$$

$$+\frac{3}{4}e[154]-\frac{9}{4}e_,[155]+\frac{9}{4}e_,[156]-\frac{21}{8}e_,[157]$$

$$+\frac{21}{8}e_,[158]+\frac{3}{8}e_,[159]-\frac{3}{8}e_,[160]+\frac{3}{8}e^2[161]$$

$$-\frac{3}{8}e^2\,[162]-\frac{15}{8}e^2\,[163]+\frac{15}{8}e^2\,[164]-\frac{3}{4}e^2\,[165]$$

$$+\frac{3}{4}e^2[166]+\frac{9}{4}ee_,[167]-\frac{9}{4}ee_,[168]+\frac{63}{8}ee_,[169]$$

$$-\frac{63}{8}ee_,[170]+\frac{3}{8}ee_,[171]-\frac{3}{8}ee_,[172]-\frac{9}{4}ee_,[173]$$

$$-\frac{9}{4}ee_,[174]\quad\frac{9}{8}ee_,[175]+\frac{9}{8}ee_,[176]\quad\frac{21}{8}ee_,[177]\Big\}$$

$$+\frac{21}{8}ee_{,}[178]-\frac{27}{8}e_{,}^{2}[179]+\frac{27}{8}e_{,}^{2}[180]-\frac{51}{8}e_{,}^{2}[181]$$

$$+\frac{51}{8}e_{,}^{2}[182]\Big\}$$

If δR denote the variation of R due to the perturbations of the moon,

$$\delta R=\left(\frac{dR}{dr'}\right)\delta r'+\left(\frac{dR}{d\lambda'}\right)\delta\lambda'+\left(\frac{dR}{ds}\right)\delta s$$

$$=-a\left(\frac{dR}{da}\right)\mathrm{r}'\delta\frac{1}{r'}+\left(\frac{dR}{d\tau}\right)\delta\lambda'+\left(\frac{dR}{ds}\right)\delta s$$

r' being the elliptic value of r'.

Let $\delta\frac{1}{r'}=r_0+r_1\cos 2\tau-e r_2\cos\xi+e r_3\cos(2\tau-\xi)+$ &c.

$$\delta\lambda'=\lambda_1\sin 2\tau+e\lambda_3\sin(2\tau-\xi)+e\lambda_4\sin(2\tau+\xi)+\text{\&c.}$$

$$\delta s=\gamma s_{146}\sin\eta+\gamma s_{147}\sin(2\tau-\eta)+\text{\&c.}$$

The multiplication of r' by $\delta\frac{1}{r'}$ may be performed by means of the Table.

$$\mathrm{r}\delta\frac{1}{r}=\left\{\left(1+\frac{e^2}{2}\right)r_1-\frac{e^2}{2}\left(1-\frac{3}{8}e^2\right)\left\{r_3+r_4\right\}-\frac{e^4}{4}\left\{r_9+r_{10}\right\}\right\}\cos 2\tau$$

[1]

$$+\left\{\left(1+\frac{e^2}{2}\right)r_2-\frac{1}{2}\left(1-\frac{3}{8}e^2\right)\left\{2r_0+e^2r_8\right\}-\frac{e^2}{4}r_2\right\}e\cos\xi$$

[2]

$+$ &c.

When the square of the eccentricity is neglected,

$$\mathrm{r}\delta\frac{1}{r}=r_1\cos 2\tau+\left\{r_2-r_0\right\}e\cos\xi+\left\{r_3-\frac{r_1}{2}\right\}e\cos(2\tau-\xi)$$

$$+\left\{r_4-\frac{r_1}{2}\right\}e\cos(2\tau+\xi)+r_5e_{,}\cos\xi_{,}$$

$$+ r_6 e_, \cos(2\tau - \xi_,) + r_7 e_, \cos(2\tau + \xi_,)$$

$$\frac{dR}{ds} = \frac{r^2}{2\,r_,^3}\,s \text{ nearly.} \qquad \delta s = \gamma s_{147} \sin(2\tau - \eta) \text{ nearly,}$$

this inequality of latitude being by far the greatest.

The development of δR may be effected with great facility by means of the Table. If

$$r\,\delta\frac{1}{r} = r_0' + r_1' \cos 2\tau + e\,r_2' \cos\xi + e\,r_3' \cos(2\tau - \xi) + \&c.$$

neglecting the squares of the eccentricities,

$$\delta R = \frac{m_,\, a^2}{a_,^3}\left\{ \frac{68}{137} r'_0 + \frac{20}{27}\left\{ r'_1 + \lambda_1 \right\} + \left\{ \frac{40}{27} r'_0 + \frac{68}{137} r_1' \right\} \cos 2\tau \right. \qquad [1]$$

$$+ \left\{ -\frac{76}{77} r_0' + \frac{20}{27}\left\{ r_1' + \lambda_1 \right\} - \frac{38}{17}\left\{ r_1' + \lambda_1 \right\} + \frac{68}{137} r_2' \right.$$

$$\left. + \frac{20}{27}\left\{ r_3' + \lambda_3 \right\} + \frac{20}{27}\left\{ r_4' + \lambda_4 \right\} \right\} e \cos\xi \qquad [2]$$

$$+ \left\{ -\frac{76}{17} r_0' - \frac{38}{77} r_1' + \frac{20}{27} r_2' + \frac{68}{137} r_3' \right\} e \cos(2\tau - \xi) \qquad [3]$$

$$+ \left\{ \frac{40}{27} r_0' - \frac{38}{77} r_1' + \frac{20}{27} r_2' + \frac{68}{137} r_4' \right\} e \cos(2\tau + \xi) \qquad [4]$$

$$+ \left\{ \frac{64}{43} r_0' - \frac{10}{27}\left\{ r_1' + \lambda_1 \right\} + \frac{70}{27}\left\{ r_1' + \lambda_1 \right\} + \frac{68}{137} r_5' \right\} e_, \cos\xi_, \qquad [5]$$

$$+ \left\{ \frac{140}{27} r_0' + \frac{32}{43} r_1' + \frac{20}{27}\left\{ r_5' + \lambda_5 \right\} + \frac{68}{137} r_6' \right\} e_, \cos(2\tau - \xi_,) \qquad [6]$$

$$\left. + \left\{ -\frac{20}{27} r_0' + \frac{32}{43} r_1' + \frac{20}{27}\left\{ r_5' - \lambda_5 \right\} + \frac{68}{137} r_7' \right\} e_, \cos(2\tau + \xi_,) \; * \right\} \qquad [7]$$

---

* The terms here given are not the only sensible: in this and the succeeding pages the expressions given are merely intended as examples to show the manner in which, by means of the Table, the developments may be carried further, as may be seen in the *Philosophical Transactions.*

$d R$ = the differential of R, supposing only $n t$ variable + the differential of R with regard to $n_{\prime} t$ only inasmuch as it is contained in the terms in r, λ and s due to the perturbations; hence $d R$ = the differential of R, supposing only $n t$ variable

$+ \frac{dR}{dr'} d.\delta r' + \frac{dR}{d\lambda'} d.\delta\lambda' + \frac{dR}{ds} d\delta s$; $d.\delta r'$, $d.\delta\lambda'$, and $d\delta s$ being restrained to mean the differentials of those quantities with regard to $n_{\prime} t$ only.

$$\left(\frac{dR}{dr}\right) d.\delta r + \left(\frac{dR}{d\lambda'}\right) d.\delta\lambda' + \left(\frac{dR}{ds}\right) d\delta s$$

$$= -a\left(\frac{dR}{da}\right) d.r\delta\frac{1}{r} + \left(\frac{dR}{d\tau}\right) d\delta\lambda' + \left(\frac{dR}{ds}\right) d.\delta s$$

Hence $\delta . d R$ = the differential of δR, supposing only $n t$ variable

$$+\frac{n_{\prime} m_{\prime} a^2}{a_{\prime}^3}\left\{\frac{2.68}{137}\sin 2\tau\right.$$

[1]

$$+\left\{-\frac{2.20}{27}\left\{r_1'+\lambda_1\right\}-\frac{2.38}{17}\left\{r_1'+\lambda_1\right\}-\frac{2.20}{27}\left\{r_3'+\lambda_3\right\}\right.$$

$$\left.+\frac{2.20}{27}\left\{r_4'+\lambda_4\right\}\right\} e\sin\xi$$

[2]

$$+\left\{-\frac{2.38}{77}r_1'+\frac{2.68}{137}r_3'\right\} e\sin(2\tau-\xi)$$

[3]

$$+\left\{-\frac{2.38}{77}r_1'+\frac{2.68}{137}r_4'\right\} e\sin(2\tau+\xi)$$

[4]

$$+\left\{\frac{2.10}{27}\left\{r_1'+\lambda_1\right\}+\frac{2.70}{27}\left\{r_1'+\lambda_1\right\}-\frac{68}{137}r_5'\right\} e_{\prime}\sin\xi_{\prime}$$

[5]

$$+\left\{\frac{2.32}{43}r_1'+\frac{20}{27}\left\{r_5'+\lambda_5\right\}+\frac{3.68}{137}r_6'\right\} e_{\prime}\sin(2\tau-\xi_{\prime})$$

[6]

$$\left.+\left\{\frac{2.32}{43}-\frac{20}{27}\left\{r_5'-\lambda_5\right\}+\frac{68}{137}r_7'\right\} e_{\prime}\sin(2\tau+\xi_{\prime})\right\}$$

[7]

Similarly

$$\delta . \frac{dR}{d\lambda'} = -a\, d . \frac{\left(\frac{dR}{d\lambda'}\right)}{da} r' \delta \frac{1}{r'} + d . \frac{\left(\frac{dR}{d\lambda'}\right)}{d\lambda'} \delta \lambda' + d . \frac{\frac{dR}{d\lambda}}{ds} \delta s$$

$$= \frac{2\, m_{\prime} a^2}{a_{\prime}^3} \Bigg\{ - \frac{40}{27} r_0' \sin 2\tau$$

[1]

$$+ \left\{ -\frac{20}{27}\left\{r_1' + \lambda_1\right\} - \frac{38}{17}\left\{r_1' + \lambda_1\right\} - \frac{20}{27}\left\{r_3' + \lambda_3\right\} \right.$$

$$\left. + \frac{20}{27}\left\{r_4' + \lambda_4\right\} \right\} e \sin \xi$$

[2]

$$+ \left\{ \frac{76}{17} r_0' - \frac{20}{27} r_2' \right\} e \sin (2\tau - \xi)$$

[3]

$$+ \left\{ -\frac{40}{27} r_0' - \frac{20}{27} r_2' \right\} e \sin (2\tau + \xi)$$

[4]

$$+ \left\{ \frac{10}{27}\left\{r_1' + \lambda_1\right\} + \frac{70}{27}\left\{r_1' + \lambda_1\right\} \right\} e_{\prime} \sin \xi_{\prime}$$

[5]

$$+ \left\{ -\frac{140}{27} r_0' - \frac{20}{27}\left\{r_5' + \lambda_5\right\} \right\} e_{\prime} \sin (2\tau - \xi_{\prime})$$

[6]

$$+ \left\{ \frac{20}{27} r_0' - \frac{20}{27}\left\{r_5' - \lambda_5\right\} \right\} e_{\prime} \sin (2\tau + \xi_{\prime}) \Bigg\}$$

[7]

In order to verify the preceding developments, suppose

$$R = \frac{38}{17} \frac{m_{\prime} a^2}{a_{\prime}^3} e \cos (2\tau - \xi)$$

$$r \delta \frac{1}{r} = r_1' \cos 2\tau \qquad \delta \lambda = \lambda_1 \sin 2\tau$$

neglecting δs

$$\delta R = \left(\frac{dR}{dr}\right) \delta r + \left(\frac{dR}{d\lambda}\right) \delta \lambda = -a \left(\frac{dR}{da}\right) r \delta \frac{1}{r} + \left(\frac{dR}{d\tau}\right) \delta \lambda$$

$$\delta R = -\frac{2 . 38}{17} \frac{m_{\prime} a^2}{a_{\prime}^3} e \left\{ \cos (2\tau - \xi)\, r_1' \cos 2\tau + \sin (2\tau - \xi)\, \lambda_1 \sin 2\tau \right\}$$

$$= \frac{m_, a^2}{a_,^3} e \left\{ -\frac{38}{17} r_1' \cos(4\tau - \xi) - \frac{38}{17} r_1' \cos\xi \right.$$

$$\left. + \frac{38}{17} \lambda_1 \cos(4\tau - \xi) - \frac{38}{17} \lambda_1 \cos\xi \right\}$$

$$= \frac{m_, a^2}{a_,^3} \left\{ -\frac{38}{17} \{r_1' + \lambda_1\} e \cos\xi - \frac{38}{17} \{r_1' - \lambda_1\} e \cos(4\tau - \xi) \right\}$$

[2] [132]

Again,

$$\delta \mathrm{d} R = \frac{4 \cdot 38}{17} \frac{n_, m_, a^2}{a_,^3} e \left\{ -\cos(2\tau - \xi) r_1' \sin 2\tau \right.$$

$$\left. + \sin(2\tau - \xi) \lambda_1 \cos 2\tau \right\}$$

$$= \frac{2 \cdot 38}{17} \frac{n_, m_, a^2}{a_,^3} e \left\{ -r_1' \sin(4\tau - \xi) - r_1' \sin\xi \right.$$

$$\left. + \lambda_1 \sin(4\tau - \xi) - \lambda_1 \sin\xi \right\}$$

$$= \frac{n_, m_, a^2}{a_,^3} \left\{ -\frac{2 \cdot 38}{17} \{r_1' + \lambda_1\} e \sin\xi \right.$$

[2]

$$\left. - \frac{2 \cdot 38}{17} \{r_1' - \lambda_1\} e \sin(4\tau - \xi) \right\}$$

[132]

Similarly

$$\delta \left(\frac{\mathrm{d} R}{\mathrm{d} \lambda} \right) = \frac{2 \cdot 2 \cdot 38}{17} \frac{m_, a^2}{a_,^3} e \left\{ \sin(2\tau - \xi) r'_1 \cos 2\tau \right.$$

$$\left. - \cos(2\tau - \xi) \lambda_1 \sin 2\tau \right\}$$

$$= \frac{2 \cdot 38}{17} \frac{m_, a^2}{a_,^3} e \left\{ r'_1 \sin(4\tau - \xi) - r'_1 \sin\xi \right.$$

$$\left. - \lambda_1 \sin(4\tau - \xi) - \lambda_1 \sin\xi \right\}$$

$$= \frac{m_, a^2}{a_,^3} \left\{ -\frac{2 \cdot 38}{17} \{r'_1 + \lambda_1\} e \sin\xi \right.$$

[2]

$$\left. + \frac{2 \cdot 38}{17} \{r'_1 - \lambda_1\} e \sin(4\tau - \xi) \right\}$$

[132]

Similarly, if δ' denote the variation due to the disturbance of the earth by the moon,

$$\delta' R = - a_{,} \mathrm{d} . \left(\frac{\mathrm{d} R}{\mathrm{d} a_{,}}\right) r_{,} \delta . \frac{1}{r_{,}} - \mathrm{d} . \left(\frac{\mathrm{d} R}{\mathrm{d} \tau}\right) \delta \lambda_{,}$$

Neglecting in R the terms multiplied by $\frac{a^3}{a^4}$ and by s^2, and omitting the factor $m_{,}$,

$$R = - \frac{r^2}{4 r_{,}^3} \left\{ 1 + 3 \cos (2 \lambda - 2 \lambda_{,}) \right\}$$

$$R = - \frac{(\mathrm{r} + \delta r)^2}{4 r_{,}^3} \left\{ 1 + 3 \cos (2 \lambda - 2 \lambda_{,} + 2 \delta \lambda) \right\}$$

$$= - \frac{\{1 + 3 \cos (2 \lambda - 2 \lambda_{,})\}}{4 r_{,}^3} \left\{ 2 \mathrm{r} \delta r + \delta r^2 \right\}$$

$$+ \frac{3 \mathrm{r}}{r_{,}^3} \sin (2 \lambda - 2 \lambda_{,}) \delta r \delta \lambda + \frac{3}{2} \frac{\mathrm{r}^2}{r_{,}^3} (\cos (2 \lambda - 2 \lambda_{,}) (\delta \lambda)^2$$

$$\delta r = - \mathrm{r}^2 \delta \frac{1}{r} + \mathrm{r}^3 \left(\delta \frac{1}{r}\right)^2$$

Neglecting the terms multiplied by $\delta \frac{1}{r}$ and $\delta \lambda$,

$$R = - 3 \mathrm{r}^2 \frac{\{1 + 3 \cos (2 \lambda - 2 \lambda_{,})\}}{4 r_{,}^3} \left(\mathrm{r} \delta \frac{1}{r}\right)^2$$

$$- \frac{3 \mathrm{r}^2}{r_{,}^3} \sin (2 \lambda - 2 \lambda_{,}) \left(\mathrm{r} \delta \frac{1}{r}\right) \delta \lambda_{,} + \frac{3}{2} \frac{\mathrm{r}^2}{r_{,}^3} \cos (2 \lambda - 2 \lambda_{,}) (\delta \lambda)^2$$

$\mathrm{d} R$ and $r \left(\frac{\mathrm{d} R}{\mathrm{d} r}\right)$ may be obtained from R as before.

$$\frac{\mathrm{d} R}{\mathrm{d} \lambda} = \frac{3}{2} \frac{r^2}{r_{,}^3} \sin (2 \lambda - 2 \lambda_{,})$$

$$\frac{\mathrm{d} R}{\mathrm{d} \lambda} = \frac{3 \{2 \mathrm{r} \delta r + \delta r^2\}}{r_{,}^3} \sin (2 \lambda - 2 \lambda_{,}) + \frac{6 \cos (2 \lambda - 2 \lambda_{,})}{r_{,}^3} \mathrm{r} \delta r \delta \lambda$$

$$- \frac{3 \mathrm{r}^2}{r_{,}^3} \sin (2 \lambda - 2 \lambda_{,}) (\delta \lambda)^2$$

Neglecting as before the terms multiplied by $\delta \frac{1}{r}$ and $\mathrm{d} \lambda$,

$$= \frac{9}{2} \frac{\mathrm{r}^2}{r_{,}^3} \sin (2 \lambda - 2 \lambda_{,}) \left(\mathrm{r} \delta \frac{1}{r}\right)^2 - \frac{6 \mathrm{r}^2 \cos (2 \lambda - 2 \lambda_{,})}{r_{,}^3} \left(\mathrm{r} \delta \frac{1}{r}\right) \delta \lambda$$

$$- \frac{3 r^2}{r_{,}^3} \sin (2 \lambda - 2 \lambda_{,}) (\delta \lambda)^2$$

The equation for determining the coefficients of the expression for the reciprocal of the radius vector is,

$$\frac{\mathrm{d}^2 . \mathrm{r}^2}{2\,\mathrm{d}t^2} - \frac{\mathrm{d}^2 . \mathrm{r}^3 \delta \frac{1}{r}}{\mathrm{d}t^2} + \frac{3\,\mathrm{d}^2 . \mathrm{r}^4 \left(\delta \frac{1}{r}\right)^2}{2\,\mathrm{d}t^2} - \frac{2\,\mathrm{d}^2 . \mathrm{r}^5 \left(\delta \frac{1}{r}\right)^3}{\mathrm{d}t^2} - \frac{\mu}{r} + \frac{\mu}{a}$$

$$+ 2 \int \mathrm{d}R + r \left(\frac{\mathrm{d}R}{\mathrm{d}r}\right) = 0$$

r being the elliptic value of r.

The development of $\mathrm{r}^3 \delta \frac{1}{r}$ is easily deduced from that of $\mathrm{r}\,\delta \frac{1}{r}$ given in the *Phil. Trans.* 1832, Part I. p. 3, and that of $\mathrm{r}^4 \left(\delta \frac{1}{r}\right)^2$ from that of $\left(\delta \frac{1}{r}\right)^2$, p. 4. If $\mathfrak{r}_n$ is that part of the coefficient of the nth argument in the development of the quantity $\mathrm{r}^3 \delta \frac{1}{r} - \frac{3}{2} \mathrm{r}^4 \left(\delta \frac{1}{r}\right)^2$ which is independent of r_n, with a contrary sign;

$$\mathfrak{r}_1 = \frac{3e^2}{2}\left(1 + \frac{3}{8}e^2\right)(r_3 + r_4)$$

$$+ \frac{3}{2}\left(1 + 5e^2\right)\left\{2 r_0 r_1 + e^2 (r_3 + r_4) r_2 + e_{\prime}^2 (r_6 + r_7) r_5\right\}$$

$$- 3e^2 \left\{2 r_1 r_2 + 2 r_0 r_3 + 2 r_0 r_4\right\}$$

$$\mathfrak{r}_2 = \frac{3}{2}\left(1 + \frac{3}{8}e^2\right)(2r_0 + e^2 r_8) + \frac{3}{2}\left\{(r_4 + r_3) r_1 + 2 r_0 r_2\right\}$$

$$- 3 . 2 \left\{r_0^2 + \frac{r_1^2}{2} + \frac{e^2 r_3^2}{2} + \&c.\right\}$$

$$\mathfrak{r}_3 = \frac{3}{2}\left(1 + \frac{3}{8}e^2\right)(e^2 r_9 + r_1) + \frac{3}{2}\left\{r_1 r_2 + 2 r_0 r_3\right\}$$

$$- 3\left\{2 r_0 r_1 + e^2 (r_3 + r_4) r_2 + e_{\prime}^2 (r_6 + r_7) r_5\right\}$$

$$\mathfrak{r}_4 = \frac{3}{2}\left(1 + \frac{3}{8}e^2\right)(r_1 + e^2 r_{10}) + \frac{3}{2}\left\{r_1 r_2 + 2 r_0 r_4\right\}$$

$$- 3\left\{2 r_0 r_1 + e^2 (r_3 + r_4) r_2 + e_{\prime}^2 (r_6 + r_7) r_5\right\}$$

Let R_n be the coefficient corresponding to the nth argument in the development of R; R'_n the coefficient corresponding to the nth argument in the development of that part of $\delta\, d\, R$ which is multiplied by $n_{\prime}$, with its sign changed; then

$$r_1 \left\{1 + 3e^2\left(1 + \frac{e^2}{8}\right)\right\} = \frac{(2-2m)^2}{(2-2m)^2-1}\mathfrak{r}_1$$

$$-\frac{2}{(2-2m)^2-1}\left\{\left\{\frac{2}{2-2m}+1\right\}\frac{m_{\prime}}{\mu}a\,R_1 + \frac{m\quad m_{\prime}}{(2-2m)\mu}a\,R_1'\right\}$$

$$c^2\left\{1 - \frac{e^2}{8} - \mathfrak{r}_2\right\}$$

$$= 1 - \frac{e^2}{8} - 2\left\{\left\{\frac{1}{c}+1\right\}\frac{m_{\prime}}{\mu}a\,R_2 + \frac{m}{c}\frac{m_{\prime}}{\mu}a\,R_2'\right\}$$

$$r^3\left\{1 + 3e^2\left(1+\frac{e^2}{8}\right)\right\} = \frac{(2-2m-c)^2}{(2-2m-c)^2-1}\mathfrak{r}_3$$

$$-\frac{2}{(2-2m-c)^2-1}\left\{\left\{\frac{2-c}{2-2m-c}+1\right\}\frac{m_{\prime}}{\mu}\,a R_2\right.$$

$$\left.+\frac{m}{(2-2m-c)}\frac{m_{\prime}}{\mu}a\,R_3'\right\}$$

$$r_4\left\{1+3e^2\left(1+\frac{e^2}{8}\right)\right\} = \frac{(2-2m+c)^2}{(2-2m+c)^2-1}\mathfrak{r}_4$$

$$-\frac{2}{(2-2m+c)^2-1}\left\{\left\{\frac{2+c}{2-2m+c}+1\right\}\frac{m_{\prime}}{\mu}a\,R_4\right.$$

$$\left.+\frac{m}{(2-2m+c)}\frac{m_{\prime}}{\mu}a\,R_4'\right\}$$

$$r_5\left\{1+3e^2\left(1+\frac{e^2}{8}\right)\right\}$$

$$= \frac{m^2}{m^2-1}\mathfrak{r}_5 - \frac{2}{m^2-1}\left\{\frac{m_{\prime}}{\mu}a\,R_5 + \frac{m_{\prime}}{\mu}a\,R_5'\right\}$$

$$r_6\left\{1+3e^2\left(1+\frac{e^2}{8}\right)\right\} = \frac{(2-3m)^2}{(2-3m)^2-1}\mathfrak{r}_6$$

$$-\frac{2}{(2-3m)^2-1}\left\{\left\{\frac{2}{2-3m}+1\right\}\frac{m_{\prime}}{\mu}a\,R_6 + \frac{m}{(2-3m)}\frac{m_{\prime}}{\mu}a\,R_6'\right\}$$

$$r_7\left\{1+3e^2\left(1+\frac{e^2}{8}\right)\right\}=\frac{(2-m)^2}{(2-m)^2-1}\,\mathfrak{r}_7$$

$$-\frac{2}{(2-m)^2-1}\left\{\left\{\frac{2}{2-m}+1\right\}\frac{m_{\prime}}{\mu}\,a\,R_7+\frac{m}{(2-m)}\frac{m_{\prime}}{\mu}\,a\,R_7'\right\}$$

The equation for determining the inequalities of latitude is

$$\frac{d^2 z}{d t^2}+\frac{\mu z}{r^3}+\frac{m_{\prime} z}{\{r^2-2 r r^{\backprime}\cos(\lambda^{\backprime}-\lambda_{\prime})+r_{\prime}^2\}^{\frac{3}{2}}}$$

$$\frac{d^2 z}{d t^2}+\frac{\mu z}{r^3}+\frac{m_{\prime} z}{r_{\prime}^3}+\frac{3 m_{\prime} z r^{\backprime} r \cos(\lambda^{\backprime}-\lambda)}{r_{\prime}^5}=0$$

$$\frac{d^2 . \delta z}{d t^2}+\frac{3\mu z \delta . \frac{1}{r}}{r^3}+\frac{\mu \delta z}{r^3}+\frac{m_{\prime} z}{r_{\prime}^3}+\frac{3 m_{\prime} z r^{\backprime} r_{\prime} \cos(\lambda^{\backprime}-\lambda)}{r_{\prime}^5}=$$

The multiplications required for the integration of this equation by the method of indeterminate coefficients may all be effected at once without difficulty by the Table.

If $\frac{z}{a}=\gamma\, z_{146}\sin\eta+\gamma\, z_{147}\sin(2\tau-\eta)+\gamma z_{148}\sin(2\tau+\eta)+$&c.

Neglecting quantities of the order of the square of the disturbing force,

$$-g^2 z_{146}+3 r_0+z_{146}+\frac{m_{\prime}}{\mu}\frac{a^3}{a_{\prime}^3}=0$$

$$-\left\{2(1-m)-g\right\}^2 z_{147}+\frac{3 r_1}{2}+z_{147}=0$$

$$-\left\{2(1-m)+g\right\}^2 z_{148}+\frac{3 r_1}{2}+z_{148}=0$$

$$s=\frac{z}{r}\text{ nearly}$$

$$=\left\{z_{146}+\frac{e^2}{2}z_{150}+\frac{e^2}{2}z_{149}\right\}\gamma\sin\eta$$

$$+\left\{z_{147}+\frac{e^2}{2}z_{151}+\frac{e^2}{2}z_{153}-\frac{r_1}{2}\right\}\gamma\sin(2\tau-\eta)$$

$$+\left\{z_{148}+\frac{e^2}{2}z_{152}+\frac{e^2}{2}z_{154}+\frac{r_1}{2}\right\}\gamma\sin(2\tau+\eta)+\text{\&c.}$$

The inequalities of longitude are determined through the equation

$$r^{\backprime 4}\frac{d\lambda^{\backprime 2}}{d t^2}=h^2-2\int r^{\backprime 2}\left(\frac{d R}{d\lambda^{\backprime}}\right)d\lambda,$$

which gives

$$\frac{d\lambda'}{dt} = \frac{h}{r'^2}\left\{1 - \frac{1}{h}\int\frac{dR}{d\lambda'}dt + \frac{1}{2h^2}\left\{\int\frac{dR}{d\lambda'}dt\right\}^2 - \frac{1.1}{2.4h^3}\left\{\int\frac{dR}{d\lambda'}dt\right\}^4\right.$$

$$\left. - \frac{1.1.3}{2.4.6h^5}\left\{\int\frac{dR}{d\lambda'}dt\right\}^6 + \&c.\right\}$$

If the coefficients corresponding to the different arguments in the quantity $\frac{a^2}{r'^2}$, be called $2\,\mathfrak{r}'_n$ and the coefficients of the different arguments in the development of the quantity

$$-na\left[\int\frac{dR}{d\lambda'}dt - \frac{1}{2h^2}\left\{\int\frac{dR}{d\lambda'}dt^2\right\}^2\right]$$ be called $\mathfrak{R}_n$, then

$$2\,\mathfrak{r}_0' = \left\{1 + \frac{\gamma^2}{2} + \frac{\gamma^2}{2}s^2_{147}{}^*\right\}\left\{1 + \frac{e^2}{2}\left(1 + \frac{3}{4}e^2\right) + 2r_0 + \frac{r_0^2}{2}\right.$$

$$\left. + \frac{r_1^2}{2} + \frac{e^2 r_2^2}{2} + \frac{e^2 r_3^2}{2} + \frac{e^2 r_4^2}{2} + \frac{e_{\prime}^2 r_5^2}{2} + \frac{e_{\prime}^2 r_6^2}{2} + \frac{e_{\prime}^2 r_7^2}{2}\right\}$$

$$\mathfrak{r}_1' = \left\{1 + \frac{\gamma^2}{2} + \frac{\gamma^2}{2}s^2_{147}\right\}\left\{r_1 - \gamma^2 s_{147} + \frac{e^2}{2}\left(1 - \frac{e^2}{8}\right)\left\{r_3 + r_4\right\}\right.$$

$$\left. + \frac{e^4}{2}\left\{r_9 + r_{10}\right\} + r_0 r_1 + \frac{e^2}{2}(r_3 + r_4)\,r_2 + \frac{e^2}{2}(r_6 + r_7)\,r_5\right\}$$

$$\mathfrak{r}_2' = \left\{1 + \frac{\gamma^2}{2} + \frac{\gamma^2}{2}s^2_{147}\right\}\left\{1 + \frac{3}{8}e^2 + r_2\right.$$

$$\left. + \frac{1}{2}\left(1 - \frac{e^2}{8}\right)\left\{2r_0 + e^2 r_8\right\} + \frac{e^2}{2}r_2 + \frac{1}{2}(r_4 + r_3)\,r_1 + r_0 r_2\right\}$$

$$\mathfrak{r}_3' = \left\{1 + \frac{\gamma^2}{2} + \frac{\gamma^2}{2}s^2_{147}\right\}\left\{r_3 + \frac{1}{2}\left(1 - \frac{e^2}{8}\right)\left\{e^2 r_9 + r_1 - \gamma^2 s_{147}\right\}\right.$$

$$\left. + \frac{e^2}{2}r_4 + \frac{1}{2}r_1 r_2 + r_0 r_3\right\}$$

* $(s_{147})^2$ is intended.

$$\mathfrak{r}_4` = \left\{1+\frac{\gamma^2}{2}+\frac{\gamma^2}{2}s^2_{147}\right\}\left\{r_4+\frac{1}{2}\left(1-\frac{e^2}{8}\right)\left\{r_1-\gamma^2 s_{147}+e^2 r_{10}\right\}\right.$$
$$\left.+\frac{e^2}{2}r_3+\frac{1}{2}r_1 r_2+r_0 r_4\right\}$$

$$\mathfrak{r}_5` = \left\{1+\frac{\gamma^2}{2}+\frac{\gamma^2}{2}s^2_{147}\right\}\left\{r_5+\frac{1}{2}\left(1-\frac{e^2}{8}\right)\left\{e^2 r_{14}+e^2 r_{11}\right\}\right.$$
$$\left.+\frac{1}{2}r_1 r_7+\frac{1}{2}r_1 r_6+r_0 r_5\right\}$$

$$\mathfrak{r}_6` = \left\{1+\frac{\gamma^2}{2}+\frac{\gamma^2}{2}s^2_{147}\right\}\left\{r_6+\frac{1}{2}\left(1-\frac{e^2}{8}\right)\left\{e^2 r_{12}+e^2 r_{16}\right\}\right.$$
$$\left.+\frac{1}{2}r_5 r_1+r_0 r_6\right\}$$

$$\mathfrak{r}_7` = \left\{1+\frac{\gamma^2}{2}+\frac{\gamma^2}{2}s^2_{147}\right\}\left\{r_7+\frac{1}{2}\left(1-\frac{e^2}{8}\right)\left\{e^2 r_{15}+e^2 r_{13}\right\}\right.$$
$$\left.+\frac{1}{2}r_5 r_1+r_0 r_7\right\}$$

$$\lambda` = \left\{\frac{2h\mathfrak{r}_0`}{a^2}+2\mathfrak{r}_0`\mathfrak{R}_0+\mathfrak{r}_1`\mathfrak{R}_1+e^2\mathfrak{r}_3`\mathfrak{R}_3+e^2\mathfrak{r}_4`\mathfrak{R}_4+e_i^2\mathfrak{r}_5`\mathfrak{R}_5+e_i^2\mathfrak{r}_6\mathfrak{R}_6\right.$$
$$\left.+e_i^2\mathfrak{r}_7\mathfrak{R}_7\right\}t$$
$$+\frac{1}{2-2m}\{2\mathfrak{r}_1`+2\mathfrak{r}_0`\mathfrak{R}_1+2\mathfrak{r}_1`\mathfrak{R}_0+e^2\mathfrak{r}_2`\mathfrak{R}_3+e^2\mathfrak{r}_2\mathfrak{R}_4+e^2\mathfrak{r}_3\mathfrak{R}_2+e^2\mathfrak{r}_3`\mathfrak{R}_2$$
$$+e_i^2\mathfrak{r}_5`\mathfrak{R}_6+e_i^2\mathfrak{r}_5`\mathfrak{R}_7+e_i^2\mathfrak{r}_6`\mathfrak{R}_5+e_i^2\mathfrak{r}_7`\mathfrak{R}_5\}\sin 2\tau$$
[1]
$$+\frac{1}{c}\{2\mathfrak{r}_2`+2\mathfrak{r}_0`\mathfrak{R}_2+2\mathfrak{r}_2`\mathfrak{R}_0+\mathfrak{r}_1`\mathfrak{R}_4+\mathfrak{r}_1`\mathfrak{R}_5+\mathfrak{r}_3`\mathfrak{R}_1+\mathfrak{r}_4`\mathfrak{R}_1\}e\sin\xi$$
[2]
$$+\frac{1}{(2-2m-c)}\{2\mathfrak{r}_3`+2\mathfrak{r}_0`\mathfrak{R}_3+2\mathfrak{r}_3`\mathfrak{R}_0+\mathfrak{r}_1\mathfrak{R}_2+\mathfrak{r}_2`\mathfrak{R}_1\}e\sin(2\tau-\xi)$$
[3]

$$+\frac{1}{(2-2m+c)}\{2\mathfrak{r}_4'+2\mathfrak{r}_0'\mathfrak{R}_4+2\mathfrak{r}_4'\mathfrak{R}_0+\mathfrak{r}_1'\mathfrak{R}_3+\mathfrak{r}_2'\mathfrak{R}_1\}\,e\sin(2\tau+\xi)$$

[4]

$$+\frac{1}{m}\{2\mathfrak{r}_5'+2\mathfrak{r}_0'\mathfrak{R}_5+2\mathfrak{r}_5'\mathfrak{R}_0+\mathfrak{r}_1'\mathfrak{R}_7+\mathfrak{r}_1'\mathfrak{R}_6+\mathfrak{r}_6'\mathfrak{R}_1+\mathfrak{r}_7'\mathfrak{R}_1\}\,e_{\prime}\sin\xi_{\prime}$$

[5]

$$+\frac{1}{(2-3m)}\{2\mathfrak{r}_6'+2\mathfrak{r}_0'\mathfrak{R}_6+2\mathfrak{r}_6'\mathfrak{R}_0+\mathfrak{r}_1'\mathfrak{R}_5+\mathfrak{r}_5'\mathfrak{R}_1\}\,e_{\prime}\sin(2\tau-\xi_{\prime})$$

[6]

$$+\frac{1}{(2-m)}\{2\mathfrak{r}_7'+2\mathfrak{r}_0'\mathfrak{R}_7+2\mathfrak{r}_7'\mathfrak{R}_0+\mathfrak{r}_1'\mathfrak{R}_5+\mathfrak{r}_5'\mathfrak{R}_1\}\,e_{\prime}\sin(2\tau+\xi_{\prime})$$

[7]

Neglecting the cubes of the eccentricities, and the quantities of the order of the square of the disturbing force,

$$R=m_{\prime}\Big\{-\frac{1}{r_{\prime}}-\frac{a^2}{4a_{\prime}^3}\Big\{1+\frac{3}{2}e^2+\frac{3}{2}e_{\prime}^2-\frac{3}{2}\gamma^2\Big\}$$

[0]

$$-\frac{3}{4}\frac{a^2}{a_{\prime}^3}\Big\{1-\frac{5}{2}e^2-\frac{5}{2}e_{\prime}^2-\frac{\gamma^2}{2}\Big\}\cos 2\tau+\frac{a^2}{2a_{\prime}^3}e\cos\xi\Big\}$$

[1] [2]

$$+\frac{9}{4}\frac{a^2}{a_{\prime}^3}e\cos(2\tau-\xi)-\frac{3}{4}\frac{a^2}{a_{\prime}^3}e\cos(2\tau+\xi)-\frac{3}{4}\frac{a^2}{a_{\prime}^3}e_{\prime}\cos\xi_{\prime}$$

[3] [4] [5]

$$-\frac{21}{8}\frac{a^2}{a_{\prime}^3}e_{\prime}\cos(2\tau-\xi_{\prime})+\frac{3}{8}\frac{a^2}{a_{\prime}^3}e_{\prime}\cos(2\tau+\xi_{\prime})$$

[6] [7]

$$-r_0-\frac{m_{\prime}a^3}{2\mu a_{\prime}^3}\Big\{1+\frac{3}{2}e^2+\frac{3}{2}e_{\prime}^2-\frac{3}{2}\gamma^2\Big\}=0$$

$$4(1-m)^2\Big\{(1+3e^2)\,r_1-\frac{3e^2}{2}\{r_3+r_4\}\Big\}-r_1$$

$$-\frac{3m_{\prime}a^3}{2\mu a_{\prime}^3}\Big\{1-\frac{5}{2}e^2-\frac{5}{2}e_{\prime}^2-\frac{\gamma^2}{2}\Big\}\Big\{\frac{1}{1-m}+1\Big\}=0$$

$$c^2\Big\{1-3r_0\Big\}-1+\frac{2m_{\prime}a^2}{\mu\,a_{\prime}^3}=0$$

$$(2-2m-c)^2\Big\{r_3-\frac{3}{2}r_1\Big\}-r_3+\frac{9}{2}\frac{m_{\prime}a^3}{\mu\,a_{\prime}^3}\Big\{\frac{1}{2-2m-c}+1\Big\}=0$$

$$(2-2m+c)^2\left\{r_4-\frac{3}{2}r_1\right\}-r_4-\frac{3}{2}\frac{m_, a^3}{\mu\, a_,^3}\left\{\frac{3}{2-2m+c}+1\right\}=0$$

$$m^2 r_5 - r_5 - \frac{3}{2}\frac{m_, a^3}{\mu\, a_,^3} = 0$$

$$(2-3m)^2 r_6 - r_6 - \frac{21 m_, a^3}{4 \mu\, a_,^3}\left\{\frac{2}{2-3m}+1\right\}=0$$

$$(2-m)^2 r_7 - r_7 + \frac{3}{4}\frac{m_,}{\mu}\frac{a^3}{a_,^3}\left\{\frac{2}{2-m}+1\right\}=0$$

$$(1-m)^2 r_{101} - r_{101} - \frac{3\, m_, a^4}{8 \mu\, a_,^4}\left\{\frac{2}{1-m}+3\right\}=0$$

$$\lambda = \frac{h}{a^2}\left\{1+\frac{e^2}{2}+\frac{\gamma^2}{2}+2\,r_0\right\}t+\frac{2\,e\,(1+r_0)}{c}\sin\xi$$

[2]

$$+\frac{5\,e^2(1+r_0)}{4\,c}\sin 2\,\xi$$

[8]

$$+\left\{2\,r_1+e^2(r_3+r_4)-\left\{-\left(1-\frac{5}{2}\,e^2-\frac{5}{2}\,e_,^2-\frac{\gamma^2}{2}\right)\frac{3}{4(1-m)}\right.\right.$$

$$\left.\left.+\frac{9\,e^2}{2\,(2-2\,m-c)}-\frac{3\,e^2}{2(2-2\,m+c)}\right\}\frac{m_,}{\mu}\frac{a^3}{a_,^3}\right\}\frac{1}{2\,(1-m)}\sin 2\,\tau$$

[1]

$$+\left\{2\,r_3+r_1-\left\{\frac{2\,.\,9}{4\,(2-2\,m-c)}-\frac{2\,.\,3}{4\,(2-2\,m)}\right\}\frac{m_,\,a^3}{\mu\,a_,^3}\right\}$$

$$\frac{e}{(2-2\,m-c)}\sin(2\,\tau-\xi)$$

[3]

$$+\left\{2\,r_4+r_1-\left\{-\frac{2\,.\,3}{4\,(2-2\,m+c)}-\frac{2\,.\,3}{4\,(2-2\,m)}\right\}\frac{m_,\,a^3}{\mu\,a_,^3}\right.$$

$$\frac{e}{(2-2\,m+c)}\sin(2\,\tau+\xi)$$

[4]

$$+\frac{2\,r_5\,e_,}{m}\sin\xi_,$$

[5]

$$+\left\{2 r_6 + \frac{2 \cdot 21\, m_, a^3}{8\,(2-3\,m)\,\mu\, a_,^3}\right\}\frac{e_,}{(2-3m)}\sin(2\,\tau - \xi_,)$$

[6]

$$+\left\{2 r_7 - \frac{2 \cdot 3\, m_, a^3}{8\,(2-m)\,\mu\, a_,^3}\right\}\frac{e_,}{(2-m)}\sin(2\,\tau + \xi_,)$$

[7]

$$+\left\{2 r_{101} - \frac{3\, m_, a^4}{8\,(1-m)\,\mu\, a_,^4}\right\}\frac{1}{(1-m)}\sin\tau$$

[101]

$$r_0 = -\frac{m_, a^3}{2\,\mu\, a_,^3}$$

$$c^2 = 1 - \frac{7\, m_, a^3}{2\,\mu\, a_,^3} \qquad c = 1 - \frac{7\, m_, a^3}{4\,\mu\, a_,^3}$$

$$g^2 = 1 - \frac{m_, a^3}{2\mu\, a_,^3} \qquad g = 1 - \frac{m_, a^3}{4\,\mu\, a_,^3}$$

If $\frac{h}{a^2}\left\{1 + 2\, r_0\right\} = \mathrm{n}$ or $n\left\{1 + 2\, r_0\right\} = \mathrm{n}$

$$c\, n = \mathrm{n}\left\{1 - \frac{3\, m_, a^3}{4\,\mu\, a_,^3}\right\}$$

$$g n = \mathrm{n}\left\{1 + \frac{3\, m_, a^3}{4\,\mu\, a_,^3}\right\}$$

If $1 - \frac{3\, m_, a^3}{4\,\mu\, a_,^3}$ be called c

and $1 + \frac{3\, m_, a^3}{4\,\mu\, a_,^3}$ be called g

the quantities c and g coincide with the quantities c and g of the *Mécanique céleste*. See the *Math. Tracts* by Prof. Airy, p. 59.

If $\sqrt{\frac{\mu}{a^3}}\left\{1 - \frac{m_, a^3}{\mu\, a_,^3}\right\} = \sqrt{\frac{\mu}{\mathrm{a}^3}}$

$$a = \mathrm{a}\left\{1 - \frac{2\, m_, a^3}{3\,\mu\, a_,^3}\right\}$$

$$\frac{1 + r_0}{a} = \frac{1 + \frac{2\, m_, a^3}{3\mu\, a_,^3} - \frac{m_, a^3}{2\, a_,^3}}{\mathrm{a}} = \frac{1 + \frac{m_, a^3}{6\,\mu\, a_,^3}}{\mathrm{a}}$$

If 2 e is the coefficient of sin ξ in the expression for the longitude,

$e = \text{e}\,(c - r_0)$ nearly

$$e = \text{e}\left\{1 - \frac{7}{4}\frac{m_{\prime}}{\mu}\frac{a^3}{a_{\prime}^3} + \frac{m_{\prime}}{2\mu}\frac{a^3}{a_{\prime}^3}\right\}$$

$$= \text{e}\left\{1 - \frac{5}{4}\frac{m_{\prime}a^3}{\mu\, a_{\prime}^3}\right\}$$

$$\frac{e}{a} = \frac{\text{e}}{\text{a}}\left\{1 - \frac{5}{4}\frac{m_{\prime}}{\mu}\frac{a^3}{a_{\prime}^3} + \frac{2}{3}\frac{m_{\prime}}{\mu}\frac{a^3}{a_{\prime}^3}\right\}$$

$$= \frac{\text{e}}{\text{a}}\left\{1 - \frac{7\,m_{\prime}}{12\,\mu}\frac{a^3}{a_{\prime}^3}\right\}$$

$$r_{105}\left\{1 - 3\,r_0 - 1\right\} - \frac{6\,m_{\prime}}{8\,\mu}\frac{a^4}{a_{\prime}^4} - \frac{9\,m_{\prime}a^4}{8\,\mu\,a_{\prime}^4} = 0$$

$$r_{105}\left\{1 - 3\,r_0 - 1\right\} = \frac{15\,m_{\prime}a^4}{8\,\mu\;\;a_{\prime}^4} = 0 \qquad r_{105} = \frac{5\,a}{4\,a_{\prime}}$$

Finally,

$$\frac{\text{a}}{r} = 1 + \frac{m_{\prime}\,a^3}{6\,\mu\,a_{\prime}^3} + \text{e}\left\{1 - \frac{7\,m_{\prime}a^3}{12\,\mu\,a_{\prime}^3}\right\}\cos\xi + \frac{5\,a}{4\,a_{\prime}}e_{\prime}\cos(\tau + \xi_{\prime}) + \&\text{c.}$$

$$\lambda = \text{n}\,t + 2\,\text{e}\sin\xi + \left\{\frac{5\,a}{2\,a_{\prime}} - \frac{3\,m_{\prime}\,a^4}{8\,\mu\,a_{\prime}^4}\right\}e_{\prime}\sin(\tau + \xi_{\prime}) + \&\text{c.}$$

This gives for the coefficient of $\sin(\tau + \xi_{\prime})$ in the expression for the longitude

$$+\left\{\frac{5\,a}{2\,a_{\prime}} - \frac{3\,m_{\prime}a^4}{8\,\mu\,a_{\prime}^4}\right\}e_{\prime}$$

which in sexagesimal seconds is $21''{\cdot}7$; according to M. Damoiseau it should be $17''{\cdot}56$. Is the difference to be attributed entirely to the quantities of the order of the square of the disturbing force neglected in the preceding approximation?

It follows from the preceding expressions that in developing R, δR, &c., e may be written instead of e, and a instead of a, provided

$$\lambda_2 = 0, \quad r_0 = \frac{m_{\prime}\,a^3}{6\,\mu\,a_{\prime}^3}, \quad r_2 = -\frac{7\,m_{\prime}a^3}{12\mu\,a_{\prime}^3}, \text{ and } \frac{m_{\prime}\,a^3}{\mu\,a_{\prime}^3} = m^2 = \frac{1}{178{\cdot}72}$$

The true longitude is thus expressed in a series of angles consisting of various combinations of the angles τ, ξ, η, and $\xi_{\prime}$, and their multiples and no others, the angle $\tau + \xi_{\prime}$, corresponding to the angle $nt + \varepsilon - \varpi_{\prime}$, in the Planetary theory, and r_{105} to the quantity f' of the *Mécanique céleste.* This being the case, the stability of the system may be inferred from the circumstance that the quantities c and g are rational, because in this case it is evident that the quantities r and s vary within given limits, none of the arguments becoming imaginary or the trigonometrical lines being converted into exponentials. The stability of the system consisting of the Earth, the Moon and the Sun, does not seem to have been specially considered by Laplace or by any who have written upon the Lunar theory. To determine the exact limits of the ratio of the distances of the sun and moon, when the quantities c and g cease to be rational, the masses $m_{\prime}$ and μ being given, is a problem which seems worthy attention, and of which I have in vain sought a satisfactory solution.

The non-sphericity of the earth introduces into the disturbing function the term

$$(M + m)\frac{\delta V}{M}$$

M being the mass of the earth, and m the mass of the moon. See *Mécanique céleste*, vol. iii. p. 183. V is the quantity denoted by Q, in my *Account of the "Traité sur le flux et réflux de la mer,"* p. 32.

$$Q = \frac{4\,\pi\,\rho\,a^3}{r}\left\{1 - \frac{e^2 + i^2}{2}\right.$$
$$\left. + \frac{a^2\left\{(e^2 + i^2)\,x^2 + (i^2 - 2\,e^2)\,y^2 + (e^2 - 2\,i^2)\,z^2\right\}}{10\,r^4}\right\}$$

When $e = i$

$$Q = \frac{4\,\pi\,\rho\,a^3}{r}\left\{1 - e^2 + \frac{a^2\,e^2\,\{2\,x^2 - y^2 - z^2\}}{10\,r^4}\right\}$$

If q is equal to the ratio of the centrifugal force to the equatorial gravity at the earth's surface, the preceding expression may be put in the form, (See *Account, &c.* p. 40,)

$$Q = \frac{M}{r}\left\{1 + \frac{(e^2 - q)}{6\,r^4}\,a^2\,\{2\,x^2 - y^2 - z^2\}\right\}$$

Substituting

for x, $r \sin f\lambda \sin \omega + r\,s \cos \omega$

for y, $r \sin f\lambda \cos \omega - r\,s \sin \omega$

and for z, $r \cos f\lambda$,

ω being the obliquity of the ecliptic, and $f\lambda$ the longitude of the moon, reckoned from the first point of Aries; the term to be added to R, is

$$-\frac{\mu\,(e^2-q)}{2\,r^3}\,a^2\left\{\sin^2 f\lambda\,\sin^2\omega+2\,s\,\sin f\lambda\,\sin\omega\cos\omega+\cos^2\omega\,s^2-\frac{1}{3}\right\}$$

which agrees with the expression of M. Damoiseau, p. 581. In his notation,

$$\frac{e^2}{2}=\alpha\,\rho,\quad q=\alpha\,\varphi,\quad a=D,\quad \omega=\lambda,\quad f\lambda=fv$$

Knowing $\frac{e^2}{2}$, (*the compression, l'aplatissement,*) the inequalities due to the quantity $\delta\,V$, may be determined; and conversely, these inequalities obtained by observation, furnish the means of determining the *compression*, which is thus found to be about $\frac{1}{305}$.

The equations

$$d\,\nu+\frac{r'^2\sin(\lambda'-\nu)}{h^2\tan\iota}\left\{(1+s^2)\left(\frac{d\,R}{d\,s}\right)-r's\left(\frac{d\,R}{d\,r'}\right)-\left(\frac{d\,R}{d\,\lambda'}\right)\left(\frac{d\,s}{d\lambda'}\right)\right\}d\,\lambda'=0$$

$$d\,\iota+\frac{r'^2\cos\iota^2\cos(\lambda'-\nu)}{h^2}\left\{(1+s^2)\right\}\left(\frac{d\,R}{d\,s}\right)-r's\left(\frac{d\,R}{d\,r'}\right)-\left(\frac{d\,R}{d\,\lambda'}\right)\left(\frac{d\,s}{d\,\lambda'}\right)\Big\}\,d\,\lambda'=0$$

(see. p. 10.), serve to verify some of the theorems of Newton in the third volume of the Principia.

In fact

$$R=-\frac{m_\prime\,r'^2}{4\,r_\prime^3}\left\{1+3\cos(2\,\lambda'-2\,\lambda_\prime)-2\,s^2\right\}$$

$$(1+s^2)\left(\frac{d\,R}{d\,s}\right)=m_\prime\,s\,(1+s^2)\,\frac{r'^2}{r_\prime^3}$$

$$r's\left(\frac{d\,R}{d\,r'}\right)=-\frac{m_\prime\,r'^2}{2\,r_\prime^3}\left\{1+3\cos(2\,\lambda'-2\,\lambda_\prime)-2\,s^2\right\}s$$

$$\left(\frac{d\,R}{d\,\lambda'}\right)=\frac{3\,m_\prime\,r'^2}{2\,r_\prime^3}\sin(2\,\lambda'-2\,\lambda_\prime)\qquad\frac{d\,s}{d\,\lambda'}=\tan\iota\cos(\lambda'-\nu)$$

$$d\,\nu+\frac{m_\prime\,r'^4\sin(\lambda'-\nu)}{h^2\,r_\prime^3}\left\{\sin(\lambda'-\nu)+\frac{\sin(\lambda'-\nu)}{2}\left\{1+3\cos(2\,\lambda'-2\lambda_\prime)\right\}\right.$$

$$-\frac{3}{2}\cos(\lambda'-\nu)\sin(2\lambda'-2\lambda_{\prime})\Big\}\,d\lambda'=0$$

$$d\nu+\frac{m_{\prime}r'^{4}\sin(\lambda'-\nu)}{h^{2}r_{\prime}^{3}}\Big\{\sin(\lambda'-\nu)+3\cos(\lambda'-\lambda_{\prime})\Big\{\cos(\lambda'-\lambda_{\prime})\sin(\lambda'-\nu)$$

$$-\sin(\lambda'-\lambda_{\prime})\cos(\lambda'-\nu)\Big\}-\sin(\lambda'-\nu)\Big\}d\lambda'=0$$

$$-d\nu=\frac{3m_{\prime}r'^{4}}{h^{2}r_{\prime}^{3}}\sin(\lambda'-\nu)\cos(\lambda'-\lambda_{\prime})\sin(\lambda'_{\prime}-\nu)\,d\lambda'$$

$$=\frac{3\,m_{\prime}a^{3}}{\mu\,a_{\prime}^{3}}\sin(\lambda-\nu)\cos(\lambda-\lambda_{\prime})\sin(\lambda_{\prime}-\nu)\,d\lambda\text{ nearly}$$

$$=\frac{1}{59{\cdot}575}\sin(\lambda-\nu)\cos(\lambda-\lambda_{\prime})\sin(\lambda_{\prime}-\nu)\,d\lambda$$

Which agrees with the result of Newton, Prop. XXX. Lib. 3. "Est igitur velocitas nodorum ut IT × PH × AZ, sive ut contentum sub sinubus trium angulorum TPI, PTN et STN. . . . . Sunto enim PK, PH et AZ prædicti tres sinus. Nempe PK sinus distantiæ lunæ a quadraturâ, PH sinus distantiæ lunæ a nodo, et AZ sinus distantiæ nodi a sole, et erit velocitas nodi ut contentum PK × PH × AZ."

Similarly

$$-d\iota=\frac{3\,m_{\prime}r^{4}\cos\iota}{h^{2}r_{\prime}^{3}}\sin\iota\cos(\lambda-\nu)\cos(\lambda-\lambda_{\prime})\sin(\lambda_{\prime}-\nu)\,d\lambda$$

"Erit angulus GPg (seu inclinationis horariæ variatio) ad angulum 33″ 16‴ 3⁗ ut IT × AZ × TG × $\frac{Pp}{PG}$ ad AT cub." Prop. XXXIV. Lib. 3.

The preceding theorems are not true for the terms multiplied by the higher powers of the ratio of a to $a_{\prime}$, and therefore they do not apply to the Planetary theory. The proof given by Wodehouse (*Treatise on Astronomy*, vol. ii. p. 442.), and noticed by Mr. Whewell, is liable to objection, because the equations which he makes use of to determine the quantities $d\iota$ and $d\nu$ (in his notation $\delta\theta$ and $\delta\phi$,) require the disturbing function to be developed in terms of the mean longitude *explicitly*. The theorems of Newton relating to the motion of the moon's node, and to the variation of the obliquity of her orbit, will ever be ranked amongst his greatest discoveries.

M. Gautier has drawn (*Essai historique sur le problème des trois corps*, p. 53.) a parallel between the methods of Clairaut, d'Alembert and Euler; and it will be seen that their methods do not resemble in any respect those which I have employed. Both Clairaut and d'Alembert, by means of the differential equation of the second order, in which the true longitude is the independent variable, obtained the expression for the reciprocal of the radius vector in terms of cosines of the true longitude. They substituted this value in the differential equation which determines the mean longitude, and obtained by integration the value of the mean motion in terms of the sines of the true longitude. By the reversion of series they then found the true longitude in terms of sines of the mean longitude. The method of Euler is not so simple, but is remarkable as introducing the employment of three rectangular coordinates and the decomposition of forces in the direction of three rectangular axes.

Although d'Alembert and Clairaut made use of the same differential equations, disguised under a different notation, yet they did not arrive at these in the same manner, nor did they employ the same method of integration. Laplace has pushed the approximations to a much greater extent; but his method coincides in all respects with that of Clairaut.

In the method of Clairaut, when the square of the disturbing force and the squares of the eccentricity and inclination are neglected, the equations employed are

$$\left\{\frac{d^2\frac{1}{r}}{d\lambda^2}+\frac{1}{r}\right\}\left\{1-\frac{2}{h^2}\int r^2\left(\frac{dR}{d\lambda}\right)d\lambda\right\}-\frac{1}{a}$$

$$-\frac{r}{h^2}\left\{r\left(\frac{dR}{dr}\right)-\frac{1}{r}\left(\frac{dR}{d\lambda}\right)\frac{dr}{d\lambda}\right\}=0$$

$$\frac{dR}{d\lambda}=\frac{3m_{\prime}r^2}{2r_{\prime}^2}\sin(2\lambda-2\lambda_{\prime})$$

$$\frac{dR}{dr}=-\frac{m_{\prime}r}{2r_{\prime}^3}\left\{1+3\cos(2\lambda-2\lambda_{\prime})\right\}$$

$$h^2=\mu a \qquad r=\frac{a}{1+e\cos(c\lambda-\varpi)} \qquad \frac{dr}{d\lambda}=ce\sin(c\lambda-\varpi)$$

$$\frac{d^2\frac{1}{r}}{d\lambda^2}+\frac{1}{r}-\frac{1}{a}-\frac{3m_{\prime}}{\mu a^2}\int\frac{r^4}{r_{\prime}^3}\sin(2\lambda-2\lambda_{\prime})\,d\lambda$$

$$+\frac{m_{\prime}}{2\mu}\frac{r^3}{a\,r_{\prime}^3}\left\{1+3\cos(2\lambda-2\lambda_{\prime})\right\}$$

$$+\frac{3m_{\prime}e\,r^2}{2\mu\,r_{\prime}^3}\sin(2\lambda-2\lambda_{\prime})\sin(\lambda-\varpi)=0$$

In order to integrate this equation, the value of $\lambda_{\prime}$ in terms of λ must be substituted, which substitution is an operation by no means simple, and therefore liable to occasion error.

$\lambda_{\prime} = m\,\lambda - 2\,m\,e \sin (c\,\lambda - \varpi) + 2\,e_{\prime} \sin (m\,\lambda - \varpi_{\prime}) + \&\text{c.}$

The equation

$$h\,\mathrm{d}\,t = \mathrm{d}\,\lambda \left\{ 1 + \frac{1}{h^2} \int \left(\frac{\mathrm{d}\,R}{\mathrm{d}\,\lambda} \right) r^2\,\mathrm{d}\,\lambda \right\}$$

gives t in terms of λ, and by the reversion of series λ may afterwards be obtained in terms of t. The equation for determining the inequalities of latitude is

$$\left\{ \frac{\mathrm{d}^2 s}{\mathrm{d}\,\lambda^2} + s \right\} \left\{ 1 - \frac{2}{h^2} \int \left(\frac{\mathrm{d}\,R}{\mathrm{d}\,\lambda} \right) r^2\,\mathrm{d}\,\lambda \right\}$$

$$+ \frac{r^2}{h^2} \left(\frac{\mathrm{d}\,R}{\mathrm{d}\,s} \right) - \frac{r^2\,s}{h^2} \left(\frac{\mathrm{d}\,R}{\mathrm{d}\,r} \right) - \frac{r^2}{h^2} \left(\frac{\mathrm{d}\,R}{\mathrm{d}\,\lambda} \right) \left(\frac{\mathrm{d}\,s}{\mathrm{d}\,\lambda} \right) = 0$$

$$\frac{\mathrm{d}\,R}{\mathrm{d}\,s} = \frac{3\,m_{\prime}}{2} \frac{r^2}{r_{\prime}^3} \left\{ 1 + \cos (2\,\lambda - 2\,\lambda_{\prime}) \right\} \qquad \frac{\mathrm{d}\,s}{\mathrm{d}\,\lambda} = g\,\gamma \cos (g\,\lambda - \nu)$$

4. The following are the numerical values of some of the quantities which occur, according to M. Damoiseau, *Théorie de la lune*, p. 502;

$e = \cdot 0548442 \qquad \log e = 8\cdot 7391307$

$e_{\prime} = \cdot 0167527 \qquad \log e_{\prime} = 8\cdot 2251205$

$\gamma = \cdot 0900684 \qquad \log \gamma = 8\cdot 9545705$

$c = \cdot 991548 \qquad g = 1\cdot 00402175$

Also $\dfrac{m_{\prime}\,a^3}{\mu\,a_{\prime}^3} = \dfrac{1}{178\cdot 72} \qquad \log \dfrac{m_{\prime}\,a^3}{\mu\,a_{\prime}^3} = 7\cdot 7478182$

The first column of the following Table gives the index of the argument; the second the approximate value of the numerical coefficient of the corresponding term in the development of R (see p. 39.); and the third the logarithm of that coefficient + $\log \dfrac{m_{\prime}\,a^3}{\mu\,a_{\prime}^3}$.

[0]	$-\frac{34}{137}$	$-7\cdot 1426047$				
[1]	$-\frac{20}{27}$	$-7\cdot 6175419$	$2-2m =$	$1\cdot 8503970$	$\log(2-2m) =$	$0\cdot 2672649$
[2]	$+\frac{38}{77}$	$+7\cdot 4411195$	$c =$	$\cdot 9915480$	$\log c =$	$9\cdot 9963137$
[3]	$+\frac{38}{17}$	$+8\cdot 0972347$	$2-2m-c =$	$\cdot 8588490$	$\log(2-2m-c) =$	$9\cdot 9339173$
[4]	$-\frac{20}{27}$	$-7\cdot 6177065$	$2-2m+c =$	$2\cdot 8419460$	$\log(2-2m+c) =$	$0\cdot 4536149$
[5]	$-\frac{32}{43}$	$-7\cdot 6196796$	$m =$	$\cdot 0748013$	$\log m =$	$8\cdot 8739091$

[6] $-\frac{70}{27}$	$-8{\cdot}1616746$	$2-3m=1{\cdot}7755960$	$\log(2-3m)=0{\cdot}2493445$
[7] $+\frac{10}{27}$	$+7{\cdot}3163403$	$2-m=1{\cdot}9251987$	$\log(2-m)=0{\cdot}2844754$
[8] $+\frac{10}{81}$	$+6{\cdot}8395953$	$2c=1{\cdot}9830960$	$\log 2c=0{\cdot}2973437$
[9] $-\frac{28}{15}$	$-8{\cdot}0187620$	$2-2m-2c=-{\cdot}1326990$	$\log(2-2m-2c)=9{\cdot}1228678$
[10] $-\frac{20}{27}$	$-7{\cdot}6175362$	$2-2m+2c=3{\cdot}8334930$	$\log(2-2m+2c)=0{\cdot}5835978$
[11] $+\frac{20}{27}$	$+7{\cdot}6175419$	$c+m=1{\cdot}0663490$	$\log(c+m)=0{\cdot}0278995$
[12] $+\frac{180}{23}$	$+8{\cdot}6413401$	$2-3m-c={\cdot}7840480$	$\log(2-3m-c)=9{\cdot}8943427$
[13] $+\frac{10}{27}$	$+7{\cdot}3169697$	$2-m+c=2{\cdot}9167470$	$\log(2-m+c)=0{\cdot}4648988$
[14] $+\frac{20}{27}$	$+7{\cdot}6175362$	$c-m={\cdot}9167470$	$\log(c-m)=9{\cdot}9622496$
[15] $-\frac{66}{59}$	$-7{\cdot}7964962$	$2-m-c={\cdot}9336500$	$\log(2-m-c)=9{\cdot}9701841$
[16] $-\frac{83}{32}$	$-8{\cdot}1618122$	$2-3m+c=2{\cdot}7671440$	$\log(2-3m+c)=0{\cdot}4420318$
[17] $-\frac{67}{70}$	$-7{\cdot}7957279$	$2m={\cdot}1496026$	$\log 2m=9{\cdot}174939$
[18] $-\frac{233}{17}$	$-8{\cdot}5469910$	$2-4m=1{\cdot}7007948$	$\log(2-4m)=0{\cdot}2306517$

Substituting in the equations of p. 49, and writing the logarithms of the coefficients between brackets instead of the coefficients themselves, we get

$$r_1=[0{\cdot}1460995]\,\mathfrak{r}_1-[0{\cdot}2308405]\frac{a\,m_{\prime}}{\mu}R_1-[8{\cdot}5192440]\frac{a\,m_{\prime}}{\mu}R_1'$$

$$r_3=-[0{\cdot}4450058]\,\mathfrak{r}_3+[1{\cdot}2154967]\frac{a\,m_{\prime}}{\mu}R_3+[9{\cdot}8181930]\frac{a\,m_{\prime}}{\mu}R_3'$$

$$r_4=[0{\cdot}0535010]\,\mathfrak{r}_4-[9{\cdot}7596140]\frac{a\,m_{\prime}}{\mu}R_4+[7{\cdot}8675954]\frac{a\,m_{\prime}}{\mu}R_4'$$

$$r_5=-[7{\cdot}7463524]\,\mathfrak{r}_5+[0{\cdot}2995642]\frac{a\,m_{\prime}}{\mu}R_5+[0{\cdot}2995642]\frac{a\,m_{\prime}}{\mu}R_5'$$

$$r_6=[0{\cdot}1617938]\,\mathfrak{r}_6-[0{\cdot}2917755]\frac{a\,m_{\prime}}{\mu}R_6+[8{\cdot}5887003]\frac{a\,m_{\prime}}{\mu}R_6'$$

$$r_7=[0{\cdot}1326574]\,\mathfrak{r}_7-[0{\cdot}1741219]\frac{a\,m_{\prime}}{\mu}R_7+[8{\cdot}4541703]\frac{a\,m_{\prime}}{\mu}R_7'$$

The quantities between brackets are essentially constant, and do not change in the successive approximations.

The terms depending upon the square of the disturbing function are very little sensible in the values of r_1 and λ_1; when these are neglected,

$$\mathfrak{r}_1=\frac{3\,e^2}{2}(r_3+r_4)\quad \mathfrak{r}_3=\frac{3}{2}r_1\quad \mathfrak{r}_4=\frac{3}{2}r_1\quad R_1'=0,\quad R_3'=0,\quad R_4'=0$$

writing the logarithms of the numbers between brackets, instead of the numbers themselves,

$$r_1 = [0{\cdot}1460995]\,\mathfrak{r}_1 - [0{\cdot}2308405]\,\frac{a\,m_,}{\mu}\,R_1$$

$$r_3 = -\,[0{\cdot}4450058]\,\mathfrak{r}_3 + [1{\cdot}2154967]\,\frac{a\,m_,}{\mu}\,R_3$$

$$r_4 = [0{\cdot}0535010]\,\mathfrak{r}_4 - [9{\cdot}7596140]\,\frac{a\,m_,}{\mu}\,R_4$$

$$r_1 = [0{\cdot}1460995]\,\mathfrak{r}_1 - [7{\cdot}8483824]$$

$$r_3 = -\,[0{\cdot}4450058]\,\mathfrak{r}_3 + [9{\cdot}3127314]$$

$$r_4 = [0{\cdot}0535010]\,\mathfrak{r}_4 + [7{\cdot}3773205]$$

whence

$$r_3 + r_4 = -\,{\cdot}24826\,r_1 + {\cdot}20831$$

$$1{\cdot}001568\,r_1 = {\cdot}001315 + {\cdot}007053 \qquad r_1 = {\cdot}008354$$

$$\lambda_1 = \left\{2r_1 + e^2(r_3 + r_4) - \left\{\frac{-\,2\,[7{\cdot}6175419]}{(2 - 2\,m)} + \frac{2\,e^2\,[8{\cdot}0972347]}{(2 - 2\,m - c)} - \frac{2\,e^2\,[7{\cdot}6177065]}{(2 - 2\,m + c)}\right\}\right\}\frac{1}{(2 - 2\,m)}$$

whence $\lambda_1 = {\cdot}011526$, which converted into sexagesimal seconds* is $2377''{\cdot}5$; the accurate value of this quantity according to M. Damoiseau is $2370''{\cdot}00$.

To the same degree of approximation

$$r_3 = {\cdot}170545 \qquad \lambda_3 = {\cdot}37817 \qquad r'_3 = r_3 - \frac{r_1}{2} = {\cdot}166368$$

The terms depending upon the square of the disturbing function are sensible in the values of r_3, λ_3: I shall now take into consideration the principal of those terms, which may be done very easily and without having recourse to elimination. For this purpose it is only necessary at first to retain in $\mathfrak{r}_3$, $\mathfrak{r}'_3$, $\delta\,R$, $\delta\,\mathrm{d}\,R$, and $\delta\,\dfrac{\mathrm{d}\,R}{\mathrm{d}\,\lambda}$, the terms multiplied by r_0, r_1, λ_1, r_2, r_3, and λ_3.

$$r_0 = \frac{m_,}{6\,\mu}\frac{a^3}{a_,^3} = {\cdot}0009325 \qquad r_2 = -\,\frac{7}{12}\frac{m_,\,a^3}{\mu\,a_,^3} = -\,{\cdot}0005439 \text{ nearly;}$$

also

$$\mathfrak{r}_3 = \frac{3}{2}\,r_1 + \frac{3}{2}\,(r_1\,r_2 + 2\,r_0\,r_3) - 3\,(2\,r_0\,r_1 + e^2\,r_2\,r_3)$$

$$= {\cdot}012531 + {\cdot}0000030 + {\cdot}0027975\,r_3 - {\cdot}0000467 + {\cdot}0000049\,r_3$$

$$= {\cdot}012487 + {\cdot}0028044\,r_3$$

---

* By adding to the logarithm of λ_1 the logarithm 5·3144251.

$$\frac{a\,m_{\prime}}{\mu} R_3 = [8{\cdot}0972347] - [8{\cdot}0972347]\, r_0 - [7{\cdot}4411195] \left\{ r_1 - \frac{e^2}{2} r_3 \right\}$$

$$+ [7{\cdot}6175419] \{ r_2 - r_0 \} + 2\,[7{\cdot}1426047] \left\{ r_3 - \frac{r_1}{2} \right\}$$

$$= {\cdot}0125093 - {\cdot}00001166 - {\cdot}00002306 + {\cdot}000004141\, r_3$$

$$- {\cdot}000006119 + {\cdot}0027773\, r_3 - {\cdot}0000116$$

$$= {\cdot}0124569 + {\cdot}0027814\, r_3$$

$$\frac{a\,m_{\prime}}{\mu} R'_3 = 2\,[7{\cdot}4411195] \left\{ r_1 - \frac{e^2}{2} r_3 \right\} + 4\,[7{\cdot}1426047] \left\{ r'_3 - \frac{r_1}{2} \right\}$$

$$= - {\cdot}00004612 + {\cdot}000008282\, r_3 + {\cdot}0053546 r_3 - {\cdot}0000232$$

$$= - {\cdot}0000693 + {\cdot}0055629\, r_3$$

$$r_3 = - [0{\cdot}4450058]\, r_3 + [1{\cdot}2154967]\, R_3 + [9{\cdot}8181930]\, R'_3$$

$$= - {\cdot}034790 - {\cdot}007807\, r_3 + {\cdot}20460 + {\cdot}045683\, r_3 - {\cdot}0000456 + {\cdot}0036601\, r_3$$

$$= {\cdot}16977 + {\cdot}041536\, r_3$$

$$r_3 = {\cdot}17712$$

$$\int \frac{d R}{d \lambda}\, d t = \Big\{ 2\,[8{\cdot}0972347] - 2\,[8{\cdot}0972347]\, r_0$$

$$+ 2\,[7{\cdot}1426047] \{ r_2 - r_0 \} \Big\} \frac{e}{(2-2m-c)} \sin (2\,\tau - \xi)$$

$$\lambda_3 = \Bigg\{ 2\, r_3 + r_1 + r_1 r_2 + 2\, r_0 r_3$$

$$- \frac{1}{(2-2m-c)} \left\{ 2\,[8{\cdot}0972347] - 2\,[8{\cdot}0972347]\, r_0 + 2\,[7{\cdot}1426047] \{ r_2 - r_0 \} \right\}$$

$$+ \frac{2\,[7{\cdot}6175419]}{(2-2m)} \Bigg\} \frac{1}{(2-2m-c)}$$

$= {\cdot}39351$, which should be ${\cdot}405714$ according to M. Damoiseau.

The magnitude of this inequality (*evection*) arises from the quantity $2-2\,m-c$, introduced in the denominator by integration, being small.

The terms depending upon the square of the disturbing force are very little sensible in the values of r_4, λ_4.

The equation

$$r_4 = [0{\cdot}0535010]\, r_4 - [9{\cdot}7596140] \frac{a\,m_{\prime}}{\mu} R_4$$

gives $r_4 = {\cdot}016557$
and the equation

$$\lambda_4 = \left\{ 2\, r_4 + r_1 + \frac{2\,[7{\cdot}6177065]}{(2-2m+c)} + \frac{2\,[7{\cdot}6175419]}{2-2m} \right\} \frac{1}{(2-2m+c)}$$

gives $\lambda_4 = \cdot 017202$, which should be $\cdot 0166992$ according to M. Damoiseau.

The preceding numerical examples serve to show how, by *false position* or gradual approximation, accurate values of the coefficients of all the inequalities may be obtained. In using, however, the method in order to perfect the Tables of the Moon, this preliminary labour is unnecessary, and it seems far better to assume the numerical values given by M. Damoiseau, which cannot differ much, if at all, from the truth, and to employ the method which I have given as one of *verification* merely; but for this purpose it is necessary first to transform the expressions of M. Damoiseau for the reciprocal of the radius vector and of the tangent of the latitude, which are given in terms of the true longitude, into others containing explicitly the mean longitude instead.

The following are the values of the quantities λ, according to M. Damoiseau, [*Mém. sur la Théorie de la Lune,*] p. 561;

$\lambda_1 = \cdot 0114901$ $\quad\lambda_3 = \cdot 405714$ $\quad\lambda_4 = \cdot 016992$

$\lambda_5 = -\cdot 194501$ $\quad\lambda_6 = \cdot 047798$ $\quad\lambda_7 = -\cdot 0071657$

$\lambda_9 = \cdot 34101$ $\quad\lambda_{10} = \cdot 023758$ $\quad\lambda_{11} = -\cdot 57521$

$\lambda_{12} = 1\cdot 090142$ $\quad\lambda_{13} = -\cdot 01595$ $\quad\lambda_{14} = \cdot 77772$

$\lambda_{15} = -\cdot 15092$ $\quad\lambda_{16} = \cdot 07733$ $\quad\lambda_{17} = -\cdot 12619$

$\lambda_{18} = \cdot 13616$ $\quad\lambda_{19} = -\cdot 0056734$ $\quad$ &c. &c.

M. Damoiseau has given, p. 348, the expression for $a\,\delta\,\frac{1}{r'}$ in terms of the true longitude. This may easily be transformed by Lagrange's theorem, into a series containing explicitly the mean longitude.

If we suppose

$$\frac{a}{r'} = A_0 + A_1 \cos(2\lambda' - 2m\lambda') + e A_2 \cos(c\lambda' - \varpi)$$

$$+ e A_3 \cos(2\lambda' - 2m\lambda' - c\lambda' + \varpi) + \&c.$$

$$s = B_{146}\gamma \sin(g\lambda' - \nu) + B_{147}\gamma \sin(2\lambda' - 2m\lambda - g\lambda' + \nu)$$

$$+ B_{148}\gamma \sin(2\lambda' - 2m\lambda' + g\lambda' - \nu) + \&c.$$

$$nt = \lambda' + C_1 \sin(2\lambda' - 2m\lambda') + e C_2 \sin(c\lambda' - \varpi)$$

$$+ e C_3 \sin(2\lambda - 2m\lambda' - c\lambda' + \varpi) + \&c.$$

in which expressions A, B, C are the same quantities as in M. Damoiseau's notation, the indices only being changed according to the remark, p. 32, in order that the Table may be applicable to the transformation required; λ' is called v, and $\delta.\frac{1}{r'}$, δu in the notation of M. Damoiseau.

$$\frac{a}{r'} = A_0 + \frac{1}{2}(2-2m) A_1 C_1 + \frac{c}{2} e^2 A_2 C_2 + \frac{1}{2}(2-2m-c) e^2 A_3 C_3$$

$$+ \frac{1}{2}(2-2m+c) e^2 A_4 C_4 + \frac{m}{2} e_i^2 A_5 C_5 + \&c.$$

$$+ \Big\{ A_1 - \frac{1}{2} c e^2 A_2 C_3 + \frac{1}{2} c e^2 A_2 C_4 - \frac{1}{2}(2-2m-c) e^2 A_3 C_2$$

$$+ \frac{1}{2}(2-2m+c) e^2 A_4 C_2 - \frac{1}{2} m e_i^2 A_5 C_6 + \frac{1}{2} m e_i^2 A_5 C_7$$

$$- \frac{1}{2}(2-3m) e_i^2 A_6 C_5 + \frac{1}{2}(2-m) e_i^2 A_7 C_5 \Big\} \cos 2\tau$$

[1]

$$+ \Big\{ A_2 + \frac{1}{2}(2-2m) A_1 C_4 + \frac{1}{2}(2-2m) A_1 C_3$$

$$+ \frac{1}{2}(2-2m-c) A_3 C_1 + \frac{1}{2}(2-2m+c) A_4 C_1 \Big\} e \cos \xi$$

[2]

$$+ \Big\{ A_3 + \frac{1}{2}(2-2m) A_1 C_2 + \frac{c}{2} A_2 C_1 \Big\} e \cos(2\tau - \xi)$$

[3]

$$+ \Big\{ A_4 - \frac{1}{2}(2-2m) A_1 C_2 - \frac{c}{2} A_2 C_1 \Big\} e \cos(2\tau + \xi)$$

[4]

+ &c. &c.

Similarly

$$s = \Big\{ B_{146} + \frac{1}{2}(2-2m+g) C_1 B_{148} - \frac{1}{2}(2-2m-g) C_1 B_{147}$$

$$+ \frac{1}{2}(c+g) e^2 C_2 B_{150} - \frac{1}{2}(c-g) e^2 C_2 B_{149}$$

$$+ \frac{1}{2}(2-2m-c+g) e^2 C_3 B_{152} - \frac{1}{2}(2-2m-c-g) e^2 C_3 B_{151}$$

$$+ \frac{1}{2}(2-2m+c+g) e^2 C_4 B_{154} - \frac{1}{2}(2-2m+c-g) e^2 C_4 B_{153}$$

$$+ \frac{1}{2}(m+g)\,e_{\prime}^{2}\,C_5 B_{156} - \frac{1}{2}(m-g)\,e_{\prime}^{2}\,C_5 B_{155}\Big\}\sin\eta$$

[146]

$$+\Big\{B_{147} - \frac{g}{2}C_1 B_{146} - \frac{1}{2}(2-2m-c-g)\,e^2\,C_2 B_{151}$$

$$+ \frac{1}{2}(2-2m+c-g)\,e^2\,C_2 B_{153} - \frac{1}{2}(c-g)\,e^2\,C_3 B_{149}$$

$$- \frac{1}{2}(c+g)\,e^2 C_4 B_{150} - \frac{1}{2}(2-3m+g)\,e_{\prime}^{2}\,C_5 B_{157}$$

$$+ \frac{1}{2}(2-m-g)\,e_{\prime}^{2}\,C_5 B_{159}\Big\}\,\gamma\sin(2\tau-\eta)$$

[147]

$$+\Big\{B_{148} - \frac{g}{2}C_1 B_{146} - \frac{1}{2}(2-2m-c+g)\,e^2\,C_2 B_{152}$$

$$+ \frac{1}{2}(2-2m+c+g)\,e^2\,C_2 B_{154} - \frac{1}{2}(c+g)\,e^2\,C_3 B_{150}$$

$$- \frac{1}{2}(c-g)\,e^2 C_4 B_{149} - \frac{1}{2}(2-3m+g)\,e_{\prime}^{2}\,C_5 B_{158}$$

$$+ \frac{1}{2}(2-m+g)\,e^2_{\prime} C_5 B_{160}\Big\}\,\gamma\sin(2\tau+\eta)$$

[148]

+ &c. &c.

In order to verify these expressions, suppose

$$\frac{a}{r} = A_2 e\cos(c\lambda' - \varpi) \qquad s = \gamma B_{146}\sin(g\lambda' - \nu)$$

$$nt = \lambda' + C_1\sin(2\lambda' - 2m\lambda')$$

Then by Lagrange's theorem, neglecting A^3, $A^2 C$, &c.

$$\frac{a}{r'} = A_2 e\cos\xi + c\,e\,A_2 C_1\sin 2\tau\sin\xi \text{ nearly}$$

$$= A_2 e\cos\xi + \frac{c A_2 C_1}{2}e\cos(2\tau-\xi) - \frac{c A_2 C_1}{2}e\cos(2\tau+\xi)$$

[2] [3] [4]

which terms are found in the expression which I have given above.

Again, by Lagrange's theorem,

$$s = \gamma\, B_{146} \sin \eta - g\,\gamma\, C_1\, B_{146} \sin 2\,\tau \cos \eta$$

$$= \gamma\, B_{146} \sin \eta - \frac{g\, C_1\, B_{146}}{2} \gamma \sin (2\,\tau - \eta) - \frac{g\, C_1\, B_{146}\,\gamma}{2} \sin (2\,\tau + \eta)$$

[146] [147] [148]

which terms are found in the expression which I have given above.

The numerical values of the quantities A, B, C, according to M. Damoiseau, are

A_0 = ?	[30] A_1 = ·00709538	[1] A_2 = ?
*[31] A_3 = ·2024622	[32] A_4 = − ·00369361	[16] A_5 = − ·0056375
[33] A_6 = ·0289158	[34] A_7 = − ·0030859	[2] A_8 = ·003183 ?
[35] A_9 = ·347942	[36] A_{10} = ·001970	[19] A_{11} = − ·19737
[41] A_{12} = ·516174	[42] A_{13} = ·0026238	[18] A_{14} = − ·286046
[39] A_{15} = − ·060625	[40] A_{17} = − ·014546	[17] A_{17} = − ·006930
[43] A_{18} = ·08125		

[0] B_{147} = ·0284942	[2] B_{149} = − ·019169	[6] B_{151} = − ·020788
[5] B_{153} = ·006113	[8] B_{155} = − ·081170	[11] B_{157} = ·071237
[10] B_{159} = − ·0033394		

[30] C_1 = − ·009216	[1] C_2 = − 2·0044055	[31] C_3 = − ·4138664
[32] C_4 = ·012939	[16] C_5 = − ·194385	[33] C_6 = − ·394172
[34] C_7 = ·0038267	[2] C_8 = ·745169	[35] C_9 = − ·286413
[36] C_{10} = − ·012575	[19] C_{11} = ·365516	[41] C_{12} = − 1·08891
[42] C_{13} = − ·008551	[18] C_{14} = − ·607534	[39] C_{15} = − ·11587
[40] C_{16} = ·055936	[17] C_{17} = ·12755	[43] C_{18} = − ·11432

5. The determination of the inequalities of the planets, and of the satellites of Jupiter, flows as an easy corollary from the method already explained of determining the inequalities of the moon.

In the theory of the planetary perturbations the terms depending upon the square of the disturbing force are much less sensible than those which occur in the theory of the moon.

In order to render the Table applicable to the planetary theory, it is only necessary to write the indeterminate i in the arguments, retaining the same indices, thus;

---

* These are the indices of the arguments in M. Damoiseau's work.

0	0	23	$2\xi+\xi_{,}$	46	$i\tau+3\xi-\xi_{,}$
1	$i\tau$	24	$i\tau-2\xi-\xi_{,}$	47	$2\xi+2\xi_{,}$
2	ξ	25	$i\tau+2\xi+\xi_{,}$	48	$i\tau-2\xi-2\xi_{,}$
3	$i\tau-\xi$	26	$2\xi-\xi_{,}$	49	$i\tau+2\xi+2\xi_{,}$
4	$i\tau+\xi$	27	$i\tau-2\xi+\xi_{,}$	50	$2\xi-2\xi_{,}$
5	$\xi_{,}$	28	$i\tau+2\xi-\xi_{,}$	51	$i\tau-2\xi+2\xi_{,}$
6	$i\tau-\xi_{,}$	29	$\xi+2\xi_{,}$	52	$i\tau+2\xi-2\xi_{,}$
7	$i\tau+\xi_{,}$	30	$i\tau-\xi+2\xi_{,}$	53	$\xi+3\xi_{,}$
8	2ξ	31	$i\tau+\xi+2\xi_{,}$	54	$i\tau-\xi-3\xi_{,}$
9	$i\tau-2\xi$	32	$\xi-2\xi_{,}$	55	$i\tau+\xi+3\xi_{,}$
10	$i\tau+2\xi$	33	$i\tau-\xi-2\xi_{,}$	56	$\xi-3\xi_{,}$
11	$\xi+\xi_{,}$	34	$i\tau+\xi+2\xi_{,}$	57	$i\tau-\xi+3\xi_{,}$
12	$i\tau-\xi-\xi_{,}$	35	$3\xi_{,}$	58	$i\tau+\xi-3\xi_{,}$
13	$i\tau+\xi+\xi_{,}$	36	$i\tau-3\xi_{,}$	59	4ξ
14	$\xi-\xi_{,}$	37	$i\tau+3\xi_{,}$	60	$i\tau-4\xi_{,}$
15	$i\tau-\xi-\xi_{,}$	38	4ξ	61	$i\tau+4\xi_{,}$
16	$i\tau+\xi-\xi_{,}$	39	$i\tau-4\xi$	62	2η
17	$2\xi_{,}$	40	$i\tau+4\xi$	63	$i\tau-2\eta$
18	$i\tau-2\xi_{,}$	41	$3\xi+\xi_{,}$	64	$i\tau+2\eta$
19	$i\tau+2\xi_{,}$	42	$i\tau-3\xi-\xi_{,}$	146	η
20	3ξ	43	$i\tau+3\xi+\xi_{,}$	147	$i\tau-\eta$
21	$i\tau-3\xi$	44	$3\xi-\xi_{,}$	148	$i\tau+\eta$
22	$i\tau+3\xi$	45	$i\tau-3\xi+\xi_{,}$		

Two inequalities only of the order of the cubes of the eccentricities have yet been considered.

If $n\,t$, $n_{,}t$ represent the mean motions of Jupiter and Saturn, the following are the arguments of one of the inequalities in question, called the great inequality of Jupiter and Saturn:

$$5\,\tau - 3\,\xi \text{ or } 2\,n\,t - 5\,n_{,}t + 3\,\varpi$$
$$4\,\tau - 2\,\xi - \xi_{,} \text{ or } 2\,n\,t - 5\,n_{,}t + 2\,\varpi + \varpi_{,}$$
$$3\,\tau - \xi - 2\,\xi_{,} \text{ or } 2\,n\,t - 5\,n_{,}t + \varpi + 2\,\varpi_{,}$$
$$3\,\tau - \xi - 2\,\eta_{,} \text{ or } 2\,n\,t - 5\,n_{,}t + \varpi + 2\,\nu_{,}$$
$$2\,\tau - \xi_{,} - 2\,\eta_{,} \text{ or } 2\,n\,t - 5\,n_{,}t + \varpi_{,} + 2\,\nu_{,}$$

If $n\,t$, $n_{,}t$ represent the mean motions of Mercury and the Earth, the other inequality arises from the arguments

$$4\,\tau - 3\,\xi \text{ or } n\,t - 4\,n_{,}t + 3\,\varpi$$
$$3\,\tau - 2\,\xi - \xi_{,} \text{ or } n\,t - 4\,n_{,}t + 2\,\varpi + \varpi_{,}$$
$$2\,\tau - \xi - 2\,\xi_{,} \text{ or } n\,t - 4\,n_{,}t + \varpi + 2\,\varpi_{,}$$
$$2\,\tau - \xi - 2\,\eta_{,} \text{ or } n\,t - 4\,n_{,}t + \varpi + 2\,\nu_{,}$$
$$\tau - \xi_{,} - 2\,\eta_{,} \text{ or } n\,t - 4\,n_{,}t + \varpi_{,} + 2\,\nu_{,}$$

The terms depending upon the fifth powers of the eccentricities are sensible in the great inequality of Jupiter. The only other inequality in which terms of this order have been found to be sensible, occurs in the theory of the Earth and Venus, and was discovered by Prof. Airy.

If $n\,t$, $n_{,}t$ represent the mean motions of Venus and the Earth, this inequality arises from the arguments,

$13\tau - 5\xi$	or $8nt - 13n_{,}t + 5\varpi$
$12\tau - 4\xi - \xi_{,}$	or $8nt - 13n_{,}t + 4\varpi + \varpi_{,}$
$11\tau - 3\xi - 2\xi_{,}$	or $8nt - 13n_{,}t + 3\varpi + 2\varpi_{,}$
$10\tau - 2\xi - 3\xi_{,}$	or $8nt - 13n_{,}t + 2\varpi + 3\varpi_{,}$
$9\tau - \xi - 4\xi_{,}$	or $8nt - 13n_{,}t + \varpi + 4\varpi_{,}$
$8\tau - 5\xi_{,}$	or $8nt - 13n_{,}t + 5\varpi_{,}$
$13\tau - \xi - 4\eta$	or $8nt - 13n_{,}t + \varpi + 4\nu$
$12\tau - \xi_{,} - 4\eta$	or $8nt - 13n_{,}t + \varpi_{,} + 4\nu$
$13\tau - 3\xi - 2\eta$	or $8nt - 13n_{,}t + 3\varpi + 2\nu$
$12\tau - 2\xi - \xi_{,} - 2\eta$	or $8nt - 13n_{,}t + 2\varpi + \varpi_{,} + 2\nu$
$11\tau - \xi - 2\xi_{,} - 2\eta$	or $8nt - 13n_{,}t + \varpi + 2\varpi_{,} + 2\nu$
$10\tau - 3\xi_{,} - 2\eta$	or $8nt - 13n_{,}t + 3\varpi_{,} + 2\nu$

The care with which Prof. Airy has determined the numerical amount of this inequality (see *Phil. Trans.* 1832, part 1.), is no less remarkable than its detection. This discovery by Professor Airy forms the only direct contribution towards the perfection of the Planetary Tables by an Englishman since the publication of the Tables of the Planets by Halley, in 1749.

These inequalities become sensible, as is well known, from the small divisors, $2n - 5n_{,}$, $n - 4n_{,}$, $8n - 13n_{,}$, respectively, introduced by integration.

Developing R according to the method explained, p. 34, we find,

$$R_2 = -\frac{a\,\mathrm{d}R_0}{\mathrm{d}a} \qquad R_3 = -\frac{a\,\mathrm{d}R_1}{2\,\mathrm{d}a} - iR_1 \qquad R_4 = -\frac{a\,\mathrm{d}R_1}{2\,\mathrm{d}a} + iR_1$$

$$2R_8 = -\frac{a\,\mathrm{d}R_2}{2\,\mathrm{d}a} - \frac{3a\,\mathrm{d}R_0}{2\,\mathrm{d}a}$$

$$2R_9 = -\frac{a\,\mathrm{d}R_3}{2\,\mathrm{d}a} - iR_3 - \frac{3a\,\mathrm{d}R_1}{4\,\mathrm{d}a} - \frac{5iR_1}{4}$$

$$2R_{10} = -\frac{a\,\mathrm{d}R_4}{2\,\mathrm{d}a} + iR_4 - \frac{3a\,\mathrm{d}R_1}{4\,\mathrm{d}a} + \frac{5iR_1}{4}$$

By successive substitutions in the expressions which have been given, it is obvious that they may be reduced so as to contain only the quantity R_1 and the differential coefficients of this quantity with respect to a and $a_{,}$.

Thus

$$R_4 = -\frac{a\,\mathrm{d}R_1}{2\,\mathrm{d}a} + iR_1$$

$$2R_{10} = -\frac{a\,\mathrm{d}R_4}{2\,\mathrm{d}a} + iR_4 - \frac{3a\,\mathrm{d}R_1}{4\,\mathrm{d}a} + \frac{5iR_1}{4}$$

$$= -\frac{1}{2}\left\{-\frac{a^2 \mathrm{d}^2 R_1}{2\,\mathrm{d}a^2} - \frac{a\,\mathrm{d}R_1}{2\,\mathrm{d}a} + i\,a\frac{\mathrm{d}R_1}{\mathrm{d}a}\right\}$$

$$-\frac{i\,a\,\mathrm{d}R_1}{2\,\mathrm{d}a} + i^2 R_1 - \frac{3\,a\,\mathrm{d}R_1}{4\,\mathrm{d}a} + \frac{5\,i R_1}{4}$$

$$R_{10} = \frac{a^2 \mathrm{d}^2 R_1}{8\,\mathrm{d}a^2} - \frac{(2i+1)}{4}\frac{a\,\mathrm{d}R_1}{\mathrm{d}a} + \frac{(4\,i^2+5\,i)\,R_1}{8}$$

Changing the sign of i, we get

$$R_9 = \frac{a^2 \mathrm{d}^2 R_1}{8\,\mathrm{d}a^2} + \frac{(2\,i-1)}{4}\frac{a\,\mathrm{d}R_1}{\mathrm{d}a} + \frac{(4\,i^2-5\,i)\,R_1}{8}$$

which accords with the expression (for $N^{(0)}$) given in the *Théor. anal.* vol. i. p. 463.

$$3\,R_{22} = -\frac{a\,\mathrm{d}R_{10}}{\mathrm{d}a} + i\,R_{10} - \frac{3}{4}\frac{a\,\mathrm{d}R_4}{\mathrm{d}a} + \frac{5\,i}{4}R_4 - \frac{17}{16}\frac{a\,\mathrm{d}R_1}{\mathrm{d}a} + \frac{13\,i\,R_1}{8}$$

$$= -\frac{1}{2}\left\{\frac{a^2 \mathrm{d}^3 R_1}{8\,\mathrm{d}a^3} + \frac{a^2 \mathrm{d}^2 R_1}{4\,\mathrm{d}a^2} - \frac{(2\,i+1)}{4}\frac{a^2 \mathrm{d}^2 R_1}{\mathrm{d}a^2} - \frac{(2\,i+1)}{4}\frac{a\,\mathrm{d}R_1}{\mathrm{d}a}\right.$$

$$\left. + \frac{(4\,i+5)}{8}\frac{i\,\mathrm{d}R_1}{\mathrm{d}a}\right\}$$

$$+ i\left\{\frac{a^2 \mathrm{d}R_1}{8\,\mathrm{d}a^2} - \frac{(2\,i+1)}{4}\frac{a\,\mathrm{d}R_1}{\mathrm{d}a} + \frac{(4\,i+5)\,i\,R_1}{8}\right\}$$

$$-\frac{3}{4}\left\{-\frac{a^2 \mathrm{d}^2 R_1}{2\,\mathrm{d}a^2} - \frac{a\,\mathrm{d}R_1}{2\,\mathrm{d}a} + \frac{i\,a\,\mathrm{d}R_1}{\mathrm{d}a}\right\}$$

$$+\frac{5\,i}{4}\left\{-\frac{a\,\mathrm{d}R_1}{2\,\mathrm{d}a} + i\,R_1\right\} - \frac{17}{16}\frac{a\,\mathrm{d}R_1}{\mathrm{d}a} + \frac{13}{8}R_1$$

$$R_{22} = \frac{1}{48}\left\{(26\,i + 30\,i^2 + 8\,i^3)\,R_1 - (9 + 27\,i + 12\,i^2)\frac{a\,\mathrm{d}R_1}{\mathrm{d}a}\right.$$

$$\left. + (6\,i + 6)\frac{a^2 \mathrm{d}^2 R_1}{\mathrm{d}a^2} - \frac{a^3 \mathrm{d}^3 R_1}{\mathrm{d}a^3}\right\}$$

Changing the sign of i, we get

$$R_{21} = -\frac{1}{48}\left\{(26\,i - 30\,i^2 + 8\,i^3)\,R_1 + (9 - 27\,i + 12\,i^2)\,a\frac{\mathrm{d}R_1}{\mathrm{d}a}\right.$$

$$\left. + (6\,i - 6)\frac{a^2 \mathrm{d}^2 R_1}{\mathrm{d}a^2} + \frac{a^3 \mathrm{d}^3 R_1}{\mathrm{d}a^3}\right\}$$

which agrees with the expression given by Burckhardt for $(M^{(0)})$, *Mémoires de l'Institut*, 1808, *Second Semestre*, p. 39.

Similarly

$$2\,R_{51} = \frac{a^2 \mathrm{d}^2 R_{19}}{4\,\mathrm{d}\,a^2} + \frac{(2\,i-1)}{2}\frac{a\,\mathrm{d}\,R_{19}}{\mathrm{d}\,a} + \frac{(4\,i^2-5\,i)\;R_{19}}{4}$$

$$2\,R_{19} = \frac{a_{\prime}^{2}\,\mathrm{d}^2 R_1}{4\,\mathrm{d}\,a_{\prime}^{2}} + \frac{(2\,i-1)}{2}\frac{a_{\prime}\,\mathrm{d}\,R_1}{\mathrm{d}\,a_{\prime}} + \frac{(4\,i^2-5\,i)\;R_1}{4}$$

If $i = 2$,

$$2\,R_{51} = \frac{a^2\,\mathrm{d}^2 R_{19}}{4\,\mathrm{d}\,a^2} + \frac{3}{2}\frac{a\,\mathrm{d}\,R_{19}}{\mathrm{d}\,a} + \frac{3}{2}R_{19}$$

$$2\,R_{19} = \frac{a_{\prime}^{2}\,\mathrm{d}^2 R_1}{4\,\mathrm{d}\,a_{\prime}^{2}} + \frac{3}{2}\frac{a_{\prime}\,\mathrm{d}\,R_1}{\mathrm{d}\,a_{\prime}} + \frac{3}{2}R_1$$

$$R_1 = -\frac{b_{1.2}}{a_{\prime}} = -\frac{3\,a^2}{4a_{\prime}^{3}} - \frac{3\,.\,5}{2\,.\,4\,.\,6}\frac{a^4}{a_{\prime}^{5}} - \frac{3\,.\,3\,.\,5\,.\,7}{2\,.\,4\,.\,4\,.\,6\,.\,8}\frac{a^6}{a_{\prime}^{7}} - \&c.$$

In the Lunar Theory, the higher terms may be neglected; and taking $R_1 = -\dfrac{3\,a^2}{4\,a_{\prime}^{3}}$, it is evident that R_{19} and R_{51} are each equal to zero. This theorem, however, cannot be extended to the other terms, and therefore in the Planetary Theory the coefficient corresponding to the argument $2\,\tau - 2\,\xi + 2\,\xi_{\prime}$ or $2\,\varpi - 2\,\varpi_{\prime}$, in the development of R, (which term is important as regards the secular inequalities,) does not vanish.

If the coefficients of the nth argument in the expressions for $\dfrac{a}{r}$ and λ be called r_n and λ_n, the Table which has been used for the preceding multiplications may also be used (when the square of the disturbing force is neglected,) for the integration of the equations

$$\frac{\mathrm{d}^2\,.\,r^2}{2\,\mathrm{d}\,t^2} - \frac{\mu}{r} + \frac{\mu}{a} + 2\int \mathrm{d}\,R + r\frac{\mathrm{d}\,R}{\mathrm{d}\,r} = 0$$

and

$$\frac{\mathrm{d}\,\lambda}{\mathrm{d}\,t} = \frac{h}{r^2} - \frac{1}{r^2}\int\frac{\mathrm{d}\,R}{\mathrm{d}\,\lambda}\mathrm{d}\,t$$

$$-\frac{\mathrm{d}^2\,\mathrm{r}^3\,\delta\dfrac{1}{r}}{\mathrm{d}\,t^2} - \mu\,\delta\frac{1}{r} + 2\int \mathrm{d}\,R + \frac{r\,\mathrm{d}\,R}{\mathrm{d}\,r} = 0$$

$$-\,r_0 + \frac{a\,m_{\prime}\,\mathrm{d}\,R_0}{\mu\,\mathrm{d}\,a} = 0$$

$$\left\{\frac{i(n-n_{\prime})}{n}\right\}^2 r_1 - r_1 + \frac{2\,n\,a}{(n-n_{\prime})}\frac{m_{\prime}}{\mu}R_1 + \frac{a^2 m_{\prime}\,\mathrm{d}\,R_1}{\mu\,\mathrm{d}\,a} = 0$$

$$\left\{\frac{i\,(n-n_{\prime})-n}{n}\right\}^2\left\{r_3 + \frac{3}{2}r_1\right\} - r_3 + \frac{2\,(i-1)\,n\,a}{(i\,(n-n_{\prime})-n)}\frac{m_{\prime}}{\mu}R_3$$

$$+\frac{a^2 m_{\prime}\,\mathrm{d}\,R_3}{\mu\,\mathrm{d}\,a} = 0$$

$$\left\{\frac{i\,(n-n_{\prime})+n}{n}\right\}^2\left\{r_4 + \frac{3}{2}r_1\right\} - r_4 + \frac{2\,(i+1)\,n\,a}{(i\,(n-n_{\prime})+n)}\frac{m_{\prime}}{\mu}R_4$$

$$+\frac{a^2 m_{\prime}\,\mathrm{d}\,R_4}{\mu\,\mathrm{d}\,a} = 0$$

$$\frac{n_{\prime}^2}{n^2}r_5 - r_5 + \frac{a^2 m_{\prime}\,\mathrm{d}\,R_5}{\mu\,\mathrm{d}\,a} = 0$$

$$\left\{\frac{i\,(n-n_{\prime})-n_{\prime}}{n}\right\}^2 r_6 - r_6 + \frac{2\,i\,n\,a}{(i\,(n-n_{\prime})-n_{\prime})}\frac{m_{\prime}}{\mu}R_6 + \frac{a^2 m_{\prime}\,\mathrm{d}\,R_6}{\mu\,\mathrm{d}\,a} = 0$$

$$\left\{\frac{i\,(n-n_{\prime})+n_{\prime}}{n}\right\}^2 r_7 - r_7 + \frac{2\,i\,n\,a}{(i\,(n-n_{\prime})+n_{\prime})}\frac{m_{\prime}}{\mu}R_7 + \frac{a^2 m_{\prime}\,\mathrm{d}\,R_7}{\mu\,\mathrm{d}\,a} = 0$$

If $\frac{\mathrm{d}\,R}{\mathrm{d}\,z} = Z$, the inequalities of latitude may be determined from the equation

$$\frac{\mathrm{d}^2 z}{\mathrm{d}\,t^2} + \frac{\mu z}{r^3} + Z = 0$$

which gives if $z = a\,\gamma \sin\eta + a\,\gamma\,z_{147}\sin(\tau-\eta) + a\,\gamma\,z_{148}\sin(i\,\tau+\eta)$

$$-g^2 + 3\,r_0 + 1 + \frac{m_{\prime}}{\mu}a^2 Z_{146} = 0$$

$$-\left\{\frac{i\,(n-n_{\prime})-n}{n}\right\}^2 z_{147} + \frac{3\,r_1}{2} + z_{147} + \frac{m_{\prime}}{\mu}a^2 Z_{147} = 0$$

$$-\left\{\frac{i\,(n-n_{\prime})+n}{n}\right\}^2 z_{148} + \frac{3\,r_1}{2} + z_{148} + \frac{m_{\prime}}{\mu}a^2 Z_{148} = 0$$

$$\frac{\mathrm{d}\,R}{\mathrm{d}\,z} = m_{\prime}\left\{\frac{z_{\prime}}{r_{\prime}^3} + \frac{(z-z_{\prime})}{\{r^2 - 2\,r\,r_{\prime}\cos(\lambda-\lambda_{\prime}) + r_{\prime}^2\}^{\frac{3}{2}}}\right\}$$

If $z_{\prime}=0$, that is, if (as in the Theory of the Moon) the longitudes are reckoned upon the orbit of $m_{\prime}$, then if $a < a_{\prime}$,

$$\frac{\mathrm{d}\,R}{\mathrm{d}\,z} = \frac{m_{\prime}\,a\,\gamma\sin\eta}{a_{\prime}^3}\left\{\frac{1}{2}b_{3,0} + b_{3,1}\cos\tau + b_{3,2}\cos 2\,\tau + \&c.\right\}$$

$$= \frac{m_{,} a \gamma}{2 a_{,}^{3}} \left\{ b_{3,0} \sin \eta - b_{3,1} \sin (\tau - \eta) - b_{3,2} \sin (2\tau - \eta) \right.$$
$$\left. + b_{3,1} \sin (\tau + \eta) + b_{3,2} \sin (2\tau + \eta) + \&c. \right\}$$

If $z = 0$, that is if the longitudes are reckoned upon the orbit of m, then if $a < a_{,}$,

$$\frac{d R}{d z} = \frac{m_{,} \gamma_{,} \sin \eta_{,}}{a_{,}^{2}} \left\{ 1 - \frac{1}{2} b_{3,0} - b_{3,1} \cos \tau - b_{3,2} \cos 2\tau - \&c. \right\}$$
$$= \frac{m_{,} \gamma_{,}}{a_{,}^{2}} \left\{ \left(1 - \frac{1}{2} b_{3,0}\right) \sin \eta_{,} + \frac{1}{2} b_{3,1} \sin (\tau - \eta_{,}) \right.$$
$$+ \frac{1}{2} b_{3,2} \sin (2\tau - \eta_{,}) - \frac{1}{2} b_{3,1} \sin (\tau + \eta_{,})$$
$$\left. - \frac{1}{2} b_{3,2} \sin (2\tau + \eta_{,}) + \&c. \right\}$$

$$\lambda = n t + \varepsilon$$
$$+ \left\{ 2 r_1 - \frac{n a}{(n - n_{,})} \frac{m_{,}}{\mu} R_1 \right\} \frac{n}{i (n - n_{,})} \sin i \tau \qquad [1]$$
$$+ \left\{ 2 r_3 + r_1 - \frac{i n a}{(i(n - n_{,}) - n)} \frac{m_{,}}{\mu} R_3 \right.$$
$$\left. - \frac{n a}{(n - n_{,})} R_1 \right\} \frac{n e}{i (n - n_{,}) - n} \sin (i \tau - \xi) \qquad [3]$$
$$+ \left\{ 2 r_4 + r_1 - \frac{i n a}{(i (n - n_{,}) + n)} \frac{m_{,}}{\mu} R_4 \right.$$
$$\left. - \frac{n a}{(n - n_{,})} \frac{m_{,}}{\mu} R_1 \right\} \frac{n e}{i (n - n_{,}) + n} \sin (i \tau + \xi) \qquad [4]$$
$$+ \frac{2 n r_5}{n_{,}} e_{,} \sin \xi_{,} \qquad [5]$$
$$+ \left\{ 2 r_6 - \frac{i n a}{(i(n - n_{,}) - n_{,})} \frac{m_{,}}{\mu} R_6 \right\} \frac{n e_{,}}{i (n - n_{,}) - n_{,}} \sin (i \tau - \xi_{,}) \qquad [6]$$
$$+ \left\{ 2 r_7 - \frac{i n a}{(i (n - n_{,}) + n_{,})} \frac{m_{,}}{\mu} R_7 \right\} \frac{n e_{,}}{i (n - n_{,}) + n_{,}} \sin (i \tau + \xi_{,}) \qquad [7]$$

i being any positive whole number; excluding the argument $\tau+\xi_{\prime}$ which must be considered separately.

If we call the quantity R_n which corresponds to $i=m$, $R_{n,m}$, then neglecting the squares of the eccentricities, when $a < a_{\prime}$

$$R_0 = -\frac{b_{1,0}}{2a_{\prime}}, \quad R_{1,1} = \frac{a}{a_{\prime}^2} - \frac{b_{1,1}}{a_{\prime}}, \quad R_{1,2} = -\frac{b_{1,2}}{a_{\prime}}, \quad R_{1,3} = -\frac{b_{1,3}}{a_{\prime}}$$

$$R_2 = -\frac{a^2}{2a_{\prime}^3} b_{3,0} + \frac{a}{2a_{\prime}^2} b_{3,1}$$

$$R_{3,1} = -\frac{3a}{2a_{\prime}^2} + \frac{3a}{4a_{\prime}^2} b_{3,0} - \frac{a^2}{2a_{\prime}^3} b_{3,1} - \frac{a}{4a_{\prime}^2} b_{3,2}$$

$$R_{3,2} = \frac{3a}{4a_{\prime}^2} b_{3,1} - \frac{a^2}{2a_{\prime}^3} b_{3,2} - \frac{a}{4a_{\prime}^2} b_{3,3}$$

$$R_{3,3} = \frac{3a}{4a_{\prime}^2} b_{3,2} - \frac{a^2}{2a_{\prime}^3} b_{3,3} - \frac{a}{4a_{\prime}^2} b_{3,4}$$

$$R_{4,1} = \frac{a}{2a_{\prime}^2} - \frac{a}{4a_{\prime}^2} b_{3,0} - \frac{a^2}{2a_{\prime}^3} b_{3,1} + \frac{3a}{4a_{\prime}^2} b_{3,2}.$$

$$R_{4,2} = -\frac{a}{4a_{\prime}^2} b_{3,1} - \frac{a^2}{2a_{\prime}^3} b_{3,2} + \frac{3a}{4a_{\prime}^2} b_{3,3}$$

$$R_{4,3} = -\frac{a}{4a_{\prime}^2} b_{3,2} - \frac{a^2}{2a_{\prime}^3} b_{3,3} + \frac{3a}{4a_{\prime}^2} b_{3,4}$$

The preceding values are deduced from the general expression, p. 21.

$$\frac{a\,\mathrm{d}\,b_{n,i}}{\mathrm{d}\,a} = -\frac{n\,a}{a_{\prime}} \left\{ \frac{a}{a_{\prime}} b_{n+2,i} - \frac{1}{2} b_{n+2,i-1} - \frac{1}{2} b_{n+2,i+1} \right\}$$

$$\frac{a\,\mathrm{d}\,R_0}{\mathrm{d}\,a} = \frac{a^2}{2a_{\prime}^3} b_{3,0} - \frac{a}{2a_{\prime}^2} b_{3,1}$$

$$\frac{a\,\mathrm{d}\,R_{1,1}}{\mathrm{d}\,a} = \frac{a}{a_{\prime}^2} + \frac{a^2}{a_{\prime}^2} \left(\frac{a}{a_{\prime}} b_{3,1} - \frac{1}{2} b_{3,0} - \frac{1}{2} b_{3,2} \right)$$

$$\frac{a\,\mathrm{d}\,R_{1,2}}{\mathrm{d}\,a} = \frac{a}{a_{\prime}^2} \left(\frac{a}{a_{\prime}} b_{3,2} - \frac{1}{2} b_{3,1} - \frac{1}{2} b_{3,2} \right)$$

$$\frac{a\,\mathrm{d}\,R_{1,3}}{\mathrm{d}\,a} = \frac{a}{a_{\prime}^2} \left(\frac{a}{a_{\prime}} b_{3,3} - \frac{1}{2} b_{3,2} - \frac{1}{2} b_{3,4} \right)$$

$$\frac{a\,\mathrm{d}\,R_2}{\mathrm{d}\,a} = \frac{a^2}{2a_{\prime}^3} b_{3,0}$$

$$\frac{a\,\mathrm{d}\,R_{3,1}}{\mathrm{d}\,a} = -\frac{3\,a}{2\,a_{\prime}^{3}} - \frac{a}{4\,a_{\prime}^{2}}b_{3,0} + \frac{3\,a^{2}}{2\,a_{\prime}^{3}}b_{3,1} - \frac{3\,a}{4\,a_{\prime}^{2}}b_{3,2}$$

$$\frac{a\,\mathrm{d}\,R_{3,2}}{\mathrm{d}\,a} = -\frac{a}{2a_{\prime}^{2}}b_{3,1} + \frac{5}{2}\frac{a^{2}}{a_{\prime}^{3}}b_{3,2} - \frac{3\,a}{2a_{\prime}^{2}}b_{3,3}$$

$$Z_{146} = \frac{a}{2a_{\prime}^{3}}b_{3,0} \qquad Z_{147,1} = -\frac{a}{2\,a_{\prime}^{3}}b_{3,1} \qquad Z_{147,2} = -\frac{a}{2a_{\prime}^{3}}b_{3,2}$$

$$Z_{148,1} = \frac{a}{2a_{\prime}^{3}}b_{3,1} \qquad Z_{148,2} = \frac{a}{2a_{\prime}^{3}}b_{3,2}$$

$$r_{1,1} = \frac{m_{\prime}}{\mu}\frac{n^{2}}{(2\,n - n_{\prime})\,n_{\prime}}\left\{\frac{2\,n}{(n-n_{\prime})}\left(\frac{a^{2}}{a_{\prime}^{2}} - \frac{a}{a_{\prime}}b_{1,1}\right)\right.$$

$$\left. + \frac{a^{2}}{a_{\prime}^{2}} + \frac{a^{2}}{a_{\prime}^{2}}\left(\frac{a}{a_{\prime}}b_{3,1} - \frac{1}{2}b_{3,0} - \frac{1}{2}b_{3,2}\right)\right\}$$

$$\lambda_{1,1} = \frac{n}{(n - n_{\prime})}\left\{2\,r_{1,1} - \frac{m_{\prime}\,n}{\mu\,(n-n_{\prime})}\left(\frac{a^{2}}{a_{\prime}^{2}} - \frac{a}{a_{\prime}}b_{1,1}\right)\right\}$$

When $a > a_{\prime}$, if

$$\frac{1}{2}b_{1,0} = 1 + \left(\frac{1}{2}\right)^{2}\frac{a_{\prime}^{2}}{a^{2}} + \left(\frac{3}{2\,.\,4}\right)^{2}\frac{a_{\prime}^{4}}{a^{4}} + \text{\&c.}$$

$$b_{1,1} = \frac{a_{\prime}}{a} + \frac{3}{2\,.\,4}\frac{a_{\prime}^{3}}{a^{3}} + \frac{3\,.\,3\,.\,5}{2\,.\,4\,.\,4\,.\,6}\frac{a_{\prime}^{5}}{a^{5}} + \text{\&c.}$$

$$\frac{a\,\mathrm{d}\,R_{0}}{\mathrm{d}\,a} = \frac{b_{1,0}}{2\,a} - \frac{a_{\prime}^{2}}{2\,a^{3}}b_{3,0} + \frac{a_{\prime}}{2\,a^{2}}b_{3,1} = \frac{1}{2}b_{3,0} - \frac{a_{\prime}}{2\,a^{2}}b_{3,1}$$

$$\frac{a\,\mathrm{d}\,b_{n,i}}{\mathrm{d}\,a} = n\,\frac{a_{\prime}}{a}\left\{\frac{a_{\prime}}{a}b_{n+2,i} - \frac{1}{2}b_{n+2,i-1} - \frac{1}{2}b_{n+2,i+1}\right\}$$

$$\frac{a\,\mathrm{d}\,R_{1,1}}{\mathrm{d}\,a} = \frac{a_{\prime}}{a^{2}} + \frac{b_{1,1}}{a} - \frac{a_{\prime}}{a^{2}}\left(\frac{a_{\prime}}{a}b_{3,1} - \frac{1}{2}\,b_{3,0} - \frac{1}{2}\,b_{3,2}\right)$$

$$\frac{a\,\mathrm{d}\,R_{1,2}}{\mathrm{d}\,a} = \frac{b_{1,2}}{a} - \frac{a_{\prime}}{a^{2}}\left(\frac{a_{\prime}}{a}\,b_{3,2} - \frac{1}{2}b_{3,1} - \frac{1}{2}b_{3,3}\right)$$

$$r_{1,1} = \frac{m_{\prime}\,n^{2}}{\mu\,(2\,n-n_{\prime})\,n_{\prime}}\left\{\frac{2\,n}{(n - n_{\prime})}\left(\frac{a^{2}}{a_{\prime}^{2}} - b_{1,1}\right)\right.$$

$$\left. + \frac{a^{2}}{a_{\prime}^{2}} + b_{1,1} - \frac{a_{\prime}}{a}\left(\frac{a_{\prime}}{a}b_{3,2} - \frac{1}{2}b_{3,0} - \frac{1}{2}b_{3,2}\right)\right\}$$

$$\lambda_{1,1} = \frac{n}{(n-n_{\prime})}\left\{2\,r_{1,1} - \frac{m_{\prime}\,n}{\mu\,(n - n_{\prime})}\left(\frac{a^{2}}{a_{\prime}^{2}} - b_{1,1}\right)\right\}$$

If $\frac{a}{r} = 1 + r_0 + e\cos\{n(1+k)t + \varepsilon - \varpi\}$

$+ e_, f_, \cos\{n(1 + k_,)t + \varepsilon - \varpi_,\}$ we find, when $a < a_,$,

$$r_0 = -\frac{a^2 \mathrm{d} R_0}{\mathrm{d} a} = \frac{m_,}{\mu}\left\{\frac{a^3}{2a_,^3} b_{3,0} - \frac{a^2}{2a_,^2} b_{3,1}\right\}$$

If the coefficient of the argument n in the development of

$$2\int \mathrm{d} R + \frac{r\,\mathrm{d} R}{\mathrm{d} r}$$

be called q_n, then the coefficient of $\cos\xi$ or $\cos(c\,n\,t - \varpi)$ may be called q_2. By the expression given p. 28, the coefficient of $\cos\xi$ in the development of R, or R_2 is

$$m_,\left\{-\frac{a^2}{2a_,^3} b_{3,0} + \frac{a}{2a_,^2} b_{3,1}\right\}$$

and by the expression of p. 29. the coefficient of $\cos\xi$ in the development of $r\frac{\mathrm{d} R}{\mathrm{d} r}$, is

$$+ m_, \frac{a^2}{2a_,^3} b_{3,0} \text{ whence } q_2 = -\frac{a^2}{2a_,^3} b_{3,0} + \frac{a}{2a_,^2} b_{3,1}$$

$$(1 + k)^2 (1 - 3r_0) - 1 + \frac{m_,}{\mu} a\,q_2 = 0$$

$$\text{whence } k = \frac{3}{2} r_0 - \frac{m_,}{2\mu} a\,q_2 = \frac{m_,}{\mu}\left\{\frac{a^3}{a_,^3} b_{3,0} - \frac{5}{4}\frac{a^2}{a_,^2} b_{3,1}\right\}$$

$1 + k$ is equal to the quantity which was called c in the Lunar Theory.

$$1 + k = 1 + \frac{m_,}{\mu}\left\{\frac{a^3}{a_,^3} b_{3,0} - \frac{5}{4}\frac{a^2}{a_,^2} b_{3,1}\right\}$$

Substituting for $b_{3,0}$ and $b_{3,1}$ their values in series,

$$1 + k = 1 - \frac{7\,m_, a^3}{4\,\mu\,a_,^3}$$

which of course agrees with the value of c obtained p. 55.

$$R_{7,1} = \frac{3}{4}\frac{a}{a_,^2} b_{3,0} - \frac{1}{2a_,} b_{3,1} - \frac{a}{4a_,^2} b_{3,2}$$

$$q_{7,1} = -\frac{a}{2a_,^2} b_{3,2}$$

$$f_{\prime}\left\{(1+k_{\prime})^2(1-3r_0)-1\right\}=\frac{m_{\prime}a^2}{2a_{\prime}^2}b_{3,2}$$

$$\lambda=n\left\{1+2r_0\right\}t+\varepsilon+\frac{2e(1+r_0)}{1+k}\sin\left(n(1+k)t+\varepsilon-\varpi\right)$$

$$+2e_{\prime}\left\{\frac{f_{\prime}}{1+k_{\prime}}-\frac{m_{\prime}a}{2\mu}R_{7,1}\right\}\sin\left(n(1+k_{\prime})t+\varepsilon-\varpi_{\prime}\right)+\&c.$$

If

$$\frac{e(1+r_0)}{1+k}=\mathrm{e},\qquad e=\mathrm{e}\left\{1+k-r_0\right\}\text{ nearly.}$$

If $n(1+2r_0)=\mathrm{n}$ and $\mathrm{n}^2=\frac{\mu}{\mathrm{a}^3}$, $\quad a=\mathrm{a}\left\{1+\frac{4}{3}r_0\right\}$

$$\frac{\mathrm{a}}{r}=1-\frac{1}{3}r_0+\mathrm{e}\left\{1+k-\frac{7}{3}r_0\right\}\cos\left(n(1+k)t+\varepsilon-\varpi\right)$$

$$+e_{\prime}f_{\prime}\cos\left(n(1+k_{\prime})t+\varepsilon-\varpi_{\prime}\right)$$

$$=1-\frac{m_{\prime}a^3}{6\mu a_{\prime}^3}b_{3,0}+\frac{m_{\prime}a^2}{6\mu a_{\prime}^2}b_{3,1}+\mathrm{e}\left\{1-\frac{m_{\prime}a^3}{2\mu a_{\prime}^3}b_{3,0}\right.$$

$$\left.-\frac{m_{\prime}a^2}{12\mu a_{\prime}^2}b_{3,1}\right\}\cos\left(\mathrm{n}\left(1-\frac{m_{\prime}a^2}{4\mu a_{\prime}^2}b_{3,1}\right)t+\varepsilon-\varpi\right)$$

$$+e_{\prime}f_{\prime}\cos\left(n(1+k_{\prime})t+\varepsilon-\varpi_{\prime}\right)$$

$$\frac{r}{\mathrm{a}}=1+\frac{1}{3}r_0-\mathrm{e}\left\{1+k-\frac{5}{3}r_0\right\}\cos\left(n(1+k)t+\varepsilon-\varpi\right)$$

$$-e_{\prime}f_{\prime}\cos\left(n(1+k_{\prime})t+\varepsilon-\varpi_{\prime}\right)$$

$$=1+\frac{m_{\prime}a^3}{6\mu a_{\prime}^3}b_{3,0}-\frac{m_{\prime}a^2}{6\mu a_{\prime}^2}b_{3,1}-\mathrm{e}\left\{1+\frac{m_{\prime}a^3}{6\mu a_{\prime}^3}b_{3,0}\right.$$

$$\left.-\frac{5m_{\prime}a^2}{12\mu a_{\prime}^2}b_{3,1}\right\}\cos\left(\mathrm{n}\left(1-\frac{m_{\prime}a^2}{4\mu a_{\prime}^2}b_{3,1}\right)t+\varepsilon-\varpi\right)$$

$$-e_{\prime}f_{\prime}\cos\left(\mathrm{n}(1+k_{\prime})t+\varepsilon-\varpi_{\prime}\right)$$

If $a < a_{\prime}$ as before, and

$$\frac{a_{\prime}}{r_{\prime}}=1+r_{\prime 0}+e_{\prime}(1+f')\cos\left(n_{\prime}(1+k')t+\varepsilon_{\prime}-\varpi_{\prime}\right)$$

$$+e_{\prime}f_{\prime}'\cos\left(n_{\prime}(1+k_{\prime}')t+\varepsilon_{\prime}-\varpi\right)$$

we find

$$r_{\prime 0}=\frac{m}{\mu_{\prime}}\left\{\frac{1}{2}b_{3,0}-\frac{a}{2\,a_{\prime}}b_{3,1}\right\}\qquad k'=\frac{m}{\mu_{\prime}}\left\{b_{3,0}-\frac{5\,a}{4\,a_{\prime}}b_{3,1}\right\}$$

$$f_{\prime}'\left\{(1+k_{\prime}')^2(1-3\,r_{\prime 0})-1\right\}=\frac{m\,a}{2\,\mu_{\prime}\,a_{\prime}}b_{3,2}$$

If $\mathrm{n}_{\prime}\{1+2\,r_0\}=\mathrm{n}_{\prime}$ and $\mathrm{n}_{\prime}^2=\frac{\mu_{\prime}}{\mathrm{a}_{\prime}^3}$, $\quad a_{\prime}=\mathrm{a}_{\prime}\left\{1+\frac{4}{3}r_{\prime 0}\right\}$

$$\frac{\mathrm{a}_{\prime}}{r_{\prime}}=1-\frac{m}{6\,\mu_{\prime}}b_{3,0}+\frac{m\,a}{6\,\mu_{\prime}a_{\prime}}b_{3,1}+\mathrm{e}_{\prime}\left\{1-\frac{m}{6\,\mu_{\prime}}b_{3,0}\right.$$

$$\left.-\frac{m\,a}{12\,\mu_{\prime}\,a_{\prime}}b_{3,1}\right\}\cos\left(\mathrm{n}_{\prime}\left(1-\frac{m\,a}{4\,\mu_{\prime}\,a_{\prime}}b_{3,1}\right)t+\varepsilon_{\prime}-\varpi_{\prime}\right)$$

$$+e\,f_{\prime}'\cos\left(n_{\prime}(1+k_{\prime}')t+\varepsilon_{\prime}-\varpi\right)$$

Neglecting k^2,

$$\frac{r}{\mathrm{a}}=1+\frac{1}{3}r_0-\mathrm{e}\left\{1+k-\frac{5}{3}r_0\right\}\cos(n\,t+\varepsilon-\varpi)$$

$$+\,e\,k\,n\,t\sin(n\,t+\varepsilon-\varpi)$$

$$=1+\frac{1}{3}r_0-\mathrm{e}\left\{1-\frac{1}{6}r_0-\frac{m_{\prime}}{2\,\mu}a\,q_2\right\}\cos(n\,t+\varepsilon-\varpi)$$

$$-e\left\{\frac{1}{2}r_0-\frac{m_{\prime}}{2\,\mu}a\,q_2\right\}n\,t\sin(n\,t+\varepsilon-\varpi)$$

It is easy to show that the preceding result is so far in accordance with that of the *Mécanique céleste*, vol. i. p. 279.

$$r_0=\frac{a^2}{2}\left(\frac{\mathrm{d}\,A^{(0)}}{\mathrm{d}\,a}\right)\qquad a\,q_2=-\frac{a^3}{2}\left(\frac{\mathrm{d}^2A^{(0)}}{\mathrm{d}\,a^2}\right)-\frac{3\,a^2}{2}\left(\frac{\mathrm{d}\,A^{(0)}}{\mathrm{d}\,a}\right)$$

$$\frac{\mu}{6\,m_{\prime}}r_0+\frac{a}{2}q_2=\frac{a^2}{12}\left(\frac{\mathrm{d}\,A^{(0)}}{\mathrm{d}\,a}\right)-\frac{a^3}{4}\left(\frac{\mathrm{d}^2A^{(0)}}{\mathrm{d}\,a^2}\right)-\frac{3}{4}a^2\left(\frac{\mathrm{d}\,A^{(0)}}{\mathrm{d}\,a}\right)$$

$$=-\frac{2\,a^2}{3}\left(\frac{\mathrm{d}\,A^{(0)}}{\mathrm{d}\,a}\right)-\frac{a^3}{4}\left(\frac{\mathrm{d}^2A^{(0)}}{\mathrm{d}\,a^2}\right)=-f$$

$$\frac{\mu}{2\,m_{\prime}}r_0+\frac{a}{2}q_2=\frac{a^2}{4}\left(\frac{\mathrm{d}\,A^{(0)}}{\mathrm{d}\,a}\right)-\frac{a^3}{4}\left(\frac{\mathrm{d}^2A^{(0)}}{\mathrm{d}\,a^2}\right)-\frac{3}{4}a^2\left(\frac{\mathrm{d}\,A^{(0)}}{\mathrm{d}\,a}\right)$$

$$= -\frac{a^2}{2}\left(\frac{\mathrm{d}\,A^{(0)}}{\mathrm{d}\,a}\right) - \frac{a^3}{4}\left(\frac{\mathrm{d}^2\,A^{(0)}}{\mathrm{d}\,a^2}\right) = -\frac{C}{2} = \frac{a^2}{4a_{\prime}^2}b\,,$$

No difficulty occurs in the determination of the quantities k and f, k being of the order of the disturbing force, and therefore k^2 may be neglected. Various difficulties however present themselves, in the Planetary theory, in the determination of the quantities $k_{\prime}$ and $f_{\prime}$. In the notation of the *Mécanique céleste*,

$$a\,q_{7,1} = -\frac{1}{2}\left\{a^2 a'\left(\frac{\mathrm{d}^2\,A^{(1)}}{\mathrm{d}\,a\,\mathrm{d}\,a'}\right) + 2a^2\left(\frac{\mathrm{d}A^{(1)}}{\mathrm{d}\,a}\right)\right.$$

$$\left. + 2\,a'\,a\left(\frac{\mathrm{d}\,A^{(1)}}{\mathrm{d}\,a'}\right) + 4\,a\,A^{(1)}\right\}$$

$$= -\frac{1}{2}\left\{2\,aA^{(1)} - 2\,a^2\left(\frac{\mathrm{d}\,A^{(1)}}{\mathrm{d}\,a}\right) - a^3\left(\frac{\mathrm{d}^2\,A^{(1)}}{\mathrm{d}\,a^2}\right)\right\}$$

$$R_{7,1} = -\frac{1}{2}\,a'\left(\frac{\mathrm{d}\,A^{(1)}}{\mathrm{d}\,a'}\right) - A^{(1)}$$

$$= \frac{a}{2}\left(\frac{\mathrm{d}\,A^{(1)}}{\mathrm{d}\,a}\right) - \frac{1}{2}\,A^{(1)}$$

$$\frac{a}{2}\,q_{7,1} - \frac{a}{2}\,R_{7,1} = -\frac{a\,A^{(1)}}{2} + \frac{a^2}{2}\left(\frac{\mathrm{d}\,A^{(1)}}{\mathrm{d}\,a}\right) + \frac{a^3}{4}\left(\frac{\mathrm{d}^2\,A^{(1)}}{\mathrm{d}\,a^2}\right)$$

$$+ \frac{a\,A^{(1)}}{4} - \frac{a^2}{4}\left(\frac{\mathrm{d}\,A^{(1)}}{\mathrm{d}\,a}\right)$$

$$= -\frac{a\,A^{(1)}}{4} + \frac{a^2}{4}\left(\frac{\mathrm{d}\,A^{(1)}}{\mathrm{d}\,a}\right) \overset{*}{+} \frac{a^3}{4}\left(\frac{\mathrm{d}^2\,A^{(1)}}{\mathrm{d}\,a^2}\right) = -f'$$

$$-\frac{1}{2}\,a\,q_{7,1} = \frac{1}{2}\left\{a\,A^{(1)} - a^2\left(\frac{\mathrm{d}\,A^{(1)}}{\mathrm{d}\,a}\right) - \frac{a^3}{2}\left(\frac{\mathrm{d}^2\,A^{(1)}}{\mathrm{d}\,a^2}\right)\right\}$$

$$= -\frac{D}{2} = \frac{a^2}{2\,a_{\prime}^2}\,b_{3,2}$$

The expression given by Laplace, *Méc. cél.* vol. i. p. 279, may be put in the form,

$$\frac{\delta\,r}{a} = \frac{m_{\prime}}{\mu}\left\{\frac{a^3}{6\,a_{\prime}^3}b_{3,0} - \frac{a^2}{6\,a_{\prime}^2}b_{3,1}\right\}$$

---

* The sign of this quantity in the *Mécanique céleste* requires alteration.

$$-\frac{m_{,}}{\mu}\left\{\frac{a^3}{6a_{,}^3}b_{3,0}-\frac{5\,a^2}{12\,a_{,}^2}b_{3,1}\right\}e\cos(nt+\epsilon-\varpi)$$

$$+\frac{m_{,}}{\mu}\left\{\frac{a^3}{a_{,}^3}b_{3,0}-\frac{5\,a^2}{4\,a_{,}^2}b_{3,1}\right\}e\,nt\sin(nt+\epsilon-\varpi)$$

$$-\frac{m_{,}}{\mu}\left\{\frac{3}{8}\frac{a^2}{a_{,}^2}b_{3,0}-\frac{a}{4\,a_{,}}b_{3,1}+\frac{a^2}{8\,a_{,}^2}b_{3,2}\right\}e_{,}\cos(nt+\epsilon-\varpi_{,})$$

$$-\frac{m_{,}}{\mu}\frac{a^2}{a_{,}^2}b_{3,2}\,e_{,}\,nt\sin(nt+\epsilon-\varpi_{,})$$

$$-r_1\cos i\tau-\{r_3-r_1\}e\cos(i\tau-\xi)-\{r_4-r_1\}e\cos(i\tau+\xi)$$

$$-r_5\,e_{,}\cos\xi_{,}-r_6\,e_{,}\cos(i\tau-\xi_{,})-r_7\,e_{,}\cos(i\tau+\xi_{,})$$

i being any positive whole number.

It may be shown that in the Planetary Theory (when $a < a_{,}$)

$$g=1+\frac{m_{,}}{\mu}\left\{\frac{a^3}{a_{,}^3}b_{3,0}-\frac{3\,a^2}{4\,a_{,}^2}b_{3,1}\right\}$$

$$gn=\mathrm{n}\left\{1+\frac{m_{,}\,a^2}{4\,\mu\,a_{,}^2}b_{3,1}\right\}$$

$$g_{,}=1+\frac{m}{\mu_{,}}\left\{b_{3,0}-\frac{3\,a}{4\,a_{,}}b_{3,1}\right\}$$

$$g_{,}n_{,}=\mathrm{n}_{,}\left\{1+\frac{m\,a}{4\,\mu_{,}\,a_{,}}b_{3,1}\right\}$$

The expressions for the variations of the elliptic constants given in a former part of this treatise, may be presented, as is well known, in another form, R being developed in terms of the mean longitudes,

$$\mathrm{d}a=-2\frac{a^2n}{\mu}\frac{\mathrm{d}R}{\mathrm{d}\epsilon}\mathrm{d}t$$

$$\mathrm{d}\epsilon=-\frac{an\sqrt{1-e^2}}{\mu e}(1-\sqrt{1-e^2})\frac{\mathrm{d}R}{\mathrm{d}e}\mathrm{d}t+2\frac{a^2n}{\mu}\frac{\mathrm{d}R}{\mathrm{d}a}\mathrm{d}t$$

$$\mathrm{d}e=\frac{an\sqrt{1-e^2}}{\mu e}(1-\sqrt{1-e^2})\frac{\mathrm{d}R}{\mathrm{d}\epsilon}\mathrm{d}t+\frac{an\sqrt{1-e^2}}{\mu e}\frac{\mathrm{d}R}{\mathrm{d}\varpi}\mathrm{d}t$$

$$\mathrm{d}\varpi=-\frac{an\sqrt{1-e^2}}{\mu e}\frac{\mathrm{d}R}{\mathrm{d}e}\mathrm{d}t$$

$$\mathrm{d}\gamma=-\frac{an}{\mu\sin\iota\sqrt{1-e^2}}\frac{\mathrm{d}R}{\mathrm{d}\iota}\mathrm{d}t$$

$$\mathrm{d}\iota=\frac{an}{\mu\sin\iota\sqrt{1-e^2}}\frac{\mathrm{d}R}{\mathrm{d}\nu}$$

See the *Théor. Anal.* vol. 1. p. 330, or *The Mechanism of the Heavens*, p. 231.

In these works R is used with a contrary sign to its acceptation in the *Mécanique céleste*, which I have followed.

When the square of the eccentricity is neglected in the value of the radius vector, the equations may be employed in the following shape:

$$\mathrm{d}a = -2a^2 n \frac{\mathrm{d}R}{\mathrm{d}\varepsilon}\mathrm{d}t \qquad \mathrm{d}\varepsilon - \mathrm{d}\varpi = \frac{an}{e}\frac{\mathrm{d}R}{\mathrm{d}e}\mathrm{d}t + 2a^2 n \frac{\mathrm{d}R}{\mathrm{d}a}\mathrm{d}t$$

$$\mathrm{d}e = \frac{ane}{2}\frac{\mathrm{d}R}{\mathrm{d}\varepsilon}\mathrm{d}t + \frac{an}{e}\frac{\mathrm{d}R}{\mathrm{d}\varpi}\mathrm{d}t$$

If $\qquad \zeta = \int n\,\mathrm{d}t \qquad \mathrm{d}\zeta = 3\iint an\,\mathrm{d}R\,\mathrm{d}t$

$$\frac{a}{r} = -\frac{\delta a}{a}\left\{1 + e\cos(nt+\varepsilon-\varpi)\right\} + \cos(nt+\varepsilon-\varpi)\,\delta e$$

$$- e\sin(nt+\varepsilon-\varpi)(\delta\varepsilon-\delta\varpi) + 2e\cos(2nt+2\varepsilon-2\varpi)\,\delta e$$

$$- 2e^2\sin(2nt+2\varepsilon-2\varpi)(\delta\varepsilon-\delta\varpi) - e\sin(nt+\varepsilon-\varpi)\,\delta\zeta$$

Considering the arguments 0, $nt+\varepsilon-\varpi$, $nt+\varepsilon-\varpi_{\prime}$, neglecting the term $\frac{m_{\prime}a}{4a_{\prime}^2}b_{3,2}\,e\,e_{\prime}\cos(\varpi-\varpi_{\prime})$ in the development of R, which requires particular attention,

$$a\,\delta\frac{1}{r} = \frac{m_{\prime}a^3}{2\mu a_{\prime}^3}b_{3,0} - \frac{m_{\prime}a^2}{2\mu a_{\prime}^2}b_{3,1}$$

$$+ \frac{m_{\prime}}{\mu}\left\{\frac{a^3}{a_{\prime}^3}b_{3,0} - \frac{3a^2}{4a_{\prime}^2}b_{3,1}\right\} e\cos(nt+\varepsilon-\varpi)$$

$$+ \frac{m_{\prime}}{\mu}\left\{\frac{a^3}{a_{\prime}^3}b_{3,0} - \frac{5}{4}\frac{a^2}{a_{\prime}^2}b_{3,2}\right\} te\sin(nt+\varepsilon-\varpi)$$

$$+ \frac{m_{\prime}}{\mu}\left\{\frac{3}{2}\frac{a^2}{a_{\prime}^2}b_{3,0} - \frac{3}{2}\frac{a^3}{a_{\prime}^3}b_{3,1} - \frac{a^2}{8a_{\prime}^2}b_{3,2}\right\} e_{\prime}\cos(nt+\varepsilon-\varpi_{\prime})$$

$$= \frac{m_{\prime}a^3}{2\mu a_{\prime}^3}b_{3,0} - \frac{m_{\prime}a^2}{2\mu a_{\prime}^2}b_{3,1} + \left\{1 + \frac{m_{\prime}}{\mu}\frac{a^3}{a_{\prime}^3}b_{3,0} - \frac{3}{4}\frac{m_{\prime}}{\mu}\frac{a^2}{a_{\prime}^2}b_{3,1}\right\}$$

$$e\cos\left(n\left(1 + \frac{m_{\prime}a^3}{\mu a_{\prime}^3}b_{3,0} - \frac{5}{4}\frac{m_{\prime}}{\mu}\frac{a^2}{a_{\prime}^2}b_{3,1}\right)t + \varepsilon - \varpi\right)$$

$$+ \frac{m_{\prime}}{\mu}\left\{\frac{3}{2}\frac{a^2}{a_{\prime}^2}b_{3,0} - \frac{3}{2}\frac{a^3}{a_{\prime}^3}b_{3,1} - \frac{a^2}{8a_{\prime}^2}b_{3,2}\right\} e_{\prime}\cos(nt+\varepsilon-\varpi_{\prime})$$

It may be shown generally (See *Théor. anal.* vol. i. p. 472, and

Phil. Trans. 1832, p. 229.) that the result obtained by the integration by the method of indeterminate coefficients of the equation

$$\frac{d^2 r^2}{2\,d t^2} - \frac{\mu}{r} + \frac{\mu}{a} + 2\int d R + r\frac{d R}{d r} = 0$$

agrees with that obtained by the integration of the expressions for the variation of the elliptic constants, which must evidently be the case à priori. Here again, however, difficulties occur in the determination of the quantities $k_{\prime}$ and $f_{\prime}$, arising from the term

$$\frac{m_{\prime}\, a}{4\, a_{\prime}^{2}}\, b_{3,2}\, e\, e_{\prime} \cos\,(\varpi - \varpi_{\prime})$$

in the development of R.

Although the results obtained by each method must be identical, nevertheless that obtained by the latter method requires many reductions before it can be made to assume the same shape as that obtained by the former, so that the former has practically very greatly the advantage, when the numerical value is required of the coefficient of any given inequality.

The part of R which is constant and independent of the angle $\varpi - \varpi_{\prime}$,

$$= m_{\prime}\left\{-\frac{b_{1,0}}{2 a_{\prime}} + \frac{2\, a}{a_{\prime}^{2}}\left(\sin^{2}\frac{1}{2} - \frac{e^{2}+e_{\prime}^{2}}{2}\right) b_{3,1}\right\}$$ See p. 27.

Whence $\dfrac{d\,\varepsilon - d\,\varpi}{d\,t} = \dfrac{m_{\prime}}{\mu}\left\{\dfrac{a^{3}}{a_{\prime}^{3}}\, b_{3,0} - \dfrac{5}{4}\dfrac{a^{2}}{a_{\prime}^{2}}\, b_{3,1}\right\} = k = c - 1$

Similarly $\dfrac{d\,\varepsilon - d\,\nu}{d\,t} = \dfrac{m_{\prime}}{\mu}\left\{\dfrac{a^{3}}{a_{\prime}^{3}}\, b_{3,0} - \dfrac{3}{4}\dfrac{a^{2}}{a_{\prime}^{2}}\, b_{3,1}\right\} = g - 1$

$$h^{2} e\, d\,\varpi + \left\{2 + e\cos(\lambda - \varpi)\right\} \sin\,(\lambda - \varpi)\, r^{2}\left(\frac{d R}{d \lambda}\right) d\,\lambda$$

$$- h^{2} r^{2} \cos\,(\lambda - \varpi)\left(\frac{d R}{d r}\right) d\,\lambda = 0$$

In order to obtain the constant term, it is sufficient to write

$$e\,\frac{d\,\varpi}{d\,\lambda} = r^{2} \cos\,(\lambda - \varpi)\left(\frac{d R}{d r}\right)$$

$$= \frac{m_{\prime}}{\mu}\{1 - e\cos\,(\lambda - \varpi)\}\cos\,(\lambda - \varpi)\left\{\frac{a^{3}}{2\, a_{\prime}^{3}}\, b_{3,0} - \frac{a^{2}}{2\, a_{\prime}^{2}}\, b_{3,1}\right.$$

$$\left. + \frac{a^{3}}{2\, a_{\prime}^{3}}\, b_{3,0}\, e \cos\,(\lambda - \varpi)\right\}$$

Retaining only the constant term,

$$-\frac{d\varpi}{d\lambda} = -\frac{m_{,}a^2}{4\mu a_{,}^2}b_{3,1} = \text{c} - 1$$

Similarly and to the same degree of approximation,

$$-\frac{d\nu}{d\lambda} = \frac{m_{,}a^2}{4\mu a_{,}^2}b_{3,1} = \text{g} - 1$$ See p. 59, and *Méc. cél.* vol. 5, p. 367.

These theorems being extended to the square of the disturbing force would perhaps furnish the readiest means of calculating the quantities c and g, c and g in the Lunar Theory.

In Laplace's notation, when $a < a_{,}$,

$$\frac{a}{2a_{,}}b_{3,0} - \frac{1}{2}b_{3,1} = -\frac{d\,b_{\frac{1}{2}}^{(0)}}{2d\alpha}, \quad \frac{a}{a_{,}}b_{3,1} - \frac{1}{2}b_{3,0} - \frac{1}{2}b_{3,2} = -\frac{d\,b_{\frac{1}{2}}^{(1)}}{d\alpha}$$

$$\frac{a}{a_{,}}b_{3,2} - \frac{1}{2}b_{3,1} - \frac{1}{2}b_{3,3} = -\frac{d\,b_{\frac{1}{2}}^{(2)}}{d\alpha},$$

$$\frac{a}{a_{,}}b_{3,3} - \frac{1}{2}b_{3,2} - \frac{1}{2}b_{3,4} = -\frac{d\,b_{\frac{1}{2}}^{(3)}}{d\alpha}$$

$$3\left\{\frac{a}{2a_{,}}b_{5,0} - \frac{1}{2}b_{5,1}\right\} = -\frac{d\,b_{\frac{3}{2}}^{(0)}}{2d\alpha}$$

$$3\left\{\frac{a}{a_{,}}b_{5,1} - \frac{1}{2}b_{5,0} - \frac{1}{2}b_{5,2}\right\} = -\frac{d\,.\,b_{\frac{3}{2}}^{(1)}}{d\alpha}$$

$$3\left\{\frac{a}{a_{,}}b_{5,2} - \frac{1}{2}b_{5,1} - \frac{1}{2}b_{5,3}\right\} = -\frac{d\,.\,b_{\frac{3}{2}}^{(2)}}{d\alpha}$$

$$3\left\{\frac{a}{a_{,}}b_{5,3} - \frac{1}{2}b_{5,2} - \frac{1}{2}b_{5,4}\right\} = -\frac{d\,.\,b_{\frac{3}{2}}^{(3)}}{d\alpha}$$

The numerical values of these quantities are given for the principal planets in third volume of the *Mécanique céleste.*

6. The following numerical examples will serve to explain the expressions given above. When $a < a_{,}$

$$\frac{r}{a} = 1 + \frac{m_{,}}{\mu}\left\{\frac{a^3}{6a_{,}^3}b_{3,0} - \frac{a^2}{6a_{,}^2}b_{3,1}\right\}$$

$$- e\left\{1+\frac{m_{\prime}}{\mu}\left\{\frac{a^3}{6\,a_{\prime}^3}b_{3,0}-\frac{5}{12}\frac{a^2}{a_{\prime}^2}b_{3,1}\right\}\right\}\cos\,(nt+\varepsilon-\varpi)$$

$- r_1 \cos\,(i(nt-n_{\prime}t)+\varepsilon-\varepsilon_{\prime})$ See p. 55.

$$r_{1,1}=\frac{m_{\prime}}{\mu}\frac{n^2}{(2\,n-n_{\prime})\,n_{\prime}}\left\{\frac{2\,n}{(n-n_{\prime})}\left(\frac{a^2}{a_{\prime}^2}-\frac{a}{a_{\prime}}b_{1,1}\right)\right.$$

$$\left.+\frac{a^2}{a_{\prime}^2}+\frac{a^2}{a_{\prime}^2}\left(\frac{a}{a_{\prime}}b_{3,1}-\frac{1}{2}\,b_{3,0}-\frac{1}{2}b_{3,2}\right)\right\}$$ See p. 57.

$$\lambda_{1,1}=\frac{n}{(n-n_{\prime})}\left\{2\,r_{1,1}-\frac{m_{\prime}\,n}{\mu\,(n-n_{\prime})}\left(\frac{a^2}{a_{\prime}^2}-\frac{a}{a_{\prime}}b_{1,1}\right)\right\}$$

In the theory of Jupiter disturbed by Saturn,

$\frac{a}{a_{\prime}} = \cdot54531725$, $b_{1,1} = \cdot6206406$, $\frac{1}{2}b_{3,0} = \frac{4\cdot358387}{2} = 2\cdot179193$

$b_{3,1} = 3\cdot185493$ $b_{3,2} = 2\cdot082131$ $\frac{m_{\prime}}{\mu} = \frac{1}{3359\cdot4}$

$n = 337210\cdot78$ $n_{\prime} = 135792\cdot34$

$a = 5\cdot20116636$ $e = \cdot0480767$

See *Mécanique céleste*, vol. iii. p. 61 & 81.

Whence

$\frac{a^3}{2\,a_{\prime}^3}b_{3,0} = \cdot353381$, $\frac{a^3}{2\,a_{\prime}^2}b_{3,1} = \cdot473636$ $\frac{a^3}{2\,a_{\prime}^3}b_{3,0} - \frac{a^2}{2\,a_{\prime}^2}b_{3,1} = -\cdot120255$

log. ·120255 = 9·0801033

log. 5·20116636 = 0·7161007

9·7962040

log. $\frac{3\,\mu}{m_{\prime}}$ = 4·0033829

5·7928211 = log. ·0000620613 minus

Laplace has . . . . . . ·0000620566 minus

See *Mécanique céleste*, vol. iii. p. 121, line 5.

log. $\left\{\frac{5}{12}\frac{a^2}{a_{\prime}^2}b_{3,1}-\frac{a^3}{6\,a_{\prime}^3}b_{3,0}\right\}$ = 9·4423277

log. e = 8·6819347

log. a = 0·7161007

8·8403631

log. $\frac{\mu}{m_{\prime}}$ = 3·5262617

5·3141014 = log. ·0000206111 +

Laplace has . . . . ·0000206111 the sign omitted, *Mécanique céleste*, vol. iii. p. 122, line 28.

Calculation of $r_{1,1}$,

$\log. n_{,} = 5{\cdot}1328751$
$\log. n \;= 5{\cdot}5279013$

$9{\cdot}6049738 = \log. {\cdot}4026928$

$1 - \frac{n_{,}}{n} = \quad {\cdot}5973072$

$\frac{a}{a_{,}} b_{3,1} - \frac{1}{2} b_{3,0} - \frac{1}{2} b_{3,2} = - \; 1{\cdot}48315$

$\log. 1{\cdot}48315 = 0{\cdot}1711851$
$\log. \frac{a^2}{a_{,}^2} \;= 9{\cdot}4732982$

$9{\cdot}6444333 = \log. {\cdot}441045$

$\log. 1{\cdot}5973072 = 0{\cdot}2033884$
$\log. \;{\cdot}4026928 = 9{\cdot}6049738$
$\log. \;3359{\cdot}4 \;= 3{\cdot}5262617$

$3{\cdot}3346239$

$r_{1,1} = - \;{\cdot}0001301383$

$\frac{a^2}{a_{,}^2} - \frac{a}{a_{,}} b_{1,1} = - {\cdot}041075$
$\log. {\cdot}041075 = 8{\cdot}6135776$
$\log. \frac{n - n_{,}}{n} \;= 0{\cdot}7761940$

$8{\cdot}8373836 = \log. {\cdot}0687676$
2

$- \;{\cdot}1375352$
$- \;{\cdot}441045$

$- \;{\cdot}5785802$
$\frac{a^2}{a_{,}^2} = {\cdot}297371$

$- \;{\cdot}281209$

$\log. {\cdot}281209 = 9{\cdot}4490293$
$3{\cdot}3346259$

$6{\cdot}1144054 = \log. {\cdot}0001301383$
$\log. a = \;{\cdot}7161007$

$6{\cdot}8305061 = \log. {\cdot}000676871 +$

Laplace has . . . . ${\cdot}000676876 +$ p. 12

Calculation of the coefficient of sin $(nt - n_{,}t + \varepsilon - \varepsilon_{,})$ in the value of λ, or $\lambda_{1,1}$. See p. 59.

$2r_{1,1} = - {\cdot}0002602766$
$\quad\quad\;\; {\cdot}0000204702$

$\log. \quad {\cdot}0002398064 = 6{\cdot}3798601$
$\log. \frac{n - n_{,}}{n} \quad\quad = 9{\cdot}7761940$

$6{\cdot}6036661$
$5{\cdot}8038801$ *

$2{\cdot}4075461 = \log. 255''{\cdot}591$ minus.

$\log. \frac{n}{n - n_{,}} \left\{ \frac{a^2}{a_{,}^2} - \frac{a}{a_{,}} b_{1,1} \right\} = 8{\cdot}8373836$
$\log. 3359{\cdot}4 = 3{\cdot}5262617$

$5{\cdot}3111219 = \log. {\cdot}0000204702$

Laplace has . . . $255{\cdot}5917$, the sign omitted, p. 120.

In the notation of the *Mécanique céleste*, vol. iii. p. 120,

$n = n^{\text{iv}}, \qquad n_{,} = n^{\text{v}}.$

---

* The logarithm 5·8038801 is added to convert the coefficient into centesimal seconds; the logarithm 5·3144251 must be added to convert it into sexagesimal seconds.

When $a_{\prime} < a$

$$r_{1,1} = \frac{m_{\prime} n^2}{\mu (2n - n_{\prime}) n_{\prime}} \left\{ \frac{2n}{(n - n_{\prime})} \left(\frac{a^2}{a_{\prime}^2} - b_{1,1} \right) + \frac{a^2}{a_{\prime}^2} + b_{1,1} - \frac{a_{\prime}}{a} \left(\frac{a}{a} b_{3,1} - \frac{1}{2} b_{3,0} - \frac{1}{2} b_{3,2} \right) \right\}$$

$$\lambda_{1,1} = \frac{n}{(n - n_{\prime})} \left\{ 2 r_{1,1} - \frac{m_{\prime} n}{\mu (n - n_{\prime})} \left(\frac{a^2}{a_{\prime}^2} - b_{1,1} \right) \right\}$$

In the theory of Saturn disturbed by Jupiter,

$\frac{a_{\prime}}{a} = \cdot 54531725$ $\quad b_{1,1} = \cdot 6206406$ $\quad \frac{m_{\prime}}{\mu} = \frac{1}{1067 \cdot 09}$

$\frac{a_{\prime}}{a} b_{3,1} - \frac{1}{2} b_{3,0} - \frac{1}{2} b_{3,2} = - 1 \cdot 483154$ $\quad a = 9 \cdot 5378709$

Calculation of $r_{1,1}$:

$\log. n_{\prime} = 5 \cdot 5279013$
$\log. n = 5 \cdot 1328751$
$0 \cdot 3950262 = \log. 2 \cdot 48328$

$1 - \frac{n_{\prime}}{n} = - 1 \cdot 48328$

$\frac{a^2}{a_{\prime}^2} - b_{1,1} = 2 \cdot 74216$
$\log. 2 \cdot 74216 = 0 \cdot 4380928$
$\log. \frac{n - n_{\prime}}{n} = 0 \cdot 1712231$
$0 \cdot 2668697 = \log. 1 \cdot 84872$
2
$3 \cdot 69744$

$\log. 1 \cdot 48315 = 0 \cdot 1711851$
$\log. \frac{a_{\prime}}{a} = 9 \cdot 7366492$
$9 \cdot 9078343 = \log. \cdot 80879$
$\frac{a_{\prime}^2}{a^2} = 3 \cdot 36280$
$b_{1,1} = \cdot 62064$
$4 \cdot 79223$
$3 \cdot 69744$

$\log. 2 - \frac{n_{\prime}}{n} = 9 \cdot 6841988$
$\log. \frac{n_{\prime}}{n} = 0 \cdot 3950262$
$\log. 1067 \cdot 09 = 3 \cdot 0282012$
$3 \cdot 1074262$

$\log. 1 \cdot 09479 = 0 \cdot 0393310$
$3 \cdot 1074262$
$\log. r_{1,1} = 6 \cdot 9319048$
$r_{1,1} = - \cdot 00085488$

$\log. r_{1,1} = 6 \cdot 9119048$
$\log. a = 0 \cdot 9794518$
$7 \cdot 9113568 = \log. \cdot 00815374$

Laplace has . . . . $\cdot 00815384$ p. 85.

Calculation of $\lambda_{1,1}$,

$$\log. \frac{n}{n-n_{\prime}}\left(\frac{a^2}{a_{\prime}^2}-b_{1,1}\right) = 0{\cdot}2668697$$
$$\log. 1067{\cdot}09 = 3{\cdot}0282012$$
$$9{\cdot}2386685 = \log. {\cdot}00173248$$

$$2\,r_{1,1} = -\,{\cdot}00170976$$
$${\cdot}00173248$$
$$\log. {\cdot}00002272 = 5{\cdot}3564083$$
$$\log. \frac{n-n_{\prime}}{n} = 0{\cdot}1712231$$
$$5{\cdot}1851852$$
$$5{\cdot}8038801$$
$$0{\cdot}9890653 = \log. 9''{\cdot}7513$$

Laplace has $9''{\cdot}742382$ (p. 134.) the sign omitted, which should be +.

CORRECTIONS.

P. v. line 6, Preface; *for* coincidences *read* coincidence.

P. 21, at foot, *dele* }

P. 24, *for* $\frac{a\,\mathrm{d}\,b_{n,i}}{\mathrm{d}^2} = -\frac{na}{a_{\prime}}\left\{\frac{a}{a_{\prime}}b_{n+2,i} - \frac{1}{2}b_{n+2,i+1} - \frac{1}{2}b_{n+2,i+1}\right\}$

read $\frac{a\,\mathrm{d}\,b_{n,i}}{\mathrm{d}\,a} = -\frac{na}{a_{\prime}}\left\{\frac{a}{a_{\prime}}b_{n+2,i} - \frac{1}{2}b_{n+2,i-1} - \frac{1}{2}b_{n+2,i+1}\right\}$

P. 77, *for* $q_2 = -\frac{a^2}{2\,a_{\prime}^3}b_{3,0} + \frac{a}{2\,a_{\prime}^2}b_{3,1}$

read $q_2 = -\frac{a^2}{2\,a_{\prime}^3}b_{3,0} + \frac{a}{a_{\prime}^2}b_{3,1}$

P. 77, line 8, *for* R_2 *read* $m_{\prime}\,R_2$

P. 78, line 9, *for* $\mathrm{e}\left\{1 - \frac{m_{\prime}\,a^3}{2\,\mu\,a_{\prime}^3}b_{3,0}\right.$ *read* $\mathrm{e}\left\{1 - \frac{m_{\prime}\,a^3}{6\,\mu\,a_{\prime}^3}b_{3,0}\right.$

TABLE

TABLE.

TABLE showing the arguments, which, by their combination with
ments. A full stop is placed after the figure when it should be

	1	2	3	4	5	6	7	8	9	10	11	12	13	
0	 1	 2	 3	 4	 5	 6	 7	 8	 9	 10	 11	 12	 13	0
1	0 131	3 4	2 132	133. − 2	6 7	5	...... − 5	9 10	8	...... − 8	12 13	11	...... −11	1
2	4. − 3	0 8	− 9. 1	10. − 1	14 11	16. − 12	− 15. 13	...... − 2	 3	...... − 4	...... − 5	 6	...... − 7	2
3	132. − 2	9 1	0	131. 8	12 15	...... − 14	...... − 11	 4	2	....	 7	5	....	3
4	2 133	1 10	8 131	0	16 13	11	14	3	....	...... − 2	6	....	...... − 5	4
5	7. − 6	−14. 11	15. − 12	− 16. 13	0 17	− 18. 1	19. − 1	....	....	....	...... − 2	 3	...... − 4	5
6	134. − 5	12 16	14	...... − 11	18 1	0	131. − 17	....	....	....	 4	2	....	6
7	5 135	15 13	11	...... − 14	1 19	17 131	0	....	....	....	3	....	...... − 2	7
8	10. − 9	2 20	− 21. 4	22. − 3	26 23	....	....	0	 1	...... − 1	14	 16	...... −15	8
9	...... − 8	21 3	...... − 2	132. − 20	24 27	....	....	 1	0	 131	 15	...... − 14	....	9
10	8	4 22	20 133	2	28 25	....	....	1	 131	0	16	....	14	10
11	13. − 12	5 23	− 24. 7	25. − 6	2 29	 4	...... − 3	...... −14	 15	...... −16	0	 1	...... − 1	11
12	...... − 11	24 6	...... − 5	134 − 23	30 3	...... − 2	....	 16	14	....	 1	0	....	12
13	11	7 25	23 135	5	4 31	....	2	15	....	...... −14	1	 131	0	13
14	16. − 15	26. − 5	− 27. 6	28. − 7	32 2	...... − 3	 4	...... −11	 12	...... −13	−17. 8	18. − 9	−19. 10	14
15	...... − 14	27 7	5	135. − 26	3 33	....	...... − 2	 13	11	....	9 19	17	...... − 8	15
16	14	6 28	26 134	...... − 5	34 4	2	....	12	....	...... −11	18 10	8	...... −17	16
17	19. − 18	−32. 29	33 − 30	− 34. 31	5 35	 7	...... − 6	....	....	....	...... −14	 15	...... −16	17
18	...... − 17	30 34	32	...... − 29	36 6	...... − 5	....	....	....	....	 16	14	....	18
19	17	33 31	29	...... − 32	7 37	....	5	....	....	....	15	....	...... −14	19
	1	2	3	4	5	6	7	8	9	10	11	12	13	

other arguments, by addition and subtraction produce given argu-
placed in the line beneath, that place being already occupied.

	14	15	16	17	18	19	20	35	146	147	148	149	150	
0	 14	 15	 16	 17	 18	 19	 20	 35	 146	 147	 148	 149	 150	0
1	15 16	14	...... −14	18 19	17	...... −17	....	....	147 148	146	...... −146	152 153	151 154	1
2	5	 7	...... − 6	....	....	....	....	....	149 150	153 −151	154 −152	146	...... −146	2
3	 6	...... − 5	....	....	....	....	....	....	151 152	...... −149	...... −150	 147	 148	3
4	7	....	5	....	....	....	....	....	153 154	150	149	148	147	4
5	 2	...... − 3	 4	...... − 5	 6	...... − 7	....	....	155 156	160 −157	160 −158	....	....	5
6	3	....	....	 7	5	....	....	....	157 158	...... −155	...... − 156	....	....	6
7	 4	2	....	6	....	...... − 5	....	....	159 160	156	155	....	....	7
8	11	 13	...... −12	....	....	....	...... − 2	....	161 162	165 −163	166. −164	150	149	8
9	 12	...... −11	....	....	....	....	 4	....	163 164	...... −161	...... −162	 151	 152	9
10	13	....	11	....	....	....	3	....	165 166	162	....	154	153	10
11	17 8	19. − 9	−18. 10	14	 16	...... −15	....	....	167 168	171 −169	172. −170	156	155	11
12	9 18	...... −17	...... − 8	 15	...... −14	....	....	....	169 170	...... −167	...... −168	 157	 158	12
13	19 10	8	17	16	....	14	....	....	171 172	168	167	160	159	13
14	0	 1	...... − 1	 11	...... −12	 13	....	....	173 174	177 −175	178. −176	...... −155	...... −156	14
15	 1	0	....	12	....	...... −11	....	....	175 176	...... −173	....	 159	 160	15
16	1	....	0	 13	11	....	....	....	177 178	174	173	158	157	16
17	 11	...... −12	 13	0	 1	...... − 1	....	...... − 5	179 180	183. −181	184. −182	....	....	17
18	12	....	...... −11	 1	0	....	....	 7	181 182	...... −179	...... −180	....	....	18
19	 13	11	....	1	....	0	....	6	183 184	180	179	....	....	19
	14	15	16	17	18	19	20	35	146	147	148	149	150	

	1	2	3	4	5	6	7	8	9	10	11	12	13	14	15	16	17	18	19	20	35	
20	22.	8						2														20
	−21		10	−9					4	−3												
21																						21
	−20	9	−8					3	−2											1		
22	20	10		8				4		2										1		22
23	25.	11			8			5			2											23
	−24		13	−12		10	−9		7	−6		4	−3									
24																						24
	−23	12	−11		9	−8		6	−5		3	−2										
25	23	13		11	10		8	7		5	4		2									25
26	28.	14												2								26
	−27		16	−15	8	−9	10	−5	6	−7					4	−3						
27					9				5						2							27
	−26	15	−14				8	−7						3								
28	26	16		14		8		6						4		2						28
					10					−5												
29	31.	17			11						5						2					29
	−30		19	−18		13	−12					7	−6					4	−3			
30			18.																			30
	−29	18	−17		12	−11					6	−5					3	−2				
31	29	19		17	13		11				7		5				4		2			31
32	34.																					32
	−33	−17	18	−19	14	−15	16							−5	6	−7	2	−3	4			
33			17		15										5		3					33
	−32	19					−14							7					−2			
34	32	18				14								6				2				34
				−17	16											−5	4					
35	37.				17												5					35
	−36					19	−18											7	−6			
86																						36
	−35				18	−17											6	−5			1	
37	35				19		17										7		5		1	37
38		20						8												2		38
			22	−21					10	−9												
39																						39
		21	−20					9	−8											3		
40		22		20				10		8										4		40
	1	2	3	4	5	6	7	8	9	10	11	12	13	14	15	16	17	18	19	20	35	

	0	1	2	3	4	5	6	7	8	9	10	11	12	13	14	15	16	17	18	19	20	35	
41			23			20			11			8									5		41
				25	−24					13	−12		10	−9									
42																							42
			24	−23		21			12	−11		9	−8								6		
43			25		23	22			13		11	10		8							7		43
44			26						14						8								44
				28	−27	20				16	−15					10	−9				−5		
45						21																	45
			27	−26					15	−14					9	−8					7		
46			28		26				16		14				10		8				6		46
						22																	
47			29			23			17			11						8					47
				31	−30					19	−18		13						10	−9			
48																							48
			30	−29		24			18	−17		12	−11					9	−8				
49			31		29	25			19		17	13		11				10		8			49
50			32												14								50
				34	−33	26			−17	18	−19					16	−15	8	−9	10			
51						27				17								9					51
			33	−32					19						15	−14				−8			
52			34		32				18						16		14		8				52
						28					−17							10					
53			35			29						17						11				2	53
				37	−36								19	−18					13	−12			
54																							54
			36	−35		30						18	−17					12	−11			3	
55			37		35	31						19		17				13		11		4	55
56																							56
			−35	36	−37	32									−17	18	−19	14	−15	16		2	
57				35		33										17		15				3	57
			37												19					−14			
58			36												18				14				58
					−35	34											−17	16				4	
59						35												17				5	59
																			19	−18			
60																							60
						36												18	−17			6	
61						37												19		17		7	61
	0	1	2	3	4	5	6	7	8	9	10	11	12	13	14	15	16	17	18	19	20	35	

	1	2	3	4	5	6	7	8	9	10	11	12	13	14	15	16	17	
62	64. −63	66. −65	68. −67	70. −69	72. −71	....	....	....	....	....	....	....	....	....	....	....	....	62
63	...... −62	67 69	65	...... −66	73 75	....	....	....	....	....	....	....	....	....	....	....	....	63
64	62	68 70	66	...... −65	74 76	....	....	....	....	....	....	....	....	....	....	....	....	64
65	69. −68	...... −62	 63	...... −64	....	....	....	....	....	....	....	....	....	....	....	....	....	65
66	70. −67	62	 64	...... −63	....	....	....	....	....	....	....	....	....	....	....	....	....	66
67	76 −66	 63	...... −62	....	....	....	....	....	....	....	....	....	....	....	....	....	....	67
68	...... −65	 64	62	....	....	....	....	....	....	....	....	....	....	....	....	....	....	68
69	65	63	....	...... −62	....	....	....	....	....	....	....	....	....	....	....	....	....	69
70	66	64	....	62	....	....	....	....	....	....	....	....	....	....	....	....	....	70
71	75. −74	....	....	....	...... −62	 63	...... −64	....	....	....	....	....	....	....	....	....	....	71
72	76. −73	....	....	....	62	 64	...... −63	....	....	....	....	....	....	....	....	....	....	72
73	...... −72	....	....	....	 63	...... −62	....	....	....	....	....	....	....	....	....	....	....	73
74	...... −71	....	....	....	 64	62	....	....	....	....	....	....	....	....	....	....	....	74
75	71	....	....	....	63	....	...... −62	....	....	....	....	....	....	....	....	....	....	75
76	72	....	....	....	64	....	62	....	....	....	....	....	....	....	....	....	....	76
77	....	65	 69	...... −68	....	....	....	...... −62	 63	...... −64	....	....	....	....	....	....	....	77
78	....	66	 70	...... −67	....	....	....	62	 64	...... −63	....	....	....	....	....	....	....	78
79	....	 67	...... −66	....	....	....	....	 63	...... −62	....	....	....	....	....	....	....	....	79
80	....	 68	...... −65	....	....	....	....	 64	62	....	....	....	....	....	....	....	....	80
81	....	69	....	65	....	....	....	63	....	...... −62	....	....	....	....	....	....	....	81
	1	2	3	4	5	6	7	8	9	10	11	12	13	14	15	16	17	

	18	19	62	65	66	77	78	146	147	148	149	150	161	162	
62								146							62
				2	−2				148	−147	150	−149			
63				3							151				63
			1		4			147	−146			153			
64			1		3			148		146		152			64
				4							154				
65															65
			2					149	−152	153	−146				
66			2					150				146			66
									154	−151					
67															67
			3		1			151	−150			147			
68			3					152							68
				1						−149	148				
69				1					149		147				69
			4					153							
70			4		1			154		150		148			70
71															71
			5					155	−158	−159					
72			5					156							72
									159	−157					
73															73
			6					157	−156						
74			6					158							74
										−155					
75									155						75
			7					159							
76			7					160		156					76
77				2							149				77
			8					161	−164	165			−146		
78			8		2			162				150		146	78
									166	−168					
79															79
			9		3		1	163	−162			151		147	
80			9					164							80
				3		1				−161	152		148		
81				4		1			161		153		147		81
			10					165							
	18	19	62	65	66	77	78	146	147	148	149	150	161	162	

	1	2	3	4	5	6	7	8	9	10	11	12	13	14	15	16	17	
82		70		66				64		62								82
83		71			65													83
			75	−74							−62	63	−64					
84		72			66						62							84
			76	−73								64	−63					
85																		85
		73	−72		67						63	−62						
86												62						86
		74	−71		68						64							
87		75		71	69						63							87
													−62					
88		76		72	70						64		62					88
89																		89
		−72	73	−76	65									−62	63	−64		
90														62				90
		−71	74	−75	66										64	63		
91			71		67													91
		75												63	−62	−62		
92			72		68										62			92
		76												64				
93		73												63				93
				−72	69													
94		74												64		62		94
				−71	70													
95					71													95
																	−62	
96					72												62	96
97																		97
					73												63	
98																		98
					74												64	
99					75												63	99
100					76												64	100
	1	2	3	4	5	6	7	8	9	10	11	12	13	14	15	16	17	

	18	19	62	65	66	77	78	146	147	148	149	150	161	162	
82			10		4		1	166		162		154		148	82
83			11	5				167	−170	171	155				83
84			11		5			168	172	−169		156			84
85			12		6			169	−168			157			85
86			12	6				170		−167	158				86
87			13	7				171	167		159				87
88			13		7			172		168		160			88
89			14	−5				173	−176	177	−156				89
90			14		−5			174	178	−175		−155			90
91			15		7			175	−174			159			91
92			15	7				176		−173	160				92
93			16	6				177	173		157				93
94			16		6			178		174		158			94
95	63	−64							−182	−183					95
96	64	−63							184	−181					96
97	−62								−180						97
98	62									−179					98
99		−62							179						99
100		62								180					100
	18	19	62	65	66	77	78	146	147	148	149	150	161	162	

	1	2	3	4	5	6	7	8	9	10	11	12	13	14	15	16	17	18	19	
101 {	116. −101	102 103	117. −102	118 −103	104 105															} 101
102 {	117. −103	101	−101	116																} 102
103 {	118. −102	101	116	−101																} 103
104 {	119. −105				101	−101	116													} 104
105 {	120. −104				101	116	−101													} 105
106 {		102	−103	117				104	−101	116										} 106
107 {		103	118	−102				101	116	−101										} 107
108 {		104	−105	119	102						101	−101	116							} 108
109 {		105	120	−104	103						101	116	−101							} 109
110 {		105	−104	120	102									101	−101	116				} 110
111 {		104	119	−105	103									101	116	−101				} 111
112 {					104												101	−101	116	} 112
113 {					105												101	116	−101	} 113
114 {																				} 114
115 {																				} 115
116 {	101	117 118	103	102	119 120															} 116
	1	2	3	4	5	6	7	8	9	10	11	12	13	14	15	16	17	18	19	

	1	2	3	4	5	6	7	8	9	10	11	12	13	14	15	16	17	18	19	
117	102	116	101																	117
118	103	116		101																118
119	104				116	101														119
120	105				116		101													120
121		117	102					116	101											121
122		118		103				116		101										122
123		119	104		117						116	101								123
124		120		105	118						116		101							124
125		120	105		117									116	101					125
126		119		104	118									116		101				126
127					119												116	101		127
128					120												116		101	128
129																				129
130																				130
131	1	132 133	4	3	134 135	7	6		10	9		13			16	15		19	18	131
132	3	131	1	9		15	12		4			7			6					132
133	4	131	10	1		13	16			3			6			7				133
134	6		16	12	131	1	18					4				3		7		134
135	7		13	15	131	19	1						3		4				6	135
136	9	132	3	21				131	1			15			12					136
137	10	133		4				131		1			16			13				137
138	12	134	6		132	3			16		131	1			18	9		15		138
139	13	135		7	133		4			15	131		1		10	19			16	139
140	15	135	7		132		3		13			19	9	131	1				12	140
141	16	134		6	133					12		10	18	131		1		13		141
142	18				134	6						16				12	131	1		142
143	19				135		7						15		13		131		1	143
144																				144
145																				145
	1	2	3	4	5	6	7	8	9	10	11	12	13	14	15	16	17	18	19	

	1	2	3	4	5	6	7	8	9	10	11	12	13	14	
146	148. −147	150. −149	152. −151	154. −153	156. −155	....	....	....	....	....	....	....	....	....	146
147	.... −146	151 153	149	.... −150	157 159	....	....	....	....	....	....	....	....	....	147
148	146	152 154	150.	.... −149	158 160	....	....	....	....	....	....	....	....	....	148
149	153. −152	.... −146	 147	.... −148	....	....	....	....	....	....	....	....	....	....	149
150	154. −151	146	148.	.... −147	....	....	....	....	....	....	....	....	....	....	150
151	.... −150	 147	.... −146	....	....	....	....	....	....	....	....	....	....	....	151
152	.... −149	 148	146	....	....	....	....	....	....	....	....	....	....	....	152
153	149	147	....	.... -146	....	....	....	....	....	....	....	....	....	....	153
154	150	148	....	146	....	....	....	....	....	....	....	....	....	....	154
155	159. −158	....	....	....	.... −146	 147	.... −148	....	....	....	....	....	....	....	155
156	160. −157	....	....	....	146	 148	.... −147	....	....	....	....	....	....	....	156
157	.... −156	....	....	....	 147	.... −146	....	....	....	....	....	....	....	....	157
158	.... −155	....	....	....	 148	146	....	....	....	....	....	....	....	....	158
159	155	....	....	....	147	....	.... −146	....	....	....	....	....	....	....	159
160	156	....	....	....	148	....	146	....	....	....	....	....	....	....	160
161	....	149	 153	.... −152	....	....	....	.... −146	 147	.... −148	....	....	....	....	161
162	....	150	 154	.... −151	....	....	....	146	 148	.... −147	....	....	....	....	162
163	....	 151	.... −150	....	....	....	....	 147	 146	....	....	....	....	....	163
164	....	 152	.... −149	....	....	....	....	147 148	146	....	....	....	....	....	164
	1	2	3	4	5	6	7	8	9	10	11	12	13	14	

	15	16	17	18	19	146	147	148	149	150	161	162
146							− 63.					
						62	1	− 1		− 2		
147						63			3			
						1				4		
148						1	62			3		
						64	131		4			
149												
						2	− 3	4				
150						2						
							4	− 3				
151												
						3	− 2			1		
152						3						
								− 2	1			
153							2		1			
						4						
154						4		2		1		
155												
						5	− 6	7				
156						5						
							7	− 6				
157												
						6	− 5					
158						6						
								5				
159							5					
						7						
160						7		5				
161									2			
						8	− 9	10				
162						8				2		
							10	9	2			
163												
						9	− 8			3		1
164						9						
								− 8	3		1	

	1	2	3	4	5	6	7	8	9	10	11	12	13	14	
165		153		149						−146					165
166		154		150				148		146					166
167		155	159	−158	149						−146	147	148		167
168		156	160	−157	150						146	148	147		168
169		157	−156		151						147	−146			169
170		158	−155		152						148	146			170
171		159			153						147		146		171
172		160		156	154						148		146		172
173		−156	157	−160	149									−146	173
174		−155	158	−159	150									146	174
175		159	155		151									147	175
176		160	156		152									148	176
177		157		−156	153									147	177
178		158		−155	154									148	178
179					155										179
180					156										180
181					157										181
182					158										182
183					159										183
184					160										184
	1	2	3	4	5	6	7	8	9	10	11	12	13	14	

	15	16	17	18	19	146	147	148	149	150	161	162	
165							8		4		1		165
						10							
166						10		8		4		1	166
167									5				167
						11	− 12	13					
168						11				5			168
							13	− 12					
169									6				169
						12	− 11			6			
170						12							170
								− 11	6				
171							11		7				171
		146				13							
172						13		11		7			172
173													173
	−147	148				14	− 15	16	− 5				
174						14							174
	−148	147					16	− 15		− 5			
175													175
	146					15	− 14			7			
176	146					15							176
								− 14	7				
177							14						177
						16							
178		146				16		14		6			178
179													179
			−146	−147	148	17	− 18	19					
180			146			17							180
				−148	147		19	− 18					
181													181
			147	146		18	− 17						
182				146		18							182
			148					− 17					
183			147				17						183
					146	19							
184			148		146	19		17					184
	15	16	17	18	19	146	147	148	149	150	161	162	

London: Printed by Richard Taylor,
Red Lion Court, Fleet Street.

PROJECTION OF THE SPHERE FOR THE LATITUDE OF THE GREENWICH OBSERVATORY.

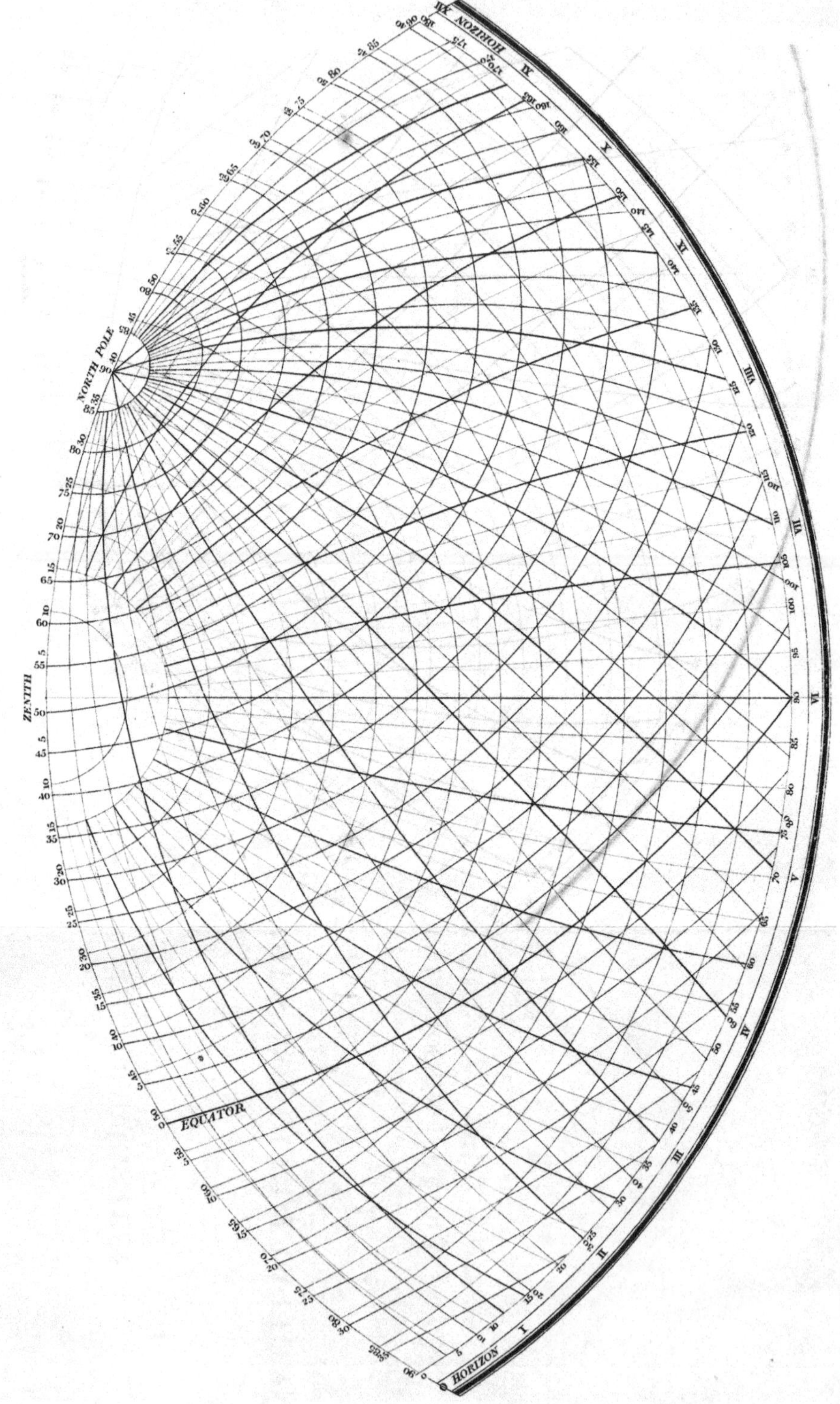

The material originally positioned here is too large for reproduction in this reissue. A PDF can be downloaded from the web address given on page iv of this book, by clicking on 'Resources Available'.

AN ELEMENTARY TREATISE ON THE COMPUTATION OF ECLIPSES AND OCCULTATIONS.

BY J. W. LUBBOCK, ESQ., F.R., R.A., and L. SS.,

MEMBER OF THE AMERICAN ACADEMY OF ARTS AND SCIENCES AND THE ACADEMY OF PALERMO.

LONDON:

CHARLES KNIGHT, 22 LUDGATE STREET.

1835.

PRINTED BY RICHARD TAYLOR,
RED LION COURT, FLEET SREET.

PREFACE.

Lagrange, in the *Astronomisches Jahrbuch* for 1782, first deduced the theory of eclipses from the general principles of analytical geometry, employing as data the true longitudes and latitudes of the heavenly bodies. This method possesses some advantages; but at present it seems better to use explicitly their right ascensions and declinations, because these quantities being given for the Moon in the Nautical Almanac, and for every hour, the operation of interpolating her place is either much simplified or altogether superseded.

The subject has lately been very fully treated by Mr. Woolhouse in the Appendix to the Nautical Almanac for 1836. I have taken the liberty of frequently referring to this very useful treatise; and I have selected my numerical examples from amongst those which Mr. Woolhouse has given, in order to facilitate the comparison of my methods and notation with those which he employs.

29, *Eaton Place*,
Sept. 1835.

ON THE

COMPUTATION

OF

ECLIPSES AND OCCULTATIONS.

If

$$y = a\,x + \alpha \qquad\qquad z = b\,x + \beta$$
$$y = a'\,x + \alpha' \qquad\qquad z = b'\,x + \beta'$$

are the equations to any straight lines, and the angle included between them be called Δ', then, by a well known expression,

$$\cos \Delta' = \frac{1 + a\,a' + b\,b'}{\sqrt{\{1 + a^2 + b^2\}}\ \sqrt{\{1 + a'^2 + b'^2\}}}$$

If these straight lines pass through given points S, S_1, S_2, of which the coordinates are

$$(x, y, z) \qquad\qquad (x_1, y_1, z_1)$$
$$(x, y, z) \qquad\qquad (x_2, y_2, z_2),$$

so that they intersect each other in the point S,

$$a = \frac{y_1 - y}{x_1 - x} \qquad\qquad b = \frac{z_1 - z}{x_1 - x}$$

$$a' = \frac{y_2 - y}{x_2 - x} \qquad\qquad b' = \frac{z_2 - z}{x_2 - x};$$

and by substituting these expressions for a, b, a', b', in the expression above, for $\cos \Delta'$,

$$\cos \Delta' = \frac{(x_1 - x)(x_2 - x) + (y_1 - y)(y_2 - y) + (z_1 - z)(z_2 - z)}{\sqrt{\{(x_1 - x)^2 + (y_1 - y)^2 + (z_1 - z)^2\}}\ \sqrt{\{(x_2 - x)^2 + (y_2 - y)^2 + (z_2 - z)^2\}}}$$

Let $x_1, y_1, z_1, x_2, y_2, z_2$ be the coordinates of the centres of any two heavenly bodies S S_2, the origin O being at the

B

earth's centre, and x, y, z the coordinates of the place of any observer; moreover, let

$$x = R\cos\phi\cos\mu \qquad y = R\cos\phi\sin\mu \qquad z = R\sin\phi$$
$$x_1 = \rho_1\cos\delta_1\cos\alpha_1 \qquad y_1 = \rho_1\cos\delta_1\sin\alpha_1 \qquad z_1 = \rho_1\sin\delta_1$$
$$x_2 = \rho_2\cos\delta_2\cos\alpha_2 \qquad y_2 = \rho_2\cos\delta_2\sin\alpha_2 \qquad z_2 = \rho_2\sin\delta_2,$$

and for abbreviation let

$$\cos\zeta_1 = \cos\delta_1\cos\phi\cos(\mu - \alpha_1) + \sin\delta_1\sin\phi$$
$$\cos\zeta_2 = \cos\delta_2\cos\phi\cos(\mu - \alpha_2) + \sin\delta_2\sin\phi$$
$$\cos\Delta = \cos\delta_1\cos\delta_2\cos(\alpha_1 - \alpha_2) + \sin\delta_1\sin\delta_2$$

$$\Pi = \frac{R}{\rho_1} \qquad \Pi' = \frac{R}{\rho_2},$$

Π and Π' being the sines of the horizontal parallaxes at S, the place of the observer.

After all these substitutions

$$\cos\Delta' = \frac{\cos\Delta - \Pi\cos\zeta_2 - \Pi'\cos\zeta_1 + \Pi\,\Pi'}{\sqrt{\{1 - 2\,\Pi\cos\zeta_1 + \Pi^2\}}\,\sqrt{\{1 - 2\,\Pi'\cos\zeta_2 + \Pi'^2\}}}.$$

So far all is general, and this equation is applicable to various systems of coordinates, the position of the coordinate axes being arbitrary. If the plane $x\,y$ be supposed to coincide with the ecliptic, the various equations first proposed by Lagrange in the *Astronomisches Jahrbuch* for 1782 may be obtained. In this case α_1, δ_1, α_2, δ_2 are the longitudes and latitudes of the bodies S_1, S_2; and if μ be the longitude, ϕ the zenith distance of the nonagesimal, or point of the ecliptic which is on the meridian of the observer, and ω the obliquity of the ecliptic, the following substitutions are required:

$$x = R\cos\phi\cos\mu$$
$$y = R\{\cos\phi\sin\mu\cos\omega - \sin\omega\sin\phi\}$$
$$z = R\{\cos\phi\sin\mu\sin\omega + \cos\omega\cos\phi\}$$
$$x_1 = \varrho_1\cos\delta_1\cos\alpha_1, \qquad y_1 = \varrho_1\cos\delta_1\sin\alpha_1, \qquad z_1 = \varrho_1\sin\delta_1.$$

But if the plane $x\,y$ coincides with the equator, and the axis $O\,x$ passes through the vernal equinox,

α_1 is the right ascension of S_1,

α_2 . . . . S_2,

μ is the sidereal time at the place of the observer, or the right ascension of an object on the meridian,

δ_1 is the declination of S_1,

δ_2 . . . . S_2,

ϕ is the geocentric latitude of the observer,

ζ_1 is the zenith distance of S_1,

ζ_2 . . . . S_2,

Δ is the true distance, or arc of great circle intercepted between $S_1\, S_2$,

Δ' is the apparent distance $S_1\, S_2$ as seen by the observer at S,

sine of the parallactic angle of $S_1 = \dfrac{\cos\phi\sin(\mu - \alpha_1)}{\sin\zeta_1}$,

sine of the azimuth of $S_1 = \dfrac{\cos\delta_1\sin(\mu - \alpha_1)}{\sin\zeta_1}$,

$\mu - \alpha_1$ being the hour angle Z P S_1.

In the examples which follow, the index 1 generally refers to the moon, so that δ_1 is the declination and $\mu - \alpha_1$ the hour angle of the moon; δ_2 is the declination of the sun or star, and $\mu - \alpha_2$ the hour angle of the star, or in eclipses the apparent time at the place of the observer.

By the equation

$$\cos\Delta' = \frac{\cos\Delta - \Pi\cos\zeta_2 - \Pi'\cos\zeta_1 + \Pi\,\Pi'}{\sqrt{\{1 - 2\,\Pi\cos\zeta_1 + \Pi^2\}}\sqrt{\{1 - 2\,\Pi'\cos\zeta_2 + \Pi'^2\}}}$$

$$\{1 - 2\,\Pi\cos\zeta_1 + \Pi^2\}\,\{1 - 2\,\Pi'\cos\zeta_2 + \Pi'^2\}\sin^2\Delta'$$
$$= \sin^2\Delta + 2\,(\Pi\cos\Delta - \Pi')\cos\zeta_2 + 2\,(\Pi'\cos\Delta - \Pi)\cos\zeta_1$$
$$+ 2\,\Pi\,\Pi'\cos\zeta_1\cos\zeta_2 + \Pi^2\sin^2\zeta_2 + \Pi'^2\sin^2\zeta_1$$
$$- 2\,\Pi\,\Pi'\,(\Pi'\cos\zeta_1 - \Pi\cos\zeta_2).$$

In an eclipse, Π' being small,

$$\{1 - 2\,\Pi\cos\zeta_1 + \Pi^2\}\sin^2\Delta'$$
$$= \sin^2\Delta + 2\,(\Pi - \Pi')\cos\Delta\cos\zeta_2 - 2\,(\Pi - \Pi')\cos\zeta_1 + (\Pi - \Pi')^2\sin^2\zeta_2$$

nearly; and when $\Pi' = 0$,

$$\cos\Delta' = \frac{\cos\Delta - \Pi\cos\zeta_2}{\sqrt{\{1 - 2\,\Pi\cos\zeta_1 + \Pi^2\}}}$$

$$\sin^2\Delta'\,\{1 - 2\,\Pi\cos\zeta_1 + \Pi^2\}$$
$$= \sin^2\Delta + 2\,\Pi\cos\Delta\cos\zeta_2 - 2\,\Pi\cos\zeta_1 + \Pi^2\sin^2\zeta_2.$$

Hence, in calculating the apparent distance of the centres of the sun and moon for an eclipse, it is evident the same expressions may be employed as for an occultation, using $\Pi - \Pi'$ instead of Π.

Conversely,

$$\cos\Delta = \cos\Delta' \sqrt{\{1 - \Pi^2 \sin^2 \zeta'_1\}} - \Pi \cos\Delta' \cos\zeta'_1 + \Pi \cos\zeta_2$$

$$\delta\Delta = -\frac{\Pi}{\sin\Delta}\cos\zeta_2 + \frac{\Pi}{\tan\Delta}\cos\zeta'_1 + \frac{\Pi^2}{2\tan\Delta}\sin^2\zeta'_1$$

nearly.

In Mr. Thompson's method of clearing a lunar distance, the *first* and *second corrections* are the quantities given by the first and second terms of the above expression for $\delta\Delta$, each increased by 5°, and the effect of the third term calculated for parallax 57′ is doubtless included with that due to the mean refraction in the *third correction* given by Mr. Thompson's Table XVIII., obtained of course empirically. It might perhaps be an improvement upon this method to include in the latter table all that is due to the mean parallax and mean refraction, which would leave only two small terms,

$$-\frac{\delta\Pi}{\sin\Delta}\cos\zeta'_2 + \frac{\delta\Pi}{\tan\Delta}\cos\zeta'_1,$$

$\delta\Pi$, being always less than 4′, to be calculated separately: but, then, perhaps the Table XVIII. would require greater extension.

From the expression

$$\cos\Delta' = \frac{\cos\Delta - \Pi\cos\zeta_2}{\sqrt{\{1 - 2\,\Pi\cos\zeta_1 + \Pi^2\}}}$$

it is easy to deduce

$$\sin^2\Delta' = \frac{\sin^2\Delta - 2\,\Pi\cos\zeta_1 + 2\,\Pi\cos\Delta\cos\zeta_2 + \Pi^2\sin^2\zeta_2}{\sqrt{\{1 - 2\,\Pi\cos\zeta_1 + \Pi^2\}}},$$

which expression may be put in the form

$$\sin^2\Delta'\{1 - 2\,\Pi\cos\zeta_1 + \Pi^2\}$$

$$= \{\cos\delta_1\sin(\alpha_1 - \alpha_2) - \Pi\cos\phi\sin(\mu - \alpha_2)\}^2$$

$$+ \left\{\sin(\delta_1 - \delta_2) + 2\cos\delta_1\sin\delta_2\sin^2\left(\frac{\alpha_1 - \alpha_2}{2}\right)\right.$$

$$\left. - \Pi\{\sin\phi\cos\delta_2 - \cos\phi\sin\delta_2\cos(\mu - \alpha_2)\}\right\}^2,$$

in which form nearly it is given by Prof. Bessel, *Ast. Nachrichten*, vol. vii. p. 3. See also Dr. Pearson's Introduction to Practical Astronomy, vol. ii. p. 634.

In eclipses and occultations, $\alpha_1 - \alpha_2$ and $\delta_1 - \delta_2$ are small angles, so that the quantity

$$2\cos\delta_1\sin\delta_2\sin^2\left(\frac{\alpha_1 - \alpha_2}{2}\right)$$

is generally insensible.

The quantities $\Pi \cos \phi \sin (\mu - \alpha_2)$

$$\Pi \{\sin \phi \cos \delta_2 - \cos \phi \sin \delta_2 \cos (\mu - \alpha_2)\}$$

are given for the latitude of Greenwich, and for $\Pi = \sin 57'$ by the Tables III. and V. subjoined.

In drawing the projection of the eclipse or occultation, the apparent orbit of the moon's centre may be considered approximately as a line passing through the series of points of which the coordinates are

$$\cos \delta_2 \sin (\alpha_1 - \alpha_2) - \Pi \cos \phi \sin (\mu - \alpha_2),$$

and

$$\sin (\delta_1 - \delta_2) - \Pi \{\sin \phi \cos \delta_2 - \cos \phi \sin \delta_2 \cos (\mu - \alpha_2)\}.$$

If α'_1, δ'_1 denote the apparent right ascension and declination of the moon,

$$\cos \delta'_1 \cos \alpha'_1 = \frac{\cos \delta_1 \cos \alpha_1 - \Pi \cos \phi \cos \mu}{\sqrt{\{1 - 2 \Pi \cos \zeta_1 + \Pi^2\}}}$$

$$\cos \delta'_1 \sin \alpha'_1 = \frac{\cos \delta_1 \sin \alpha_1 - \Pi \cos \phi \sin \mu}{\sqrt{\{1 - 2 \Pi \cos \zeta_1 + \Pi^2\}}}$$

$$\sin \delta'_1 = \frac{\sin \delta_1 - \Pi \sin \phi}{\sqrt{\{1 - 2 \Pi \cos \zeta_1 + \Pi^2\}}}$$

$$\cos \zeta'_1 = \frac{\cos \zeta_1 - \Pi}{\sqrt{\{1 - 2 \Pi \cos \zeta_1 + \Pi^2\}}}$$

$$\sin \zeta'_1 = \frac{\sin \zeta_1}{\sqrt{\{1 - 2 \Pi \cos \zeta_1 + \Pi^2\}}},$$

ζ'_1 being the apparent zenith distance of the moon.

The moon's semidiameter given in the Nautical Almanac is the angle under which her semidiameter would appear if viewed from the centre of the earth, that is, at the distance ϱ_1; let this quantity be called σ_1, and let σ'_1 be her apparent semidiameter as seen at the place of the observer and at any instant, then, since the distance of the centre of the moon from the observer

$$= \sqrt{\{(x_1 - x)^2 + (y_1 - y)^2 + (z_1 - z)^2\}}$$

$$= \varrho_1 \sqrt{\{1 - 2 \Pi \cos \zeta_1 + \Pi^2\}},$$

and as the apparent semidiameters are inversely as the distances from which the object is viewed,

$$\sin \sigma'_1 = \frac{\sin \sigma_1}{\sqrt{\{1 - 2 \Pi \cos \zeta_1 + \Pi^2\}}}$$

If Δ' refers to the instant of the immersion or emersion of a star,

$$\Delta' = \sigma'_1;$$

therefore when this phenomenon takes place,

$$\sin^2 \sigma_1 = \{\cos \delta_1 \sin (\alpha_1 - \alpha_2) - \Pi \cos \phi \sin (\mu - \alpha_2)\}^2$$
$$+ \left\{\sin (\delta_1 - \delta_2) + 2 \cos \delta_1 \sin \delta_2 \sin^2 \left(\frac{\alpha_1 - \alpha_2}{2}\right) - \Pi \{\sin \phi \cos \delta_2 - \cos \phi \sin \delta_2 \cos (\mu - \alpha_2)\}\right\}^2.$$

If P be the sine of the moon's horizontal equatorial parallax given in the Nautical Almanac, $\sin \sigma_1 = k P$; k is a constant of which the logarithm, according to Burckhardt, *Tables de la Lune*, page 73, $= 9{\cdot}4353665$. $k = {\cdot}2725$. If the equatorial radius of the earth be taken equal to unity, $P = \frac{1}{\rho_1}$, hence $\Pi = R P$.

When the edge of the sun is in contact with that of the moon,

$$\sin \Delta' = \sin (\sigma'_1 + \sigma'_2).$$

In calculating eclipses, σ'_1 the apparent semidiameter of the moon, must be found if accuracy is required,

$$\sin \sigma'_1 = \frac{\sin \sigma_1}{\surd \{1 - 2 \Pi \cos \zeta_1 + \Pi^2\}};$$

or ζ_2 may be substituted for ζ_1 without risk of any sensible error, so that

$$\sin \sigma'_1 = \frac{\sin \sigma_1}{\surd \{1 - 2 \Pi \cos \zeta_2 + \Pi^2\}}$$

If

$$2 \Pi \cos \zeta_2 - \Pi^2 = \sin^2 \theta,$$

$$\sin \sigma'_1 = \frac{\sin \sigma_1}{\cos \theta}.$$

If, moreover, the quantity on the left-hand side of the mark of equality in the equation at the foot of p. 4 be called $\sin^2 \Sigma$, and the sun's semidiameter be called σ_2,

$$\text{For} \left\{\begin{matrix}\text{partial} \\ \text{total or annular}\end{matrix}\right\} \textit{phase}, \sin \Sigma = \left\{\begin{matrix}\sin (\sigma'_1 + \sigma_2) \cos \theta \\ \sin (\sigma'_1 \backsim \sigma_2) \cos \theta\end{matrix}\right.$$

The augmentation of the moon's semidiameter may be taken from a table given by Mr. Woolhouse in the Appendix to the Nautical Almanac, 1836, p. 83, or from a similar table in other works.

Let

$$\cos\delta_1 \sin(\alpha_1 - \alpha_2) = p + p' t$$

$$\sin(\delta_1 - \delta_2) + 2\cos\delta_1 \sin\delta_2 \sin^2\left(\frac{\alpha_1 - \alpha_2}{2}\right) = q + q' t$$

$$\Pi \cos\phi \sin(\mu - \alpha_2) = u + u' t$$

$$\Pi \{\sin\phi \cos\delta_2 - \cos\phi \sin\delta_2 \cos(\mu - \alpha_2)\} = v + v' t,$$

p, q, u, and v are the values of these quantities at some given epoch T, and if t be expressed in hours, the hourly variations of p, q, u, and v may be taken for p', q', u', and v'.

In occultations α_2 is constant,

$$u' = \Pi \cos\phi \cos(\mu - \alpha_2) \frac{d\mu}{dt} \qquad v' = \Pi \cos\phi \sin\delta_2 \sin(\mu - \alpha_2) \frac{d\mu}{dt}.$$

The hourly diurnal motion of the earth is 15° 2′ 28″, or 54148″, and this quantity must be multiplied by sin 1″ to reduce it to the radius of the tables, so that in the calculation of occultations of stars by the moon, $\frac{d\mu}{dt}$ is a constant of which the logarithm is 9·41916.

In the calculation of eclipses, $\mu - \alpha_2$ is the apparent time at the place of the observer, and the logarithm of the corresponding quantity is 9·41797. Table VII. subjoined contains the value for Greenwich of the quantity [9·41916] sin 57′ $\cos\phi \cos(\mu - \alpha_2)$.

$\mu - \alpha_2 =$ Apparent solar time at the place of the observer,

$=$ Apparent solar time at Greenwich $\mp$ longitude $\begin{matrix}\text{west}\\ \text{east.}\end{matrix}$

If

$$p - u = m \sin M \qquad p' - u' = n \sin N$$

$$q - v = m \cos M \qquad q' - v' = n \cos N,$$

and if $\sin^2\Sigma$ always denote the quantity on the left-hand side of the equation at the foot of p. 4, the solution of this quadratic gives generally

$$t = -\frac{m}{n}\cos(M - N)$$

$$\mp \frac{1}{n} \sqrt{\{\sin\Sigma - m\sin(M-N)\}\{\sin\Sigma + m\sin(M-N)\}}.$$

If

$$\frac{m}{\sin\Sigma}\sin(M - N) = \cos\Psi$$

$$t = -\frac{m}{n}\cos(M-N) \mp \frac{\sin\Sigma}{n}\sin\Psi$$

$$= -\frac{m}{n}\frac{\cos(M-N\mp\Psi)}{\cos\Psi}$$

When employed for occultations, the upper sign gives the time t of the immersion reckoned from the time T, the lower that of the emersion reckoned from the same epoch, "provided Ψ has been taken $< 180°$, which may always be done. If we find, however, $\frac{m}{\sin\Sigma}\sin(M-N) > 1$, there will be no occultation, but the moon will pass by the star without occultation. It is evident, however, that this is only necessarily the case after the approximation has been pushed far enough, and that an error in N may produce the appearance of the impossibility of an occultation which really will take place, and *vice versâ*. If $\cos\Psi$ be found > 1, t is notwithstanding to be calculated by the formula

$$t = -\frac{m}{n}\cos(M-N),$$

and with this value the approximation is to be continued; it will then appear whether $\cos\Psi$ is really greater than 1. In like manner a value of Ψ which a rough approximation would show to be possible, might prove impossible by a greater approximation. These cases, however, if T is not too distant from the time of occultation, will only occur when the star remains very near the limb of the moon."—(Philosophical Magazine, vol. vi. p. 339. Prof. Bessel on the Calculations requisite for predicting Occultations of Stars by the Moon; translated from the *Astronomische Nachrichten.*)

Prof. Bessel also shows, that at the immersion and emersion of a star, the angle from north towards the east, for the *direct* image

$$= N \pm \Psi - 90°,$$

For the *inverted* image this angle

$$= N \pm \Psi + 90°.$$

For the same angles from the vertex the parallactic angle must be deducted.

This approximate solution rests upon the hypothesis that the quantities of which p, q, u, and v are the particular values at the epoch T, vary as the time, during a certain interval.

When the star is in the equator $\sin\delta_2 = 0$

$$v = \Pi\sin\phi \qquad v' = 0.$$

Generally v' is small, the error of the hypothesis is greatest with respect to the quantity

$$\Pi \cos \phi \sin (\mu - \alpha_2),$$

and this method cannot be relied on, except as a tolerably near approximation, unless the epoch from which t is reckoned is very near the time of occurrence of the *phase* required, that is, unless t is small.

At true conjunction

$$p = 0.$$

The time (t) of apparent conjunction reckoned from true conjunction

$$= \frac{u}{p' - u'}.$$

If

$$p = m' \sin M' \qquad q = m' \cos M'$$

$$\cos \Psi = \frac{(m' \sin M' - u)\, n \cos N - (m' \cos M' - v)\, n \sin N}{n \sin \Sigma}$$

$$= \frac{m' \sin (M' - N)}{\sin \Sigma} - \frac{\cos N u}{\sin \Sigma} + \frac{\sin N v}{\sin \Sigma}$$

$$= \frac{m' \sin (M' - N)}{\sin \Sigma} + \frac{\Pi \cos \delta_2 \sin N}{\sin \Sigma} \sin \phi$$

$$- \frac{\Pi \{\sin \delta_2 \sin N \cos (\mu - \alpha_2) + \cos N \sin (\mu - \alpha_2)\}}{\sin \Sigma} \cos \phi.$$

If $\sin N \sin \delta_2 = A \cos \beta \quad \cos N = A \sin \beta \quad \cot \beta = \sin \delta_2 \tan N$, and if H be the apparent time at Greenwich, $\mu - \alpha_2$ the hour angle at any place of which λ is the longitude east of Greenwich,

$$\mu - \alpha_2 = \lambda + H$$

$$\cos \Psi = \frac{m' \sin (M' - N)}{\sin \Sigma} + \frac{\Pi \cos \delta_2 \sin N}{\sin \Sigma} \sin \phi$$

$$- \frac{A \Pi}{\sin \Sigma} \cos (\lambda + H - \beta) \cos \phi$$

$$= \frac{m' \sin (M' - N)}{\sin \Sigma} + \frac{\Pi \cos \delta_2 \sin N}{\sin \Sigma} \sin \phi$$

$$- \frac{\Pi \sin N \sin \delta_2}{\sin \Sigma \cos \beta} \cos \phi \cos (\lambda + H - \beta).$$

Similarly, if $\tan \beta' = \dfrac{\tan N}{\sin \delta_2}$

$$t = -\frac{m}{n}\cos(M - N) \mp \frac{\sin\Sigma}{n}\sin\Psi$$

$$= -\frac{m'\cos(M' - N)}{n} \mp \frac{\sin\Sigma}{n}\sin\Psi$$

$$+ \frac{\Pi\cos\delta_2\cos N}{n}\sin\phi - \frac{\Pi\sin\delta_2\cos N}{n\cos\beta'}\cos\phi\cos(\lambda + H + \beta').$$

If, therefore, the epoch from which t is reckoned be the approximate time at which a phase of an eclipse or an occultation will happen at any given place, the last equations serve to obtain the time at which the phase or occultation will happen at any place not far distant, of which the latitude is ϕ and the longitude east of Greenwich λ, and they agree with the equations which Mr. Woolhouse employs for that purpose.

If

$$p = m'\sin M' \qquad p' = n'\sin N'$$

$$q = m'\cos M' \qquad q' = n'\cos N'$$

$$\cos\Psi' = \frac{m'\sin(M' - N')}{\sin\Sigma}.$$

In the problem of the transit of Venus or Mercury over the sun's disc, the times of the ingress and egress of the planet as seen from the earth's centre are given by the equation

$$t = -\frac{m'}{n'}\cos(M' - N') \mp \frac{\sin\Sigma}{n'}\sin\Psi'$$

Σ for external contact being the sum of the semidiameters of the sun and planet, α_2, δ_2 being the right ascension and declination of the sun, α_1, δ_1 of the planet.

$$-\sin\Psi' d\Psi' = \frac{m}{\sin\Sigma}\cos(M' - N')\, d M' + \frac{\sin(M' - N')}{\sin\Sigma}\, d m'$$

$$-u = m'\cos M'\, d M' + \sin M'\, d m'$$

$$-v = -m'\sin M'\, d M' + \cos M'\, d m'.$$

Neglecting altogether u' and v',

$$d t = \frac{m'}{n'}\sin(M' - N')\, d M' - \frac{\cos(M' - N')}{n'}\, d M \mp \frac{\sin\Sigma}{n'}\cos\Psi'\, d\Psi'$$

$$= \frac{\sin(N' + \Psi')}{n'\sin\Psi'}v - \frac{\cos(N' + \Psi')}{n'\sin\Psi'}u$$

$$= \frac{\sin (N' + \Psi') \cos \delta_2 \, \Pi \sin \phi}{n' \sin \Psi'}$$

$$- \frac{\{\sin (N' + \Psi') \sin \delta_2 \cos (\mu - \alpha_2) + \cos (N' + \Psi') \sin (\mu - \alpha_2)\} \Pi \cos \phi}{n' \sin \Psi'}.$$

If H be the apparent time at Greenwich, α_2 being the sun's right ascension, λ the longitude east,

$$\mu - \alpha_2 = \lambda + H; \text{ and if } \cot \beta'' = \tan (N' + \Psi') \sin \delta_2$$

$$\mathrm{d} t = \frac{\sin (N' + \Psi') \cos \delta_2 \, \Pi \sin \phi}{n' \sin \Psi'}$$

$$- \frac{\Pi \sin (N' + \Psi')}{n' \sin \Psi' \cos \beta''} \sin \delta_2 \cos (\lambda + H - \beta'') \cos \phi.$$

So that if T be the time of ingress or egress of the planet as seen from the centre of the earth, $T + \mathrm{d} t$ is the time of ingress or egress for any place of which the latitude is ϕ, and λ the longitude east of Greenwich. This last equation agrees with that given by Mr. Woolhouse for the same purpose. In this approximate method the quantities u' and v' are altogether neglected *.

If the roots of the equation

$$t = - \frac{m}{n} \cos (M - N)$$

$$\mp \frac{1}{n} \sqrt{\{\sin \Sigma - m \sin (M - N)\} \{\sin \Sigma + m \sin (M - N)\}}$$

are equal, in the occultation of the star, the times of immersion and emersion coincide, and the star grazes the edge of the moon. The observer is situated on the boundary of the country within which the occultation is visible. In an eclipse, the phase, determined by the value of Σ, is just visible:

$$\sin^2 \Sigma - m^2 \sin^2 (M - N) = 0.$$

If, moreover, $t = 0$,

---

* The transits of Mercury and Venus over the disc of the sun "are in many respects analogous to the annular eclipse of the sun, and admit of a similar calculation; the principal distinction consists in the negative sign of the relative motion of the planet in right ascension, which will make the inclination of the orbit always obtuse. . . . . . As the relative parallax is always very small, the ingress and egress of the planet will be seen at all places on the earth at nearly the same absolute time. It will, for this reason, be best to compute first the circumstances for the centre of the earth, and then to ascertain the small variations produced by parallax for any assumed place on the surface." (Appendix to Nautical Almanac, 1836, p. 121.)

$$(p-u)(p'-u')+(q-v)(q'-v')=0$$

$$(p-v)\sin N+(q-v)\cos N=0$$

$$(p-u)^2+(q-v)^2=\sin^2\Sigma$$

$$p-u=\frac{\sin\Sigma\,(q'-v')}{\sqrt{\{(p'-u')^2+(q'-v')^2\}}}=\pm\sin\Sigma\cos N$$

$$q-v=\frac{\sin\Sigma\,(p'-u')}{\sqrt{\{(p'-u')^2+(q'-v')^2\}}}=\mp\sin\Sigma\sin N.$$

Hence

$$P\cos\phi\sin(\mu-\alpha_2)=p\pm\sin\Sigma\cos N$$

$$P\{\sin\phi\cos\delta_2-\cos\phi\sin\delta_2\cos(\mu-\alpha_2)\}=q\mp\sin\Sigma\sin N.$$

If ζ be the zenith distance of the star or sun's centre, γ the parallactic angle, then, by spherical trigonometry,

$$\cos\phi\sin(\mu-\alpha_2)=\sin\zeta\sin\gamma$$

$$\sin\phi\cos\delta_2-\cos\phi\sin\delta_2\cos(\mu-\alpha_2)=\sin\zeta\cos\gamma$$

$$\sin\phi=\sin\zeta\cos\delta_2\cos\gamma+\sin\delta_2\cos\zeta.$$

If θ be an auxiliary angle, such that

$$\tan\theta=\tan\zeta\cos\gamma,$$

then

$$\sin\phi=\sin(\theta+\delta_2)\frac{\cos\zeta}{\cos\theta}$$

$$\tan(\mu-\alpha_2)=\frac{\sin\theta\tan\gamma}{\cos(\theta+\delta_2)}$$

$$\tan\phi=\tan(\theta+\delta_2)\cos(\mu-\alpha_2):$$

γ is M in Mr. Woolhouse's notation. The last equations are analogous to those which Mr. Woolhouse employs to determine the extreme latitudinal limits of a phase *.

* "The determination of the extreme latitudinal limits of a phase, or of the terrestrial lines whereon that phase will appear, as the middle of the local eclipse, is the most complex and unmanageable of all operations which relate to a general eclipse. For any given phase, at different places on the earth, the moon must be so reduced by parallax as to touch a given concentric circle on the solar disc; and if we consider this circle, by way of illustration, to represent, instead of the sun, the disc of the luminous body, the places on the earth which severally see the given phase must be situated in the surface of the penumbral or umbral cone, according as the interposing limb of the moon only approaches or projects over the centre of the sun; that is, the places must all be found in the intersection of this cone with the surface of the earth. This intersection will assume

Although the quantity N does in fact depend implicitly upon the angles φ and $\mu - \alpha_2$, which are sought, a first approximation may be obtained by substituting N' for N.

$$P \sin \zeta \sin \gamma = p \pm \sin \Sigma \cos N'$$

$$P \sin \zeta \cos \gamma = q \mp \sin \Sigma \sin N'$$

$$\tan N' = \frac{p'}{q'}$$

$$u' = [9{\cdot}41797] \frac{P \cos \zeta \cos (\theta + \delta_2)}{\cos \theta} \qquad v' = [9{\cdot}41797] P \sin \zeta \sin \gamma \sin \delta_2.$$

A nearer approximation may then be obtained by employing for N the value from the expression

$$\tan N = \frac{p' - u'}{q' - v'}.$$

If the constant of which the logarithm is [9·41797] be called A, the following equations may be found:

$$\cos^2 \zeta = \frac{\left\{\cos N \left(\frac{p'}{A} + q \sin \delta_2\right) - \sin N \left(\frac{q'}{A} - p \sin \delta_2\right)\right\}^2}{P^2 \cos^2 N \cos^2 \delta_2}$$

$$\sin^2 \zeta = \{p \pm \sin \Sigma \cos N\}^2 + \{q \mp \sin \Sigma \sin N\}^2.$$

The equation which results between $\sin N$, $\cos N$, and known quantities, by adding together the two last equations, is too complicated to be used with advantage.

If ϕ is a maximum or minimum, $d\phi = 0$,

$$P \cos \phi \cos (\mu - \alpha_2) \frac{d\mu}{dt} = \frac{dp}{dt} = p'$$

$$P \cos \phi \sin \delta_2 \sin (\mu - \alpha_2) \frac{d\mu}{dt} = \frac{dq}{dt} = q'.$$

---

a complete or partial oval form, according as the cone falls wholly or partially on the earth's illuminated disc. When it falls only partially on the earth, the extreme points will evidently see the sun in the horizon, and be therefore two points belonging to the horizon limits, but in the other case the phase cannot at that instant be seen in the horizon. .... If the rising and setting limits of any phase do not extend throughout the general partial eclipse, there will be both a northern and southern limit to that phase; but, on the contrary, when the rising and setting limits continue throughout the eclipse, there will be only one of these limits to the phase, viz. a southern limit when the difference of declination at conjunction is positive, and a northern one when that difference is negative." (Appendix to Nautical Almanac, 1836, p. 65.)

Eliminating $\frac{d\mu}{dt}$,

$$p' \sin \delta_2 \sin (\mu - \alpha_2) - q' \cos (\mu - \alpha_2) = 0$$

$$p' \sin^2 \delta_2 \{p + \sin \Sigma \cos N\} + q' \{q - \sin \Sigma \sin N - P \sin \phi \cos \delta_2\} = 0$$

$$\{p + \sin \Sigma \cos N\}^2 \sin^2 \delta_2 + \{q - \sin \Sigma \sin N - P \sin \phi \cos \delta_2\}^2$$
$$= P^2 \cos^2 \phi \sin^2 \delta_2.$$

The terms multiplied by $\sin \Sigma$ which accompany p and q may be neglected at first and afterwards replaced, in which case

$$p' \sin^2 \delta_2 \, p + q' (q - P \sin \phi \cos \delta_2) = 0$$

$$p^2 \sin^2 \delta_2 + (q - P \sin \phi \cos \delta_2)^2 = P^2 \cos^2 \phi \sin^2 \delta_2 .$$

Let $p = (p) + p' t$ $\qquad\qquad q = (q) + q' t$

so that (p) and (q) are the values of p and q at the epoch from which t is reckoned,

$$q = (q) + q' t = (q) + \frac{q'}{p'} \left(p - (p)\right)$$

$$p' \sin^2 \delta_2 \, p + q' \left\{(q) + \frac{q'}{p'} \left(p - (p)\right) - P \sin \phi \cos \delta_2 \right\} = o$$

$$p = - \frac{\frac{q'}{p'} \left\{(q) - \frac{q'}{p'} (p) - P \sin \phi \cos \delta_2\right\}}{\sin^2 \delta_2 + \frac{q'^2}{p'^2}}$$

$$q - P \sin \phi \cos \delta_2 = \frac{\sin^2 \delta_2 \left\{(q) - \frac{q'}{p'} (p) - P \sin \phi \cos \delta_2\right\}}{\sin^2 \delta_2 + \frac{q'^2}{p'^2}}$$

Substituting these values of p and $q - P \sin \phi \cos \delta_2$ in the equation of line

$$\left\{(q) - \frac{q'}{p'} (p) - P \sin \phi \cos \delta_2\right\}^2 = P^2 \cos^2 \phi \left\{\sin^2 \delta_2 + \frac{q'^2}{p'^2}\right\}.$$

The expression for $\sin \phi$ may be obtained in a very simple form by the introduction of two auxiliary variables η and ϵ, such that

$$\frac{(q) - \frac{q'}{p'}(p)}{\sqrt{1 + \frac{q'^2}{p'^2}}} = P \sin \eta \qquad \frac{\cos \delta_2}{\sqrt{1 + \frac{q'^2}{p'^2}}} = \cos \varepsilon$$

$$\{\sin \eta - \sin \phi \cos \varepsilon\}^2 = \cos^2 \phi \sin^2 \varepsilon,$$

whence

$$\sin \phi = \sin (\eta \pm \varepsilon).$$

Now, $(p) \pm \sin \Sigma \cos N$

and $(q) \mp \sin \Sigma \sin N$

may be written instead of (p) and (q), and the equation

$$\sin \phi = \sin (\eta \pm \varepsilon)$$

will serve to give the boundary beyond which any given phase of an eclipse, determined by the value of Σ, cannot possibly be visible. So in ascertaining the limits on the earth's surface beyond which the occultation of a star is not visible, $\Sigma = \sigma_1$, and if the time be reckoned from true conjunction

$$(p) = 0, \qquad (q) = \sin (\delta_1 - \delta_2)$$

$$\frac{(q) \mp \sin \sigma_1 \sin N \mp \frac{q'}{p'} \sin \sigma_1 \cos N}{\sqrt{1 + \frac{q'^2}{p'^2}}} = P \sin \eta.$$

Neglecting u' and v', $\qquad q' = n \cos N, \qquad p' = n \sin N,$

$$\sin (\delta_1 - \delta_2) \sin N \mp \sin \sigma_1 = P \sin \eta$$

$$\frac{\sin (\delta_1 - \delta_2) \sin N}{P} \mp k = \sin \eta$$

$$\cos \delta_2 \sin N = \cos \varepsilon.$$

The last equations agree with those employed by Mr. Woolhouse.

The eclipses of the moon by the earth's shadow may be resolved in the same way as those of the sun. The absolute positions of the moon and shadow being independent of the position of the spectator on the earth, the determination of parallaxes will be here unnecessary, which much simplifies the calculation of these eclipses. The considerations requisite to be attended to by way of distinction are the following:

Semidiameter of shadow $= \frac{61}{60} (P' + \pi - \sigma)$,

Semidiameter of the penumbra $= \frac{61}{60} (P' + \pi - \sigma) + 2\sigma$,

(P' being the moon's horizontal parallax for latitude 45°, π the sun's parallax, and σ the sun's true semidiameter.)

Right ascension of centre of shadow = that of sun $\pm$ 12^h.

Declination of centre of shadow = that of the sun with a contrary name." See Appendix to Nautical Almanac, 1836, p. 128. This subject is also fully treated in Woodhouse's Astronomy, vol. i. chap. xxxv.

The phænomena which take place upon the earth generally in a solar eclipse or the occultation of a star can be more readily determined by considering the appearances which would present themselves to an observer situate at the star or sun's centre, and having the moon between him and the earth.

I therefore now change the direction of the coordinate axes, taking for the new axis O z', the line joining the centre of the earth with that of the sun or star. Let x', y', z' be the new coordinates, and let

$$x' = a\,x + b\,y + c\,z$$
$$y' = a'\,x + b'\,y + c'\,z$$
$$z' = a''\,x + b''\,y + c''\,z$$

By well known equations of condition, the new axes being rectangular,

$$a^2 + b^2 + c^2 = 1 \qquad a\,a' + b\,b' + c\,c' = 0$$
$$a'^2 + b'^2 + c'^2 = 1 \qquad a'\,a'' + b'\,b'' + c'\,c'' = 0$$
$$a''^2 + b''^2 + c''^2 = 1 \qquad a\,a'' + b\,b'' + c\,c'' = 0.$$

Since the axis O z' coincides with O S_2,

$$a'' = \cos\alpha_2 \cos\delta_2 \qquad b'' = -\sin\alpha_2 \cos\delta_2 \qquad c'' = -\sin\delta_2.$$

Making $c = 0$, which amounts to supposing the axis O x' to be in the plane of the equator, and the plane $z'\,y'$ to coincide with the meridian of the star,

$$a = -\sin\alpha_2 \qquad b = -\cos\alpha_2 \qquad c = 0$$
$$a' = -\cos\alpha_2 \sin\delta_2 \qquad b' = \sin\alpha_2 \sin\delta_2 \qquad c' = -\cos\delta_2.$$

The coordinates of the moon's centre are

$$x'_1 = \rho_1 \cos\delta_1 \sin(\alpha_1 - \alpha_2)$$
$$y'_1 = \rho_1 \{\sin\delta_1 \cos\delta_2 - \cos\delta_1 \sin\delta_2 \cos(\alpha_2 - \alpha_1)\},$$

and those of any point on the earth's surface are

$$x' = R \cos \varphi \sin (\mu - \alpha_2)$$
$$y' = R \{\sin \phi \cos \delta_2 - \cos \phi \sin \delta_2 \cos (\mu - \alpha_2)\},$$

in which equations if ρ_1 be made equal to unity, $R = \Pi$. These equations are sufficient to determine the orthographic projection of any point upon the plane $x' y'$; and since the distance of the sun or star is considerable in comparison with that of the moon and with the diameter of the earth, the orthographic projection of the moon upon the plane $x' y'$ may be taken to define the shadow of the star or the points on the earth's surface where the *occultation* has taken place. This supposition amounts to substituting the *cylinder* which envelops the moon, and of which the axis is parallel to the line joining the star and the centre of the earth, for the *cone* which envelops the moon, and of which the apex coincides with the star. So also in an eclipse of the sun, the limits of the penumbra upon the earth's surface, at any moment, may be determined by describing a circle, with distance $\sigma_1 + \sigma_2$, the centre being the projection of the moon's centre; and similarly the limits within which the centre of the sun is eclipsed, and of the total ellipse by circles whose radii are σ_1 and σ_2. If $\sigma_1 < \sigma_2$, the last-mentioned circle determines the limits of the annular eclipse.

Since

$$x' = \Pi \cos \phi \sin (\mu - \alpha_2)$$
$$y' = \Pi \{\sin \phi \cos \delta_2 - \cos \phi \sin \delta_2 \cos (\mu - \alpha_2)\}$$
$$x'^2 \sin^2 \delta_2 + \{y' - \Pi \sin \varphi \cos \delta_2\}^2 = \Pi^2 \cos^2 \varphi \sin^2 \delta_2,$$

which equation is rigorous.

If the earth be considered as a sphere, R is constant; P, the sine of the moon's horizontal equatorial parallax, may be substituted for Π; and

$$x'^2 \sin^2 \delta_2 + \{y' - P \sin \varphi \cos \delta_2\}^2 = P^2 \cos^2 \varphi \sin^2 \delta_2$$

is the equation to the orthographic projection of the parallel of geocentric latitude ϕ. A nearer approach to accuracy may be had by employing, instead of the sine of the moon s horizontal equatorial parallax, that for latitude 45°.

For latitude 45°, log R = 9·99929
sin 57′ = 8·21958

8·21887 = log. sin 56′ 55″.

Therefore, in order to have the value of P for latitude 45°, it is sufficient to diminish the horizontal equatorial parallax by 5″.

In future the accents may be suppressed; and unless the contrary is stated, the axis $O z$ is to be understood as drawn from the centre of the earth to the star, the axis $O x$ perpendicular to $O z$ in the plane of the equator, and the plane $z y$ as coinciding with the meridian of the star. In eclipses of the sun, P is to be taken equal to the difference of the horizontal equatorial parallaxes, agreeably to the remark in p. 3.

Hence in this projection any given place on the earth's surface describes an ellipse which is easily constructed by means of the last equation. The centres of all the ellipses are situated on the axis $O y$, at a distance from the origin O equal to $P \sin \phi \cos \delta_2$. Their principal axes coincide in direction with the axes $O x$, $O y$. Their foci are all situated at a distance from the axis $O x$ equal to $P \cos \phi \cos \delta_2$, and upon a circle whose centre coincides with the origin, described with radius $P \cos \delta_2$. Therefore, to find the ellipse described by a place of which the latitude is ϕ as seen from the occulted star, draw the circle E N Q with radius P and centre O, and the circle A P C with radius $\Pi \cos \delta_2$. Take the angle $F O A = \phi$, produce O F, cutting the circle E N Q in K. Draw K G perpendicular to F F′. G G′ is the axis major of the ellipse required, and F F′ the foci. In fig. 1 the declination of the star is supposed south, in which case P, the south pole, is visible; in fig. 2 the declination of the star is north, and the north pole, P, is visible.

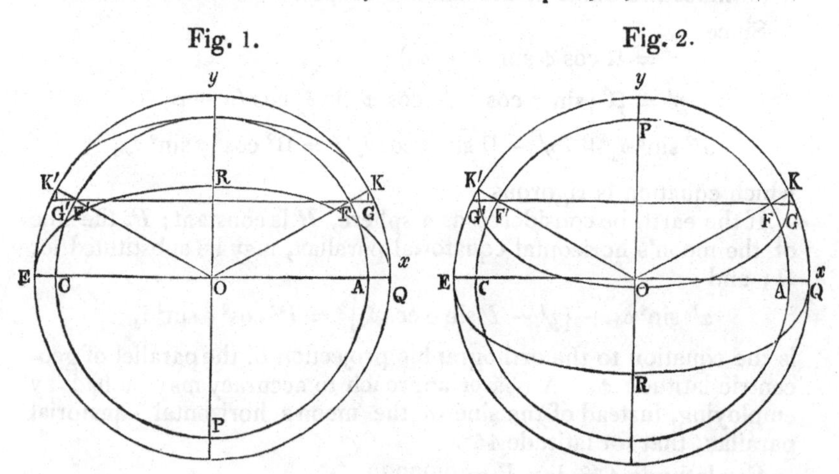

Let $y = a x + b$ be the equation to the apparent orbit of the moon's centre, or the path of the edge or any other *phase*, and that it be required to find the parallel of latitude which touches this straight line, or in other words the geographical latitude for which this phase is just visible:

$$x \sin^2 \delta_2 \, d x + \{y - P \sin \varphi \cos \delta_2\} \, d y = 0$$

$$x \sin^2 \delta_2 + a \{a x + b - P \sin \varphi \cos \delta_2\} = 0$$

$$x = -\frac{a \{b - P \sin \varphi \cos \delta_2\}}{\sin^2 \delta_2 + a^2}$$

$$y - P \sin \varphi \cos \delta_2 = \frac{\sin^2 \delta_2 \{b - P \sin \varphi \cos \delta_2\}}{\sin^2 \delta_2 + a^2}$$

Substituting these values of x and $y - P \sin \varphi \cos \delta_2$ in the equation to the ellipse,

$$\frac{\{b - P \sin \varphi \cos \delta_2\}^2}{\sin^2 \delta_2 + a^2} \{a^2 \sin^2 \delta_2 + \sin^4 \delta_2\} = P^2 \cos^2 \varphi \sin^2 \delta_2$$

$$\{b - P \sin \phi \cos \delta_2\}^2 = P^2 \cos^2 \varphi \{\sin^2 \delta_2 + a^2\}$$

$$\sin \varphi = \frac{b \cos \delta_2}{P (1 + a^2)} \pm \frac{\sqrt{\sin^2 \delta_2 + a^2}}{\sqrt{1 + a^2}} \sqrt{1 - \frac{b^2}{P^2 (1 + a^2)}}.$$

The value of φ from this equation may be conveniently obtained by the introduction of two auxiliary angles η and ε.

Let

$$\frac{b}{P \sqrt{1 + a^2}} = \sin \eta, \text{ then } \sqrt{1 - \frac{b^2}{P^2 (1 + a^2)}} = \cos \eta;$$

also let

$$\frac{\cos \delta_2}{\sqrt{1 + a^2}} = \cos \varepsilon, \text{ then } \frac{\sqrt{\sin^2 \delta_2 + a^2}}{\sqrt{1 + a^2}} = \sin \varepsilon$$

$$\sin \varphi = \sin (\eta \pm \varepsilon).$$

If $y = a x + b$ be the equation to the projection of the moon's centre as seen from the star, a is the tangent of the inclination of this line with the axis x;

$$\frac{1}{\sqrt{1 + a^2}}$$

is the cosine of this angle, b is the difference of declination of the moon and star at the time of conjunction, or $\sin (\delta_1 - \delta_2)$;

$$\frac{b}{\sqrt{1 + a^2}}$$

is the shortest distance from the origin to the projection of the moon's orbit, or n in Mr. Woolhouse's notation.

It is obvious that in considering any other phase it is sufficient to use the preceding expression for $P \sin \phi$, retaining the same quantity a, and substituting for b some other quantity B.

If, for instance, in the problem of ascertaining the limits of geocentric latitude within which the occultation of a star will be visible, σ_1 is the moon's semidiameter,

$$B = b \pm \sin \sigma_1 \sqrt{1 + a^2}$$

$$= b \pm k P \sqrt{1 + a^2} \qquad\qquad k = \cdot 2725$$

then will

$$\sin \phi = \frac{B \cos \delta_2}{P(1 + a^2)} \pm \frac{\sqrt{\sin^2 \delta_2 + a^2}}{\sqrt{1 + a^2}} \sqrt{1 - \frac{B^2}{P^2(1 + a^2)}}$$

$$\sin \eta = \frac{b}{P\sqrt{1 + a^2}} \pm k \qquad\qquad \cos \varepsilon = \frac{\cos \delta_2}{\sqrt{1 + a^2}}$$

$$\sin \phi = \sin (\eta \pm \varepsilon).$$

One of these values of $\sin \phi$ determines a parallel of geocentric latitude beyond which an occultation cannot possibly be visible; the other has reference to the projection of a parallel of latitude on the opposite hemisphere, invisible from the star. Generally in order to have the former value take

$$\frac{\sin (\delta_1 - \delta_2)}{P \cos \iota} \pm \cdot 2725 = \sin \eta \qquad\qquad \cos \iota = + \sqrt{1 + a^2}$$

$$\cos \delta_2 \cos \iota = \cos \varepsilon \qquad\qquad \sin \phi = \sin (\eta \pm \varepsilon).$$

The upper sign to be used if δ_2 is North or $+$.

lower . . . . South or $-$.

Fig. 3.

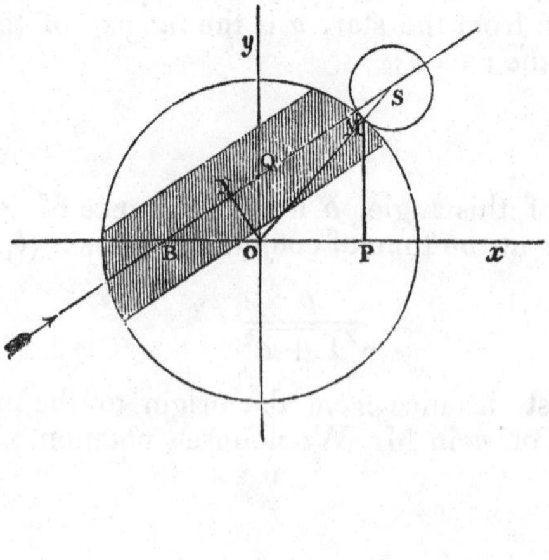

Let S be the centre of the moon, B S the path of the moon, and let O N be drawn perpendicular to B S, O N = $(q) \cos \iota$. The centre of the moon is at the point Q at true conjunction. The middle of the eclipse is when the centre of the moon is at N; the time in N Q (or the time of the middle of the eclipse reckoned from true conjunction, expressed in hours)

$$= \frac{(q) \cos \iota \sin \iota}{p'} = \frac{(q) \sin^2 \iota}{q'}$$

$$\text{S B O} = \iota,$$

(q) being as before, p. 14, the value of $\sin (\delta_1 - \delta_2)$ at true conjunction.

Let the distance O S corresponding to any phase be called Δ,

$$\text{for a} \begin{Bmatrix} \text{partial} \\ \text{central} \\ \text{total} \\ \text{annular} \end{Bmatrix} \text{eclipse } \Delta = \begin{cases} \Pi + \sigma_1 + \sigma_2 \\ \Pi \\ \Pi + \sigma_1 - \sigma_2 \\ \Pi - \sigma_1 + \sigma_2. \end{cases}$$

Let the angle N O S be called ω, the parallactic angle Q O S, as before, be called γ, then

$$(q) \cos \iota = \Delta \cos \omega \qquad \cos \omega = \frac{(q) \cos \iota}{\Delta}$$

$$\text{N S} = \Delta \sin \omega = (q) \cos \iota \tan \omega.$$

$$\text{The time in N S} = \text{time in N Q} \times \frac{\text{N S}}{\text{N Q}}$$

$$= \frac{(q) \cos \iota \sin \iota \tan \omega}{q'}$$

$$\text{O P} = \Pi \sin \gamma = \Pi \cos \phi \sin (\mu - \alpha_2)$$

$$\text{P M} = \Pi \cos \gamma = \Pi \{\sin \phi \cos \delta_2 - \cos \phi \sin \delta_2 \cos (\mu - \alpha_2)\}$$

$$\sin \phi = \cos \gamma \cos \delta_2 \qquad \sin (\mu - \alpha_2) = \frac{\sin \gamma}{\cos \phi}$$

$$\tan (\mu - \alpha_2) = - \frac{\tan \gamma}{\sin \delta_2} \qquad \zeta_2 = 90^\circ.$$

If H is the apparent Greenwich time, the geographical longitude east of Greenwich of M

$$= \mu - \alpha_2 - H.$$

In the diagram, fig. 3, the moon is leaving, and the star is seen *setting* by an observer on the earth's surface at M.

"The place on the surface of the earth where the limbs of the sun and moon first appear in contact will be where the penumbra first touches the earth, and consequently at this place the apparent contact will be in the horizon, the disc of the moon being wholly above the horizon, and that of the sun below it. The point of contact will be in the same vertical with the two centres; and therefore the real as well as the apparent places will be in the same vertical circle; and the lower limb of the moon, being in the horizon, will be depressed by the whole amount of the horizontal parallax which belongs at that time to the latitude of the place. Similarly, the place which first has a central eclipse will be where the straight line through the centres of the sun and moon comes first in contact with the earth and at this place the centres of both objects will be in the horizon, that of the moon experiencing the whole effect of the horizontal parallax. The same circumstances will have place where the phænomena finally quit the earth." (Appendix to Nautical Almanac, 1836, p. 59.)

The equation of any straight line passing through points of which the coordinates are $x_{(1)}, y_{(1)}, x_{(2)}, y_{(2)}$, is

$$y - y_{(1)} = \frac{y_{(1)} - y_{(2)}}{x_{(1)} - x_{(2)}} (x - x_1),$$

also $y_1 = \sin(\delta_1 - \delta_2)$ nearly

$$x_1 = \cos \delta_2 \sin(\alpha_1 - \alpha_2) \text{ nearly.}$$

Hence if $D_1 =$ the ☽'s relative motion in declination, or ☽'s, motion in declination — ☉'s motion in declination, $\alpha_1 =$ the ☽'s relative motion in right ascension, or the motion of the ☽ — that of the ☉

$$= \tan \iota = \frac{D_1}{\alpha_1 \cos \delta_2} = \frac{q'}{p'},$$

D_1 and α_1 in the last equation having the same signification as in Mr. Woolhouse's treatise.

"Suppose an observer situate at a star, and having the moon between him and the earth, and that he could see the moon projected on the earth's disc, he would see it moving across the disc from west to east, covering a zone whose breadth would be equal to the apparent diameter of the moon. Now it is only within the limits of this zone that the occultation of a star by the moon can be visible. To all the places through which the boundary lines pass, the star will appear just to touch the moon's limb; and that projected parallel of latitude, to which one of the boundary lines is a tangent, is one of the limiting parallels, while the intersection of the other boundary line with the circumference of the earth's disc determines the other limiting parallel." (Nautical Almanac.)

"The moon's dark shadow covers only a spot on the earth's surface, about 180 English miles broad, when the moon's diameter appears largest and the sun's least, and the total darkness can extend no further than the dark shadow covers. Yet the moon's partial shadow or penumbra may then cover a circular space 4900 miles in diameter, within all which the sun is more or less eclipsed, as the places are less or more distant from the centre of the penumbra. When the moon changes exactly in the node, the penumbra is circular on the earth at the middle of the general eclipse, because at that time it falls perpendicularly on the earth's surface; but at every other moment it falls obliquely, and will therefore be elliptical, and the more so as the time is longer before or after the middle of the general eclipse, and then much greater portions of the earth's surface are involved in the penumbra. When the penumbra first touches the earth the general eclipse begins; when it leaves the earth the general eclipse ends; from the beginning to the end the sun appears eclipsed in some part of the earth or other. When the penumbra touches any place the eclipse begins at that place, and ends when the penumbra leaves it." (Ferguson's Astronomy, 2nd edition, p. 238.)

The following numerical examples will serve to illustrate the methods which have been given:

Occultation of *i* Lionis, January 7, 1832.

I find in the Nautical Almanac for 1836, page 445, the following "Elements for facilitating the computation."

	Greenwich *Mean Time* of *Apparent* † ☌ in R.A. of ☾ and ✱.	At Greenwich Mean Time of ☌.		
		Apparent R.A. of ☾ and ✱. α_2.	*Apparent* Declination of ✱. δ_2.	Diff. of *Apparent* Dec. of ☾ and ✱. $\delta_1 - \delta_2$.
	h m s	h m s	° ′ ″	″′
Jan. 7.	12 12 7	10 23 26·29	N. 14 58 38·8	34 24

† This is with reference to an observer at the earth's centre.

	The Moon's				
			Hourly variation of		
	Right Ascension. α_1	Declination. δ_1	Right Ascension.		Declination.
			In Time.	In Arc.	
Hour.	h m s	° ′ ″	m s	′ ″	′ ″
10	10 18 55·52	N. 15 58 50·1	+2 2·95	+30 44	−11 39·1
11	10 20 58·47	15 47 11·0	+2 2·80	+30 42	−11 43·9
12	10 23 1·27	15 35 27·1	+2 2·65	+30 40	−11 48·7
13	10 25 3·92	15 23 38·4			

	The Moon's equatorial				Sidereal Time at Mean Noon.
	Semidiameter.		Horizontal Parallax.		
	Noon.	Midnight.	Noon.	Midnight.	
	′ ″	′ ″	′ ″	′ ″	h m s
Jan. 7.	15 12·4	15 16·6	55 48·2	56 3·5	19 4 22·41
8.	15 21·0	15 25·6	56 29·8	56 36·8	19 8 18·96

Hence at $12^h\ 12^m$ the time of conjunction,

$$q' = -11'48'' \qquad \frac{p'}{\cos\delta_2} = 30'40'' \qquad P = 56'\ 4''$$

$$\begin{array}{rl@{\qquad\qquad}rl} \sin P + & 8{\cdot}21241 & \sin P + & 8{\cdot}21241 \\ R + & 9{\cdot}99913 & R + & 9{\cdot}99913 \\ \cos\phi + & 9{\cdot}79610 & \sin\phi + & 9{\cdot}89230 \\ \hline & 8{\cdot}00764 & & 8{\cdot}10384 \end{array}$$

The values of log R and log cos ϕ are taken from the Table given by Mr. Woolhouse, p. 130, Appendix to Nautical Almanac, 1836.

$$\begin{array}{rl@{\qquad}rl@{\qquad}rl} \sin(\mu-\alpha_2) - & 9{\cdot}85876 & \cos(\mu-\alpha_2) + & 9{\cdot}83980 & & 7{\cdot}95062 \\ \Pi\cos\phi + & 8{\cdot}00764 & \Pi\cos\phi + & 8{\cdot}00764 & \cos\delta_2 + & 9{\cdot}98499 \\ & & \text{const} + & 9{\cdot}41916 & & \\ - & 7{\cdot}86640 & & & & 7{\cdot}93561 \\ \sin 23'\ 17'' + & 7{\cdot}83077 & & 7{\cdot}26660 & p' = & 29'\ 39'' \\ & & & 6'\ 22'' & & 6\ \ 22 \\ - & 0{\cdot}03563 & & & & \\ & & & & & 23\ \ 17 \end{array}$$

$$-1{\cdot}085^h = -(1^h\ 5^m)$$

Time of apparent conjunction at Greenwich $= 12^h\ 12^m - (1^h\ 5^m)$, or $11^h\ 7^m$.

In order that the epoch may be the same as that taken by Mr. Woolhouse, who has worked out the same example in the Appendix to the Nautical Almanac for 1836, I take the mean solar time of the apparent conjunction at Greenwich to be $11^h\ 6^m$, and this time for the epoch T, from which t is reckoned, proceeding to determine approximately the times of immersion and emersion according to the method explained at p. 7.

$\cos\delta_2 + 9{\cdot}98499$
$\Pi\sin\phi + 8{\cdot}10384$

$8{\cdot}08883$
$42'\ 11''$
$4\ \ 8$

$v = +38\ \ 3$
$q = +47\ \ 22$

$q - v = 9\ \ 19$
$\alpha_1 - \alpha_2 = -33\ \ 54$
$-7{\cdot}99392$
$\cos\delta_2 + 9{\cdot}98499$

$-7{\cdot}97891$
$p = -32'\ 45''$
$\Pi\cos\phi\sin(\mu - \alpha_2) - 7{\cdot}95676$
$u = -31'\ 7''$
$p - u = -\ 1\ \ 38$
$p - u\quad -6{\cdot}67680$
$q - v\quad +7{\cdot}43298$

$\tan M - 9{\cdot}24382$
$\sin M - 9{\cdot}23725$

$m + 7{\cdot}43955$
$\sin(M - N) - 9{\cdot}93508$

$-7{\cdot}37463$
$\sin\sigma_1\quad 7{\cdot}64778$

$\cos\Psi - 9{\cdot}72685$

$\cos(M - N + \Psi) - 9{\cdot}65931$
$-1{\cdot}59125$

$+1{\cdot}25056$

$\cos(\mu - \alpha_2) + 9{\cdot}65983$
$\Pi\cos\phi + 8{\cdot}00764$

$7{\cdot}67747$
$\sin\delta_2 + 9{\cdot}41236$

$7{\cdot}07983$
$4'\ 8''$

$P + 8{\cdot}21241$
$\text{const.} + 9{\cdot}43537$

$\sin\sigma_1 + 7{\cdot}64778$

$M = 350^\circ\ \ 3'\ 20''$
$N = 110\ \ 36\ \ 30$

$M - N = 239\ \ 26\ \ 50$
$\sin(M - N) - 9{\cdot}93508$
$m + 7{\cdot}43955$

$-7{\cdot}37463$
$\sin\sigma_1 + 7{\cdot}64778$

$\cos\Psi - 9{\cdot}72685$
$\Psi = 122^\circ\ 13'\ \ 0''$
$M - N = 239\ \ 26\ \ 50$

$M - N - \Psi = 117\ \ 13\ \ 50$
$M - N + \Psi = 1\ \ 39\ \ 50$

$\sin(\mu - \alpha_2) - 9{\cdot}94915$
$\Pi\cos\phi + 8{\cdot}00764$

$-7{\cdot}95679$
$\text{const.} + 9{\cdot}41916$
$\sin\delta_2 + 9{\cdot}41236$

$6{\cdot}78831$
$v' = -\ \ 2'\ \ 7''$
$q' = -\ 11\ \ 42$

$q' - v' = -\ \ 9'\ 35''$
$\dfrac{p'}{\cos\delta_2} = +30\ \ 44$
$+7{\cdot}95133$
$\cos\delta_2 + 9{\cdot}98499$

$7{\cdot}93632$
$p' = 29'\ 41''$
$\Pi\cos\phi\cos(\mu - \alpha_2) + 7{\cdot}66744$
$\text{const.} + 9{\cdot}41916$

$7{\cdot}08660$
$u' = 4'\ 12''$
$p' - u' + 25\ \ 29$
$p' - u' + 7{\cdot}86998$
$q' - v' - 7{\cdot}44524$

$\tan N - 0{\cdot}42474$
$\sin N + 9{\cdot}97128$
$n + 7{\cdot}89870$
$\cos\Psi - 9{\cdot}72685$

$n\cos\Psi - 7{\cdot}62555$
$m + 7{\cdot}43955$

$-9{\cdot}81400$

$\cos(M - N - \Psi) - 9{\cdot}66046$
$-9{\cdot}81400$

$+9{\cdot}47446$

$\cos(M - N + \Psi) + 9{\cdot}99982$
$-9{\cdot}81400$

$-9{\cdot}81382$

$t_1 = -\ 0^h\ 17{\cdot}9^m$	$t_2 = +\ 0^h\ 39{\cdot}1^m$
$T = \ \ 11\ \ 6$	$T = \ \ 11\ \ 6$
Immersion 10 48·1	Emersion 11 45·1
$N + \Psi - 90° = 142°\ 49'\ 30''$	$N - \Psi - 90° = 258°\ 36'\ 30''$
$N + \Psi + 90 = 322\ \ 49\ \ 30$	$N - \Psi + 90 = 78\ \ 36\ \ 30$

ECLIPSE OF THE SUN.

Solar Eclipse of May 15, 1836, as it will be seen at the Observatory of Edinburgh.

The elements of this eclipse are stated in the Appendix to the Nautical Almanac, 1836, page 86.

Greenwich mean time of true ☌	2^h	21^m	22^s
Longitude of Edinburgh		12	44 W.
Edinburgh mean time of true ☌	2	8	38
Equation of time		3	56
At true conjunction $\mu - \alpha_2$ in time . .	2	12	34
At true conjunction $\mu - \alpha_2$ in arc . .	33°	8′	30″

CONSTANTS.

☽'s equatorial horizontal parallax . . .	54′	23″·4
☉'s equatorial horizontal parallax . . .		8·5
Difference	54	14·9
☽'s true semidiameter	14′	49″·5
☉'s true semidiameter	15	49·9
Sum	30	39·4

sin 54′ 15″	+ 8·19811	log Π	+ 8·19713
log R	9·99902	sin φ	+ 9·91745
log Π	+ 8·19713		8·11458
cos φ	+ 9·75001		
Π cos φ	+ 7·94714		

Computation of the Time of Apparent Conjunction.

$\mu - \alpha_2 = 33° \ 8' \ 30''$ $\delta_1 = 19° \ 25' \ 10''$ N.

☽'s hourly motion in R. A. . .	30′ 8″
☉'s	2 28
Difference	27 40

$\sin(\mu - \alpha_2) + 9{\cdot}73776$
$\Pi \cos \phi + 7{\cdot}94714$
$u + 7{\cdot}68490$
$\sin 19' \ 26'' + 7{\cdot}75227$
$9{\cdot}93263$

$\cos(\mu - \alpha_2) + 9{\cdot}92289$
$\Pi \cos \phi + 7{\cdot}94714$
const. $+ 9{\cdot}41797$
$u' + 7{\cdot}28800$
$u' = 6' \ 41''$

$\sin 27' \ 40'' + 7{\cdot}90568$
$\cos \delta_1 + 9{\cdot}97456$
$p' + 7{\cdot}88024$
$p' = 26' \ 7''$
$u' = 6 \ 41$
$p' - u' = 19 \ 26$

$\cdot 8563^h = 0^h \ 51^m$ nearly
2 21
3 12

Mr. Woolhouse takes $3^h \ 13^m$ for the Greenwich time of apparent conjunction.

Computation for $3^h \ 13^m$, Greenwich Time.

$\delta_1 + 19° \ 33' \ 43''$ $\delta_2 + 18° \ 58' \ 29''$ $\alpha_1 - \alpha_2 + 23' \ 49''$ $\delta_1 - \delta_2 + 35' \ 14''$

☽'s hourly motion in declination — ☉'s ditto + 9′ 19″
☉'s . . . . R. A. — ☉'s ditto + 27 43

Greenwich mean time .	3^h	13^m	0^s	
Longitude		12	44	W.
	3	0	16	
Equation of time . . .		3	56	
$\mu - \alpha_2$ in time	3	4	12	
$\mu - \alpha_2$ in arc	46°	3′		

$\Pi \sin \phi + 8{\cdot}11458$
$\cos \delta_2 + 9{\cdot}97573$
$8{\cdot}09031$
42′ 22″

$\Pi \cos \phi + 7{\cdot}94714$
$\cos(\mu - \alpha_2) + 9{\cdot}84138$
$7{\cdot}78852$
$\sin \delta_2 + 9{\cdot}51209$
$7{\cdot}30061$
6′ 52″

$\Pi \cos \phi + 7{\cdot}94714$
$\sin(\mu - \alpha_2) + 9{\cdot}85730$
$+ 7{\cdot}80444$
const. $+ 9{\cdot}41797$
$\sin \delta_2 + 9{\cdot}51209$
$6{\cdot}73450$

$$42'\ 22''$$
$$6\ \ 52$$

$v = +\ 35\ \ 30$

$q = +\ 35\ \ 14$

$q - v = -\ \ 16$

$\sin(\alpha_1 - \alpha_2) + 7{\cdot}84060$

$\cos\delta_1 + 9{\cdot}97418$

$7{\cdot}81478$

$p = +\ 22'\ 26''$

$\Pi \cos\phi \sin(\mu - \alpha_2) + 7{\cdot}80444$

$u = +\ 21'\ 55''$

$p - u = \ \ 31$

$p - u + 6{\cdot}17694$

$q - v - 5{\cdot}88969$

$\tan M - 0{\cdot}28725$

$\sin M - 9{\cdot}94871$

$m + 6{\cdot}22823$

$v' = +\ 1'\ 52''$

$q' = +\ 9\ \ 19$

$q' - v' = 7\ \ 27$

☽'s hourly motion in R. A. — ☉'s ditto . . $+ 7{\cdot}90646$

$\cos\delta_1 + 9{\cdot}97418$

$p' + 7{\cdot}88064$

$p' +\ 26'\ 7''$

$\Pi \cos\phi (\mu - \alpha_2) + 7{\cdot}78852$

const. $+ 9{\cdot}41797$

$7{\cdot}20649$

$u' = +\ \ 5'\ 32''$

$p' - u' = \ \ 20\ \ 35$

$p' - u' + 7{\cdot}77724$

$q' - v' + 7{\cdot}33588$

$\tan N + 0{\cdot}44136$

$\sin N + 9{\cdot}97327$

$n + 7{\cdot}80397$

$\cos\Psi + 8{\cdot}14360$

$+ 5{\cdot}94757$

$m + 6{\cdot}22823$

$+ 0{\cdot}28066$

$M = 117^\circ\ 18'\ \ 0''$

$N = \ \ 70\ \ \ 6\ \ 10$

$M - N = \ \ 47\ \ 11\ \ 50$

$\sin(M - N) + 9{\cdot}86552$

$m + 6{\cdot}22823$

$6{\cdot}09375$

$\sin\Sigma + 7{\cdot}95015$

$\cos\Psi + 8{\cdot}14360$

$\Psi = \ \ 89^\circ\ 12'\ 10''$

$M - N - \Psi = 317\ \ 59\ \ 40$

$M - N + \Psi = 136\ \ 24\ \ \ 0$

$\cos(M - N - \Psi) + 9{\cdot}87103$

$+ 0{\cdot}28066$

$+ 0{\cdot}15169$

$\cos(M - N + \Psi) - 9{\cdot}85986$

$+ 0{\cdot}28066$

$- 0{\cdot}14052$

	h	m	s		h	m	s	
$t_1 = -1{\cdot}4181 =$	-1	25	5	$t_2 = +1{\cdot}382 =$	$+1$	22	55	
Assumed time .	3	13	0		3	13	0	
Beginning . .	1	47	55	Ending	4	35	55	Greenwich mean times.
Longitude . .		12	44 W.			12	44 W.	
PARTIAL beginning	1	35	11	Ending	4	23	11	Edinburgh mean times.

COMPUTATION FOR $1^h\ 48^m$ FOR AN ACCURATE DETERMINATION OF PARTIAL BEGINNING.

$\delta_1 + 19^\circ\ 19'\ 35''{\cdot}9$ $\qquad$ $\delta_2 + 18^\circ\ 57'\ 39''{\cdot}3$

$\alpha_1 - \alpha_2 - 15'\ 23''{\cdot}2$ $\qquad$ $\delta_1 - \delta_2 + 35'\ 14''$

☽'s hourly motion in declination − ☉'s ditto + 9′ 26″
☽'s . . . . R. A. − ☉'s ditto + 27 38

Greenwich mean time .	1^h	48^m	0^s
Longitude		12	44 W.
	1	35	16
Equation of time . . .		3	56

$\mu - \alpha_2$ in $\begin{cases} \text{time} & 1 \quad 39 \quad 12 \\ \text{arc} & 24° \; 48' \end{cases}$

$\cos\delta_2 + 9{\cdot}97577$
$\Pi \sin\phi + 8{\cdot}11458$

8·09035

42′ 20″
8 59

$v = +33 \; 21$
$q = 21 \; 56$

$q - v = -11 \; 25$

$\cos(\mu - \alpha_2) + 9{\cdot}95798$
$\Pi \cos\phi + 7{\cdot}94714$

7·90512
$\sin\delta_2 + 9{\cdot}51178$

7·41690

8′ 59″

$\sin(\mu - \alpha_2) + 9{\cdot}62268$
$\Pi \cos\phi + 7{\cdot}94714$

7·56982
const. + 9·41797
$\sin\delta_2 + 9{\cdot}51178$

6·49957

$v' = + \; 1' \; 5''$
$q' = + \; 9 \; 26$

$q' - v' = + \; 8 \; 21$

$\sin(\alpha_1 - \alpha_2) - 7{\cdot}65077$
$\cos\delta_1 + 9{\cdot}97481$

−7·62558

$p = -14' \; 31''$

$\alpha_1 - \alpha_2 = -15' \; 23''{\cdot}2$
7 41·6

7·34927
2

$\sin^2 \dfrac{(\alpha_1 - \alpha_2)}{2} + 4{\cdot}6985$

7·90515
$\cos\delta_1 + 9{\cdot}97481$

7·8799

$p' = + 26' \; 5''$

$\Pi \cos\phi \sin(\mu - \alpha_2) + 7{\cdot}56982$

$u = + 12' \; 46''$
$p - u = -27 \; 17$

$p - u - \; 7{\cdot}89961$
$q - v - \; 7{\cdot}52126$

$\tan M + 0{\cdot}37835$
$\sin M - 9{\cdot}96496$
$m + 7{\cdot}93465$

$\cos\delta_2 + 9{\cdot}97577$
$\Pi \cos\phi \cos(\mu - \alpha_2) + 7{\cdot}90512$

7·88089

$\sin\delta_2 + 9{\cdot}51178$
$\cos\delta_1 + 9{\cdot}97481$
0·30103

4·48616
(insensible)

$M = 247° \; 17' \; 40''$
$N = 66 \; 6 \; 30$

$M - N = 181 \; 11 \; 10$

$\sin(M - N) + 8{\cdot}31597$
$m + 7{\cdot}93465$

6·25062

$\Pi \cos\phi \cos(\mu - \alpha_2) + 7{\cdot}90512$
const. + 9·41797

7·32309

$u' = 7' \; 14''$
$p' - u' = 18 \; 51$

$p' - u' + 7{\cdot}73903$
$q' - v' + 7{\cdot}38541$

$\tan N + 0{\cdot}35362$
$\sin N + 9{\cdot}96109$
$n + 7{\cdot}77794$

$\sin\sigma_1 + 7{\cdot}63447$
$\cos\theta + 9{\cdot}99485$

$\sin\sigma'_1 + 7{\cdot}63962$

26′ 8″
14 33

40 41
2

1° 21′ 22″
51

1 20 31 = $2\Pi\cos\zeta_2 - \Pi^2$

$\sin\delta_2 + 9{\cdot}51178$
$\Pi\sin\phi + 8{\cdot}11458$

7·62636

14′ 33″

$\Pi + 8{\cdot}19713$
2

6·39426

51″

$\sin^2\theta + 8{\cdot}36957$
$\sin\theta + 9{\cdot}18479$
$\cos\theta + 9{\cdot}99485$

$\sigma'_1 = 15'\ 0''$
$\sigma_2 = 15\ 50$

$\sigma'_1 + \sigma_2 = 30\ 50$

7·95274
$\cos\theta + 9{\cdot}99485$

$\sin\Sigma + 7{\cdot}94759$
$- 6{\cdot}25062$

$\cos\Psi - 8{\cdot}30302$

$\Psi = 91°\ 9'\ 5''$
$M - N - \Psi = 90\ 2\ 5$

$\cos(M - N - \Psi) - 6{\cdot}78248$
$- 1{\cdot}85369$

8·63617

$t_1 = -{\cdot}04326^h = 0^h\ 2^m\ 36^s$

$\cos\Psi - 8{\cdot}30302$
$n + 7{\cdot}77794$

$- 6{\cdot}08096$
$m + 7{\cdot}93465$

$- 1{\cdot}85369$

	h	m	s	
t_1	0	2	36	
Assumed time	1	48	0	
Beginning . .	1	45	24	Greenwich time.
Longitude . .		12	44	W.
PARTIAL beginning .	1	32	40	Edinburgh mean time.

EQUATIONS FOR THE REDUCTION OF PARTIAL BEGINNING.

Greenwich mean time . $1^h\ 48^m\ 0^s$
Equation of time . . + 3 56

H in { time 1 51 56 ; arc 28° 5′ }

$p - 7{\cdot}62558$
$q + 7{\cdot}80483$

$\tan M' - 9{\cdot}82075$
$\sin M' - 9{\cdot}74186$

$M'\ 326°\ 30'\ 10''$
$N = 66\ 6\ 30$

$M' - N = 260\ 23\ 40$

$m' + 7{\cdot}81372$
$\cos(M' - N) - 9{\cdot}22336$

$- 7{\cdot}10608$
$n + 7{\cdot}77794$

$- 9{\cdot}32814$

$m' + 7{\cdot}88372$
$\sin(M' - N) - 9{\cdot}99386$

$- 7{\cdot}87758$
$\sin\Sigma + 7{\cdot}94760$

$- 9{\cdot}92998$
$- {\cdot}8511$

$+ 0^h\ 12^m\ 46^s$
1 48

2 0 46

Π + 8·19713	Π + 8·19713
sin Σ + 7·94760	3·55630
0·24953	1·75343
	n + 7·77794
	3·97549

cos δ_2 + 9·97481	cos δ_2 + 9·97481
sin N + 9·96109	cos N + 9·60747
0·24953	3·97549
0·18543	3·55777

	sin δ_2 + 9·51178		sin δ_2 + 9·51178
	tan N + 0·35362		tan N + 0·35362
β = 53° 44′	cot β + 9·86540	β' = 81° 49′	tan β' + 0·84184
H = 28 5		H = 28 5	
	sin δ_2 + 9·51178		sin δ_2 + 9·51178
$H-\beta$ = − 25 39	sin N + 9·96109	$H+\beta'$ = 109 54	cos N + 9·60747
	0·24953		3·97549
	9·72240		3·09474
	cos β + 9·77193		cos β' + 9·15362
	9·95047		3·94112

Hence for the Greenwich time t (reckoned in seconds) of beginning at any place whose latitude is ϕ, and longitude east of Greenwich λ,

$$\cos\Psi = -\cdot 8511 + [0\cdot 18543]\sin\phi - [9\cdot 95047]\cos\phi\cos(\lambda - 25^\circ\ 39')$$

$$t = 2^h\ 0^m\ 46^s - [3\cdot 72596]\sin\Psi - [3\cdot 55777]\sin\phi - [3\cdot 94112]\cos\phi\cos[\lambda + 109^\circ\ 54'].$$

Calculation of the Transit of Mercury,

November 7, 1835.

The data are taken from the Appendix to the Nautical Almanac, 1836, p. 143, and are for November 7, $7^h\ 40^m$ Greenwich mean time.

Planet's equatorial horizontal parallax	12″·66
Sun's equatorial horizontal parallax .	8 ·66
Difference	4 ·00

Planet's semidiameter . .	16′ 10″·4
Sun's semidiameter . .	4 ·8
Σ	16 15 ·2

δ_1 (declination of planet) $= -16^\circ\ 22'\ 4''\cdot 2$ $\alpha_1 - \alpha_2 = -10''\cdot 95$

δ_2 (declination of sun) $= -16\ \ 15\ \ 58\cdot 2$

$\delta_1 - \delta_2 = -\ \ 6\ \ 6$

Planet's hourly motion in R. A. — ⊙'s ditto = — 5′ 32″·7

$\sin(\alpha_1-\alpha_2) - 5\cdot 72697$	Planet's H. M. in R. A. — ⊙'s ditto	$7\cdot 20802$
$\cos\delta_1 + 9\cdot 98203$		$\cos\delta_1 + 9\cdot 98203$
	$M' = 181^\circ\ 39'\ 6''$	
$p - 5\cdot 70900$	$N' = 295\ \ 26\ \ 30$	$p' - 7\cdot 19005$
$q - 7\cdot 24905$		$q' + 6\cdot 86742$
	$M' - N' = 248\ \ 12\ \ 36$	
$\tan M' + 8\cdot 45995$		$\tan N' - 0\cdot 32263$
$\sin M' - 8\cdot 45974$	$\sin(M'-N') - 9\cdot 96144$	$\sin N' - 9\cdot 95570$
$m + 7\cdot 24926$	$m + 7\cdot 24926$	
		$n + 7\cdot 23435$
	$-7\cdot 21070$	$\cos\Psi' - 9\cdot 53612$
	$\sin\Sigma + 7\cdot 67458$	
		$-6\cdot 77047$
	$-9\cdot 53612$	$m + 7\cdot 24926$
	$\Psi' = 110^\circ\ 6'\ 0''$	$-1\cdot 47879$
	$M' - N' - \Psi' = 136\ \ 6\ \ 36$	
	$M' - N' + \Psi' = 356\ \ 18\ \ 36$	

$\cos(M'-N'-\Psi') - 9\cdot 85775$	$\cos(M'-N'+\Psi') + 9\cdot 99910$
$-1\cdot 47879$	$-1\cdot 47879$
$0\cdot 33654$	$-0\cdot 47779$
$t_1 = -2^h\ 10^m\ 13\cdot 8^s$	$t_2 = +\ 3^h\ 0^m\ 19\cdot 8^s$
Assumed time . . 7 40	7 40
Ingress . . . . 5 29 46·2	Egress . . . 10 40 19·8
Equation of time 16 10	

H in $\begin{cases}\text{time} . . 5\ \ 45\ \ 56\\ \text{arc}\ \ . .\ 86^\circ\ 30'\end{cases}$

$\sin\Psi' + 9\cdot 97271$	$\sin 4'' + 5\cdot 28763$
$n + 7\cdot 23435$	$3\cdot 55630$
$7\cdot 20706$	$8\cdot 84393$
	$7\cdot 20706$
	$1\cdot 63687$

$$\begin{array}{rl} \cos\delta_2 & +\ 9{\cdot}98226 \\ \sin(N' + \Psi') & +\ 9{\cdot}85355 \\ & 1{\cdot}63687 \\ \hline & 1{\cdot}47268 \end{array}$$

$$\begin{array}{rl} \tan(N' + \Psi') & +\ 0{\cdot}00821 \\ \sin\delta_2 & -\ 9{\cdot}44732 \\ \hline \cot\beta'' & -\ 9{\cdot}45553 \\ \beta'' & = 105\ \ 56 \\ H & = \ \ 86\ \ 30 \\ \hline H - \beta'' & = -\ \ 19\ \ 26 \end{array}$$

$$\begin{array}{rl} \sin\delta_2 & -\ 9{\cdot}44732 \\ \sin(N' + \Psi') & +\ 9{\cdot}85355 \\ & 1{\cdot}63687 \\ \hline & -\ 0{\cdot}93774 \\ \cos\beta'' & -\ 9{\cdot}43835 \\ \hline & 1{\cdot}49939 \end{array}$$

Ingress, Nov. $7^d\ 5^h\ 29^m\ 46^s + [1{\cdot}473]\ R\sin\phi - [1{\cdot}499]\ R\cos\phi\cos(\lambda - 19^\circ\ 26')$.

In this method, which is employed by Mr. Woolhouse, the quantities u' and v' are neglected altogether.

Table I.

Showing the Zenith Distance and Azimuth for every five degrees of Declination and ten degrees (or forty minutes in time) of Hour Angle, and for the latitude of Greenwich.

H. A.	0		40m		1h 20m		2h 0m	
N. Dec.	Z. D.	Az.	Z. D.	Az.	Z. D.	Az.	Z. D.	Az.
°	° ′	°	° ′	° ′	° ′	° ′	° ′	° ′
90	38 42	180	38 42	180 0	38 42	180 0	38 42	180 0
85	33 42	180	33 47	178 27	34 3	176 57	34 27	175 35
80	28 42	180	28 54	175 30	29 29	173 4	30 24	170 7
75	23 42	180	24 3	173 40	25 4	167 56	26 38	163 13
70	18 42	180	19 17	169 38	20 53	160 51	23 17	154 23
65	13 42	180	14 39	163 8	17 8	150 38	20 35	143 4
60	8 42	180	10 21	151 7	14 8	135 35	18 49	129 11
55	3 42	180	7 2	125 40	12 31	115 6	18 13	113 30
50	1 17	0	6 28	81 59	12 42	92 3	18 56	97 54
45	6 17	0	9 9	50 29	14 41	72 28	20 49	84 11
40	11 17	0	13 15	35 27	17 52	58 36	23 35	73 11
35	16 17	0	17 49	27 42	21 44	49 10	27 2	64 20
30	21 17	0	22 33	23 5	25 58	42 34	30 46	57 48
25	26 17	0	27 23	20 0	30 25	37 45	34 51	52 27
20	31 17	0	32 16	17 48	35 0	34 4	39 7	48 8
15	36 17	0	37 10	16 7	39 41	31 9	43 32	44 32
10	41 17	0	42 6	14 47	45 34	28 46	48 1	41 29
5	46 17	0	47 2	13 40	49 12	26 45	52 35	38 50
0	51 17	0	51 59	12 44	54 1	25 0	57 13	36 30

H. A.	2h 40m		3h 20m		4h 0m		4h 40m	
N. Dec.	Z. D.	Az.	Z. D.	Az.	Z. D.	Az.	Z. D.	Az.
°	° ′	° ′	° ′	° ′	° ′	° ′	° ′	° ′
90	38 42	180 0	38 42	180 0	38 42	180 0	38 42	180 0
85	35 10	174 24	35 40	173 26	36 25	172 42	37 15	172 13
80	31 36	167 42	33 2	165 53	34 38	164 39	36 20	164 1
75	28 38	159 41	30 56	157 19	33 13	155 51	35 59	155 33
70	26 13	150 7	29 27	147 48	32 50	146 53	36 15	147 5
65	24 32	139 9	28 42	137 37	32 55	137 41	37 6	138 49
60	23 45	127 3	28 44	127 12	33 41	128 40	38 29	130 59
55	23 56	114 40	29 34	117 4	35 3	120 9	40 22	123 41
50	25 5	102 56	31 7	107 37	36 59	112 18	42 40	116 59
45	27 4	87 32	33 17	99 11	39 23	105 12	45 20	110 53
40	29 42	83 30	35 57	91 49	42 10	98 50	48 17	105 21
35	32 52	76 1	39 1	85 16	45 16	93 10	51 29	100 19
30	36 23	69 46	42 25	79 36	48 38	88 8	54 52	95 43
25	40 11	64 31	46 3	74 40	52 11	83 29	58 25	91 32
20	44 11	60 4	49 52	70 19	55 54	79 22	62 6	92 23
15	48 21	56 12	53 50	66 26	59 44	75 35	65 53	83 58
10	52 37	52 49	57 54	62 56	63 41	72 5	69 45	80 32
5	56 58	49 48	62 4	59 44	67 42	68 49	73 40	77 17
0	61 23	47 5	66 18	56 47	71 47	65 45	77 39	74 9

H. A.	5h 20m		6h 0m		6h 40m		7h 20m	
N. Dec.	Z. D.	Az.	Z. D.	Az.	Z. D.	Az.	Z. D.	Az.
°	° ′	° ′	° ′	° ′	° ′	° ′	° ′	° ′
90	38 42	180 0	38 42	180 0	38 42	180 0	38 42	180 0
85	38 6	172 0	38 59	172 2	39 50	172 18	40 39	172 47
80	38 4	163 54	39 47	164 15	41 26	165 2	43 0	166 9
75	38 34	155 52	41 5	156 48	43 28	158 15	45 42	160 8
70	39 36	148 6	42 50	149 48	45 53	152 1	48 41	154 40
65	41 8	140 45	44 59	143 17	48 36	146 18	51 55	149 42
60	43 6	133 54	47 29	137 17	51 34	141 3	55 20	145 10
55	45 27	127 34	50 16	131 46	54 46	136 15	58 54	140 59
50	48 7	121 46	53 17	126 42	58 8	131 49	62 36	137 8
45	51 3	116 27	56 30	122 1	61 38	127 41	66 23	133 31
40	54 13	111 34	59 54	117 41	65 16	123 50	70 15	130 7
35	57 33	107 4	63 25	113 39	68 59	120 13	74 11	126 53
30	61 2	102 54	67 2	109 51	72 46	116 45	78 10	123 45
25	64 39	99 1	70 45	106 16	76 37	113 27	82 11	120 44
20	68 21	95 21	74 31	102 49	80 31	110 14	86 13	117 45
15	72 8	91 32	78 21	99 31	84 27	102 3		
10	75 58	91 27	82 13	96 18	88 22	104 1		
5	79 51	85 18	86 6	93 8				
0	83 46	82 10	90 0	90 0				

H. A.	8h 0m		8h 40m		9h 20m		10h 0m	
N. Dec.	Z. D.	Az.	Z. D.	Az.	Z. D.	Az.	Z. D.	Az.
°	° ′	° ′	° ′	° ′	° ′	° ′	° ′	° ′
90	38 42	180 0	38 42	180 0	38 42	180 0	38 42	180 0
85	41 24	173 27	42 4	174 17	42 38	175 15	43 6	176 21
80	44 25	167 36	45 40	169 17	46 44	171 11	47 35	173 15
75	47 51	162 24	49 29	164 53	50 58	167 38	52 9	170 34
70	51 13	157 40	53 26	160 58	55 17	164 29	56 46	168 12
65	54 53	153 25	57 30	157 26	59 41	161 40	61 25	166 5
60	58 42	149 33	61 39	154 12	64 8	159 5	66 6	164 8
55	62 33	145 58	65 53	151 13	68 37	156 41	70 49	162 19
50	66 37	142 40	70 9	148 26	73 9	154 26	75 32	160 37
45	70 41	139 33	74 29	145 48	77 42	152 17	80 17	158 59
40	74 48	136 34	78 50	143 16	82 16	150 12	85 1	157 23
35	78 58	133 43	83 12	140 48	86 50	148 11	89 45	155 49
30	83 8	130 56	87 35	138 24				
25	87 20	128 13						

H. A.	10h 40m		11h 20m		12h 0m	
N. Dec.	Z. D.	Az.	Z. D.	Az.	Z. D.	Az.
°	° ′	° ′	° ′	° ′	° ′	°
90	38 42	180 0	38 42	180 0	38 42	180
85	43 26	177 31	43 38	178 45	43 42	180
80	48 12	175 26	48 35	177 6	48 42	180
75	53 0	173 38	53 32	176 48	53 42	180
70	57 50	172 4	58 29	176 0	58 42	180
65	62 41	170 38	63 27	175 18	63 42	180
60	67 32	169 20	68 25	174 39	68 42	180
55	72 25	168 8	73 23	174 2	73 42	180
50	77 17	166 57	78 21	173 27	78 42	180
45	82 10	165 52	83 19	172 54	83 42	180
40	87 3	164 48	88 17	172 21	88 42	180
35						

H. A.	0h 0m		0h 40m		1h 20m		2h 0m	
S. Dec.	Z. D.	Az.	Z. D.	Az.	Z. D.	Az.	Z. D.	Az.
°	° ′	°	° ′	° ′	° ′	° ′	° ′	° ′
5	56 17	0	56 57	11 55	58 51	23 28	61 52	34 23
10	61 17	0	61 54	11 11	63 41	22 4	66 33	32 28
15	66 17	0	66 52	10 31	68 33	20 47	71 16	30 34
20	71 17	0	71 50	9 53	73 25	19 35	76 0	28 58
25	76 17	0	76 48	9 18	78 18	18 27	80 44	27 20
30	81 17	0	81 46	8 44	83 11	17 21	85 29	25 45
35	86 17	0	86 40	8 11	88 4	16 17		

H. A.	2h 40m		3h 20m		4h 0m		4h 40m	
S. Dec.	Z. D.	Az.	Z. D.	Az.	Z. D.	Az.	Z. D.	Az.
°	° ′	° ′	° ′	° ′	° ′	° ′	° ′	° ′
5	65 51	44 34	70 35	54 1	75 54	62 49	81 39	71 6
10	70 21	42 14	74 54	51 23	80 4	59 59	85 41	68 8
15	74 53	40 1	79 16	48 52	84 15	57 13	87 21	65 19
20	79 26	37 55	83 38	46 25	88 27	54 30		
25	84 1	35 51	88 1	44 0				
30	88 35	33 50						

H. A.	5h 20m	
S. Dec.	Z. D.	Az.
° 5	° ′ 87 42	° ′ 79 4

Table II.

Showing the *parallactic* angle for Greenwich.

H. A.	0h	40m	1h 20m	2h 0m	2h 40m	3h 20m	4h 0m
N. Dec.		° ′	° ′	° ′	° ′	° ′	° ′
30°	0	16 26	29 14	37 40	42 39	45 15	46 11
25	0	13 39	24 59	33 10	38 32	41 43	43 17
20	0	11 44	21 53	29 42	35 13	38 48	40 51
15	0	10 21	19 34	27 0	32 33	36 24	38 50
10	0	9 19	17 47	24 52	30 23	34 26	37 10
5	0	8 32	16 25	23 11	28 39	32 50	35 49
0	0	7 55	15 20	21 50	27 15	31 33	34 45

H. A.	4h 40m	5h 20m	6h 0m	6h 40m	7h 20m	8h 0m	8h 40m
N. Dec.	° ′	° ′	° ′	° ′	° ′	° ′	° ′
30°	45 56	44 44	42 47	40 9	36 54	33 3	28 39
25	43 36	42 57	41 29	39 16	36 23	32 50	
20	41 40	41 30	40 27	38 38	36 5		
15	40 4	40 19	39 41	38 13			
10	38 47	39 24	39 8	38 2			
5	37 45	38 44	38 49				
0	36 59	38 17					

H. A.	0h	40m	1h 20m	2h 0m	2h 40m	3h 20m	4h 0m	4h 40m	5h 20m
S. Dec.									
5°	0	7° 27′	14° 28′	20° 46′	26° 8′	30° 31′	33° 56′	36° 26′	38° 3′
10	0	7 4	13 48	19 55	25 16	29 45	33 21	36 6	
15	0	6 47	13 17	19 17	24 36	29 11	32 58	36 2	
20	0	6 33	12 54	18 48	24 8	28 49	32 48		
25	0	6 24	12 37	18 28	23 50	28 38			
30	0	6 18	12 26	18 17	23 42				

TABLE III.

Showing the value for Greenwich of the quantity

$$\sin 57' \cos \phi \sin (\mu - \alpha_2). \qquad (u)$$

H. A.	h m 0 10 11 50	h m 0 20 11 40	h m 0 30 11 30	h m 0 40 11 20	h m 0 50 11 10	h m 1 0 11 0	h m 1 10 10 50	h m 1 20 10 40	h m 1 30 10 30
	1′·5	3′·1	4′·6	6′·2	7′·7	9′·2	10′·7	12′·2	13′·6
H. A.	h m 1 40 10 20	h m 1 50 10 10	h m 2 0 10 0	h m 2 10 9 50	h m 2 20 9 40	h m 2 30 9 30	h m 2 40 9 20	h m 2 50 9 10	h m 3 0 9 0
	15′·1	16′·4	17′·8	19′·1	20′·4	21′·7	22′·9	24′·1	25′·2
H. A.	h m 3 10 8 50	h m 3 20 8 40	h m 3 30 8 30	h m 3 40 8 20	h m 3 50 8 10	h m 4 0 8 0	h m 4 10 7 50	h m 4 20 7 40	h m 4 30 7 30
	26′·3	27′·3	28′·3	29′·2	30′·1	30′·9	31′·6	32′·3	32′·9
H. A.	h m 4 40 7 20	h m 4 50 7 10	h m 5 0 7 0	h m 5 10 6 50	h m 5 20 6 40	h m 5 30 6 30	h m 5 40 6 20	h m 5 50 6 10	h m 6 0
	33′·5	34′·0	34′·4	34′·8	35′·1	35′·3	35′·5	35′·6	35′·6

TABLE IV.

Showing the difference or decrease of the quantities in the preceding table for 10 degrees of geographical latitude.

H. A.	20^m	40^m	$1^h\,0^m$	$1^h\,20^m$	$1^h\,40^m$	$2^h\,0^m$	$2^h\,20^m$	$2^h\,40^m$	$3^h\,0^m$
	′·7	1′·4	2′·1	2′·9	3′·5	4′·1	4′·8	5′·3	5′·8

H. A.	$3^h\,20^m$	$3^h\,40^m$	$4^h\,0^m$	$4^h\,20^m$	$4^h\,40^m$	$5^h\,0^m$	$5^h\,20^m$	$5^h\,40^m$	$6^h\,0^m$
	6′·3	6′·8	7′·2	7′·5	7′·8	8′·0	8′·1	8′·2	8′·3

TABLE V.

Showing the value for Greenwich of the quantity

$$\sin 57' \{\sin \phi \cos \delta_2 - \cos \phi \sin \delta_2 \cos (\mu - \alpha_2)\}. \qquad (v)$$

H. A.	0^h	40^m	$1^h\,20^m$	$2^h\,0^m$	$2^h\,40^m$	$3^h\,20^m$	$4^h\,0^m$
N. Dec.							
30°	20′·7	21′·0	21′·8	23′·1	24′·9	27′·1	29′·6
25	25·2	25·5	26.1	27·3	28·8	30·6	32·8
20	29·6	29·8	30·3	31·2	32·5	34·0	35·7
15	33·7	33·9	34·3	35·0	35·9	37·0	38·2
10	37·6	37·7	38·0	38·4	39·1	39·8	40·7
5	41·2	41·2	41·4	41·6	41·9	42·3	42·7
0	44.5	44·5	44·5	44·5	44·5	44·5	44·5

H. A.	$4^h\,40^m$	$5^h\,20^m$	$6^h\,0^m$	$6^h\,40^m$	$7^h\,20^m$	$8^h\,0^m$	$8^h\,40^m$
N. Dec.							
30°	32′·4	35′·4	38′·5	41′·6	44′·6	47′·4	50′·0
25	35·2	37·7	40·3	42·9	45·5	47·8	
20	37·6	39·7	41·8	43·9	46·0		
15	39·8	41·4	43·0	44·6			
10	41·7	42·7	43·8	44·9			
5	43·2	43·8	44·3				
0	44·5	44·5	44·5				

H. A.	0h	40m	1h 20m	2h 0m	2h 40m	3h 20m	4h 0m	4h 40m	5h 20m
S. Dec.									
5°	47′·4	47′·4	47′·2	47′·0	46′·7	46′·3	45′·9	45′·4	44′·8
10	50·0	49·9	49·6	49·2	48·5	47·8	46·9	45·6	
15	52·2	52·0	51·6	50·9	50·0	48·9	47·7	46·1	
20	54·0	53·8	53·2	52·3	51·1	49·5	47·9		
25	55·4	55·1	54·4	53·4	51·8	50·0			
30	56·3	56·0	55·3	53·9	52·1				

Table VI.

Showing the difference or increase of the quantities in the preceding table for 10 degrees of geographical latitude.

H. A.	0h	40m	1h 20m	2h 0m	2h 40m	3h 20m	4h 0m
N. Dec.							
30°	8′·9	8′·8	8′·7	8′·4	7′·9	7′·4	6′·8
25	8·5	8·4	8·3	7·9	7·7	7·2	6·7
20	8·0	8·0	7·9	7·6	7·3	7·0	6·6
15	7·4	7·4	7·3	7·2	7·0	6·7	6·5
10	6·9	8·8	6·8	6·7	6·5	6·3	6·2
5	6·2	6·2	6·2	6·1	6·0	6·0	5·9
0	5·5	5·5	5·5	5·5	5·5	5·5	5·5

H. A.	4h 40m	5h 20m	6h 0m	6h 40m	7h 20m	8h 0m	8h 40m
N. Dec.							
30°	6′·2	5′·5	4′·8	4′·0	3′·3	2′·7	2′·1
25	6·2	5·6	5·0	4·4	3·8	3·2	
20	6·1	5·7	5·2	4·7	4·2		
15	6·0	5·7	5·3	4·9			
10	5·9	5·7	5·4				
5	5·7	5·6	5·5				
0	5·5	5·5	5·5				

H. A.	0h	40m	1h 20m	2h 0m	2h 40m	3h 20m	4h 0m	4h 40m	5h 20m
S. Dec.									
5°	4′·8	4′·8	4′·8	4′·9	4′·9	5′·0	5′·1	5′·2	5′·4
10	4·0	4·0	4·1	4·2	4·3	4·5	4·7	4·9	
15	3·2	3·2	3·3	3·5	3·7	3·9	4·1	4·6	
20	2·3	2·5	2·5	2·7	3·0	3·5	3·8		
25	1·5	1·5	1·7	2·0	2·3	2·7			
30	·6	·7	·9	1·2	1·6				

Table VII.

Showing the value for Greenwich of the quantity

$$[9{\cdot}41916] \sin 57' \cos \varphi \cos (\mu - \alpha_{☽}). \qquad (u')$$

H. A.	h m 0 0 12 0	h m 0 10 11 50	h m 0 20 11 40	h m 0 30 11 30	h m 0 40 11 20	h m 0 50 11 10	h m 1 0 11 0	h m 1 10 10 50	h m 1 20 10 40
	±9'·3	±9'·3	±9'·3	±9'·3	±9'·2	±9'·1	±9'·0	±8'·9	±8'·8

H. A.	h m 1 30 10 30	h m 1 40 10 20	h m 1 50 10 10	h m 2 0 10 0	h m 2 10 9 50	h m 2 20 9 40	h m 2 30 9 30	h m 2 40 9 20	h m 2 50 9 10
	±8'·7	±8'·5	±8'·3	±8'·1	±7'·9	±7'·7	±7'·5	±7'·2	±6'·9

H. A.	h m 3 0 9 0	h m 3 10 8 50	h m 3 20 8 40	h m 3 30 8 30	h m 3 40 8 20	h m 3 50 8 10	h m 4 0 8 0	h m 4 10 7 50	h m 4 20 7 40
	±6'·6	±6'·3	±6'·0	±5'·7	±5'·4	±5'·0	±4'·6	±4'·3	±4'·0

H. A.	h m 4 30 7 30	h m 4 40 7 20	h m 4 50 7 10	h m 5 0 7 0	h m 5 10 6 50	h m 5 20 6 40	h m 5 30 6 30	h m 5 40 6 20	h m 5 50 6 10
	±3'·6	±3'·2	±2'·8	±2'·4	±2'·0	±1'·6	±1'·2	±'·8	±'·4

Table VIII.

Showing the increase in any of the quantities in the last five tables for 1' increase in the moon's horizontal parallax.

1'	'·02	4'	'·07	7'	'·12	10'	'·17	40'	'·70
2	·03	5	·09	8	·14	20	·35	50	·87
3	·05	6	·10	9	·16	30	·52	60	1·05

Printed by Richard Taylor, Red Lion Court, Fleet Street.

AN

ELEMENTARY TREATISE

ON

THE TIDES.

BY

J. W. LUBBOCK, Esq., Treas. R.S., F.R.A.S. and F.L.S.,

VICE-CHANCELLOR OF THE UNIVERSITY OF LONDON,
MEMBER OF THE AMERICAN ACADEMY OF ARTS AND SCIENCES,
AND OF THE ACADEMY OF PALERMO.

"Les marées ne sont pas moins intéressantes à connoître, que les inégalités des mouvemens célestes. On a negligé pendant longtemps de les suivre avec une exactitude convenable, à cause des irrégularités qu'elles présentent; mais ces irrégularités disparaissent en multipliant les observations."

LONDON:
CHARLES KNIGHT AND CO., 22 LUDGATE STREET.
1839.

PRINTED BY RICHARD AND JOHN E. TAYLOR,
RED LION COURT, FLEET STREET.

PREFACE.

In consequence of the laborious discussions of observations of the Tides, made for me by Mr. Dessiou, by Mr. Jones, and by Mr. Russell, I have endeavoured to form tables which may be employed in predicting the time and height of high water upon our coasts.

My first examination of the London Dock observations was undertaken in the year 1829, at the instance of the Committee of the Society for the Diffusion of Useful Knowledge, with a view of obtaining correct tables for predicting the time and height of high water in the British Almanac. I found that the *establishment* of the port given in the *Annuaire du Bureau des Longitudes* and in other books was erroneous, and that the tables given in works in which this question is specially treated could not be safely trusted.

But when the *semi-menstrual inequality* had been determined with sufficient accuracy for practical purposes, it was evidently inconsistent with the objects for which that Society was founded, that their funds should be employed in the encouragement of researches in this difficult problem. The discussions therefore, the utility of which I endeavoured to demonstrate in the Companion to the British Almanac for 1830, could not have been so soon undertaken but for the interest felt in the subject by some individuals distinguished in science, particularly by Mr. Whewell, and but for the pecuniary grants which were in consequence devoted to it by the British Association. I was thus enabled to procure the valuable assistance of Mr. E. Russell and Mr. Jones. The results have been published in the Philosophical Transactions, but as I am not aware that any detached elementary Treatise on the Tides exists in the English or in any other language, I trust that a more connected view of the subject may not be uninteresting.

The Hydrographer-Royal, Captain Beaufort, has already published *Tide Tables for the English and Irish Channels and the River Thames*, which are to be continued annually, and which give the times of High Water for Brest, Plymouth, Portsmouth, Ramsgate, Sheerness, London, Pembroke, Bristol, Howth, Liverpool and Leith, and the heights of High Water for Plymouth, Portsmouth, Sheerness, London, Pembroke, Bristol, Liverpool and Leith. The comparison of such calculations, carried on upon a uniform system, with the observations made at those places, cannot fail to suggest improvements in the methods employed, if hereafter they should be found necessary. It seems better at all events at present to adhere to a uniform method, and, as far as possible, to tables founded upon theory, even at a slight disadvantage. A similar work has recently been undertaken in France by M. Chazallon.

Much confusion arises from the manner in which the word *Tables* has been used in this country in connexion with the Tides. In other astronomical problems, the use of this word has been confined to such general conversions of the algebraical formulæ as may be employed in calculating the place of a star or planet for any instant, but the Right Ascensions and Declinations of the Moon, for example, given in the Nautical Almanac, have never been confounded with the *Lunar Tables*. Unfortunately, calculations of the times and heights of high water have been styled *Tide-tables*, which should rather be termed *Tide-calculations* or *predictions*.

The branch of the subject which appears at present to stand in need of the most laborious exertions, is the accurate determination of the *semi-menstrual inequality* and the local constants (from which the quantities (A), D, (E) may be immediately obtained) for other places than Liverpool and London. Until this is done, our knowledge of the progress of the tide-wave and of the circumstances which attend it must be very defective. This question is not uninteresting to geologists, for, until it has been taken up carefully and extensively, it will be impossible to detect slight changes in the relative level of sea and land, except in narrow seas, such as the Baltic, which are exempt from tides. Even to trace the gradual extinction of the *semi-menstrual inequality* between London and Teddington, or in any similar locality, would not be unprofitable.

Our knowledge of the *diurnal inequality* is also very imperfect. Mr. Whewell's researches on this subject have opened a wide field, and have shown us how much remains to be done.

Hitherto no care has been bestowed in specifying the transit to which the tide at any place is referred, and various mistakes might be adduced which have originated in this want of precision. It seems desirable that some conventional agreement should be adopted as soon as possible upon this point. I have employed in this work the transit which precedes a given high water at London by about 2 days 3 hours, and which I term transit *B*. If the prior transit, or that which I term transit *A*, and which occurs 12 hours and 25 minutes sooner be preferred, I have no objection to offer. This latter transit corresponds in syzygy to the average *interval*. At this period of the moon's age the *interval* changes rapidly, but the height of high water on the contrary is stationary, so that it is impossible to determine directly with accuracy how long after a given transit the highest tide takes place. Laplace defines the *establishment* to be the interval between the time of high water and the transit at $0^h\ 0^m$ *immediately preceding*. But this interval changes slightly with the parallaxes and declinations of the luminaries.

The phenomena of the tides are actually composite, they originate in the attractions of the luminaries, but the laws which would obtain in the perfect sphere, are liable to modification in the progress of the tide-wave, so that if we find a deviation from theory, such as that in the *position* of the semi-menstrual inequality in the height, it is difficult to determine whether it be due to the omission of any of the small corrections which would be required in the perfect sphere, or whether it be produced by the different velocities with which spring and neap tides travel in narrow channels, or to any other cause peculiar to the *derived* tide-wave. The hydrodynamical researches and experiments of Mr. Russell of Edinburgh may be of great service in determining points of this nature.

The great real obstacle to perfection in calculations or *predictions* of the tides consists in the fluctuations of the *establishment*. Suppose the establishment (or the constant to be added to ψ) changes a minute *per annum*, and that having determined it from all the observations of twenty-one years, we proceed to employ it in calculations of the time of high water. It is obvious that if we calculate for the *eleventh* year, and compare the calculated times of high water with the observed times, they will not be affected with any *constant* error, but if we calculate for the *twenty-first* year the calculated times will have a constant error of 10 minutes. Similar remarks apply to the heights, that is, to the constant D. If the channel becomes deeper, the tide-wave travels with greater velocity, and the high water happens sooner. Thus again an indication arises which may hereafter be useful to the geologist.

Difficulties unfortunately occur to prevent our obtaining accurate observations. In our great commercial docks such objects can only be subordinate, and the circumstance of the high water happening on succeeding days at different hours interferes with arrangements which could otherwise be made. Lately, at the instigation of the Royal Society, observations have been instituted at Her Majesty's Dock-yards, and are printed from time to time at the public expense.

Much has been accomplished, but the calculations and discussions which are still wanted to place the subject on a satisfactory footing, and which will be required from time to time to watch the fluctuations of the local constants are so extensive, that until some Officer is appointed by Government, well versed in astronomy, specially to superintend these investigations, we cannot expect THE TIDES to be placed on a par with the other provinces of physical astronomy.

29, Eaton Place, Jan. 1839.

CONTENTS.

THE TIDES.

THE ocean is often agitated by winds and tempests which disturb its equilibrium; but these causes are inadequate to produce the phenomena of its motion, for in the calmest weather it rises and falls alternately. It is *high water* at any given place a certain time after the moon has passed the meridian of that place, after which it falls until it is *low water*, when it again gradually rises. These phenomena recur after nearly the same intervals of time, and are called the *tides*.

The interval of time between low water and the following high water is called *flood tide*, and that between high water and the following low water is called *ebb tide*. If the interval of time between two successive passages of the moon over the same semi-meridian be called a lunar day, there are generally two high tides in one lunar day.

The time of the moon's synodic revolution, or period of her mean conjunctions, is 29.530588716 days. Hence the mean interval between successive high tides is 12 h. 24 m. 22 s.

The interval, however, which elapses between the moon's transit and the time of high water is subject to irregularities, which depend principally upon her angular distance from the sun. These variations depend also upon the distances of both luminaries from the earth, and upon their declinations. The heights of the tide at high water also vary; they are greatest soon after the moon is in syzygy, and least soon after she is in quadrature; the one are called *spring*, the other *neap* tides. These heights depend also upon the declinations and distances of the luminaries. The time of high water at any port when the moon is in syzygy (new or full) has been called the *establishment* of the port.

The evident connexion between the period of the tides and that of the phases of the moon led philosophers to attribute these phenomena to her action long before their true theory was understood. A passage in Pliny, lib. ii. c. 97, is quoted by Lalande in the fourth volume of his Astronomy, p. 8. It begins thus: "Æstus maris accedere et reciprocare maxime mirum est, pluribus quidem modis accidit, *verum causa in sole lunaque.*" Pliny proceeds to describe the principal phenomena of the tides. Kepler says, in his *Epist. Ast.* p. 555, "quemadmodum igitur ut magnes magnetem aut ferrum trahat cognatio corporum efficit, sic etiam de luna non est incredibile ut illa moveatur a terra cognato corpore, licet nec hic nec illic intercedat aliquis contactus corporum. Adeoque quid mirum lunam a terra moveri, cum videamus vicissim et lunam transitu suo super vertices locorum causare fluxum Oceani reciprocum in tellure? Nonne satis evidens hoc est documentum communicationis motuum inter hæc duo corpora?" The principle of gravitation is here clearly indicated, but the law is wanting. There is also a passage in his work *De Stellá Martis*, relating to the same phenomena, which is written in the peculiar style of Kepler's fancy. "Dissolvitur discessu lunæ concilium aquarum seu exercitus qui est in itinere versus torridam, &c." Galileo, in his Dialogues upon the System of the World, expresses his regret that so ingenious a philosopher as Kepler should have given such an explanation. He, on the contrary, attributed them to the rotation of the earth combined with its revolution about the sun.—See the Life of Galileo by Mr. Drinkwater Bethune, p. 71. Wallis, in 1666, in letters to Mr. Boyle, derives the phenomena of the tides from the consideration that it is the common centre of gravity of the moon and the earth which describes an orbit about the sun, while they revolve about this common centre. These letters are published in his works, and in the Phil. Trans. vol. i. p. 263. Wallis appears to have conjectured the principle of gravitation. In answer to an objection which was made to him in the form of a query, How two bodies which have no tie can have one common centre of gravity? Wallis says, "It is harder to show how they have, than that they have it—that the loadstone and iron have somewhat equivalent to a tie, though we see it not, yet by the effects we know. And it would be easy to show that two loadstones at once applied in different positions to the same needle at a convenient distance, will draw it not to a point directly to either of them, but to some point between both. . . . As to the present case, how the earth and moon are connected, I will not undertake to show, nor is it necessary to my purpose; but that there is somewhat that does connect them, *as much as what connects the loadstone and the iron which it draws,* is past doubt to those who allow them to be carried about the sun, as one aggregate or body, whose parts keep a respective position to one another, like as Jupiter with his four satellites, and Saturn with his one. Some tie there is that makes those satellites attend their lords and move in a body, though we do not see that tie nor hear the words of command."

Bacon's treatise *De fluxu et refluxu maris* does not contain an accurate solution of the question; he conceives the phenomena to have been erroneously attributed to the influence of the moon, "vulgo levi conjectura refertur ad lunam." Bacon was aware of the remarkable circumstance that high water takes place at the same instant on the opposite shores of the Atlantic: "de industria inquisitum atque repertum, aquas ad littora adversa Europæ et Floridæ iisdem horis ab utroque littore refluere, neque deserere littus Europæ cum advolvantur ad littora Floridæ, more aquæ (ut supra diximus) agitatæ in pelvi, sed plane simul ad utrumque littus attolli et demitti." Bacon's treatise appears to have been written before the year 1619.

In the Philosophical Transactions, vol. xiii. p. 10, Flamsteed gives "a correct tide-table, showing the true times of the high waters at London Bridge to every day in the year 1683." It appears from the remarks of Flamsteed, that the earliest tide-tables which had been calculated for the port of London had been made on the supposition that the tide always followed three hours after the time of the moon's southing. Mr. Philips, a writer on navigation, appears to have been the first who introduced an empirical correction, varying as the sine of twice the moon's angular distance from the sun, by which the tables were made to agree better with observation. Flamsteed's attention was turned to the subject from his having frequently occasion to go to Greenwich by water. Mr. Philips had placed the greatest and least differences between the moon's true southing and the high waters at the full or new and quarter moons, the error of which Flamsteed detected. Flamsteed was the first who gave tide tables, in which the time of high water is calculated, corresponding to the moon's second passage of the same meridian. A more detailed notice of the authors who endeavoured to explain the phenomena of the tides may be found in the 4th volume of Lalande's Astronomy, in the *Encyclopædia Brit.*, Art. Tides; or in the life of Galileo by Mr. Bethune, which forms part of the Library of Useful Knowledge.

Before Sir Isaac Newton, these explanations were, at best, but vague surmises; to him was reserved the honor of discovering the true theory of these remarkable phenomena, and of tracing the operation of the cause which produces them. This theory is contained in the *Principia,* Prop. 66. Lib. i. Cor. 19 and 20, and Lib. iii. Prop. 24, 36 and 37, not 26 and 27, as is said in the *Méc. Cél.* vol. v. p. 146. In Prop. 66. Lib. i. Newton shows that in a system of three bodies S, P, T, the motion of P about T, relative to T, is disturbed by the difference of the attractive forces of S upon P and T, and he infers that if a sphere be covered by a fluid, and this system be attracted by a body (S) the equilibrium of the fluid will be disturbed by the difference of the attractive forces of S upon the centre of the sphere, and upon each particle of which the fluid is composed. So that if S revolve about the sphere, the fluid will be constantly in motion relatively to the centre of the sphere.

In the 24th Prop. Newton describes the different phenomena of the tides; assuming without demonstration that the ocean takes the form of a spheroid, in consequence of the attraction of the luminary.

In the 36th Prop. he finds the force which the luminary exerts upon a particle of the ocean; and in the 37th Prop. he determines, from observations made by Sturm, at the mouth of the river Avon, near Bristol, of the heights of the tide when the moon was in conjunction and opposition, the ratio of the force of the moon to that of the sun $\left(\frac{m'}{r'^3} : \frac{m}{r^3}\right)$ to be 4·4815 : 1; and from this determination he deduced the mass of the earth to be to that of the moon as 39·788 : 1, the density of the sun to be to the density of the earth as 1 : 4, and the density of the moon to that of the earth as 11 : 9.

About the beginning of the eighteenth century many observations of the tides were communicated to the Academy of Sciences at Paris from different parts of France. See *Mém. de l'Acad.* 1710, 1713, &c. In the volume for 1713, Cassini discusses at great length the observations which had been sent to him from Brest; but although they agree with the theory of Newton, he never mentions his name, but only gives empirical tables of the heights of the tide for the port of Brest, when the moon is in syzygy, or in quadrature, the arguments of which are the declination of the moon and her distance from the earth.

Halley wrote a paper on the subject of the tides, which is given in the Philosophical Transactions, vol. xix. p. 445: it contains an explanation at length of Sir Isaac Newton's theory, but nothing original; it purports to have been drawn up "for the late King James's use."

In 1738, the Academy of Sciences proposed the theory of the tides as a prize question. The prize was divided between Daniel Bernoulli, professor of Anatomy and Botany at Basle; Maclaurin, professor of Mathematics in Edinburgh; Euler, professor of Mathematics at Petersburg, and the Jesuit Cavalleri: the three former founded their theory on the principle of universal gravitation; the latter adopted the system of vortices, "tourbillons." There is something, however, very like this in Euler. "Explosis hoc saltem tempore qualitatibus occultis missaque *Anglorum quorundam* renovata attractione quæ cum saniori philosophandi modo nullatenus consistere potest," &c. Euler then proceeds to show how a force, varying inversely as the square of the distance, might arise from the revolution of a thin medium about the sun and moon, "Materia subtilis in gyrum acta ac vorticem formans."

The treatises of Bernoulli, Maclaurin, and Euler, are printed in the Jesuits' edition of the *Principia*.

I shall now endeavour to explain Bernoulli's solution of the problem. Let us allow, that were the earth a perfect sphere covered throughout by a fluid, the fluid would assume the same form at any given instant as it would do if the forces then acting upon each particle were invariable in magnitude and direction. The actual approximation to this state of things is greater in the southern hemisphere than in northern latitudes, and on our coasts. Moreover, let us suppose that the tide-wave is subject to this law at the Cape of Good Hope, or in some region still more remote, and that it is propagated along the Atlantic Ocean and round our island, "according to the stamp first set upon it by the moon's pressure." Upon these suppositions, which are virtually those of Bernoulli, and which may be said to constitute the *equilibrium-theory*, it is easy to calculate the variations in the time and height of high water at any given place, if the time in which the tide-wave is propagated does not vary.

Bernoulli calculated tables for some of the corrections, but he did not explain with sufficient precision the manner in which these tables must be used. I allude here particularly to Bernoulli's parallax correction for the interval, p. 165. He says, "Pour se servir de cette table, il ne faudra plus qu'ajouter aux nombres des six dernières colonnes l'heure moyenne du port." But it is not sufficient to increase the argument of the table, which is the angular distance between the luminaries, by twenty degrees, as Bernoulli supposes, in order to accommodate the table to the reasoning in p. 161, where he says, "Et enfin on trouve une conformité exacte entre les deux points en question, en donnant un jour et demi au retardement des marées, c'est-à-dire, en supposant que l'état des marées est tel qu'il devroit être naturellement, un jour et demi plutôt." Although this supposition is admissible, at Brest, for example, and although Bernoulli's table would I think afford the true correction in the interval between the moon's transit and the time of high water, in the case of a perfect sphere covered by an ocean, by applying it to the transit of the moon *immediately preceding*, still this is not the case actually, and no approximation even would be obtained to the true parallax or calendar-month inequality in this manner. Nor did Bernoulli indicate the great difference in the *retard*, or *age of the tide*, at different places; and he appears to have attributed this retard* to the inertia of the

* "Nous avons encore fait voir, que sans le concours des causes secondes les plus grandes marées devroient se faire dans les syzygies et les plus petites dans les quadratures. Cependant on a observé, que les unes et les autres se font un ou deux jours plus tard. Ce retardement est encore produit, si non pour le tout, au

water, an error which Laplace pointed out. The difficulty to which I have alluded in ascertaining the correct interval between a given transit of the moon and the time of high water does not influence so much the calculation of the heights, because the parallax and declination corrections for the height change very little with the moon's age.

In 1774, Laplace took up this subject. Availing himself of the discoveries of d'Alembert in the Theory of Hydrodynamics, he attempted to deduce from the equations of motion of fluids, the principal phenomena of the tides. The integration of these equations presents, however, great difficulties, even in the case in which the depth of the ocean is constant, and the solid nucleus but little different from a sphere.

Finally, Laplace had recourse to the following indirect consideration, namely, "*that the state of any system in which the primitive conditions have disappeared through the resistances which its motion encounters, is periodical with the forces which act upon it.*" Hence he concludes, that if the system is disturbed by a periodic force expressed by a series of cosines of variable angles, the height of the tide is represented by a similar series in which the *periods* of the *arguments* are the same, but the *epochs* and the *coefficients* are different. This amounts to representing the height of the tide (or depth of the ocean) by an expression in which every thing is arbitrary except the *periods* of the arguments. Sometimes Laplace appears to treat the phenomena at Brest as *derived*, but in other places he appears to consider that they must coincide with those which would obtain at that geographical latitude in the perfect sphere.

The problem of the Tides in which the motions of the fluid are due to the action of a force of which the intensity and direction are continually changing, presents serious difficulties, which are further increased by the circumstance that the bed of the ocean is far too irregular to be represented even approximately by any algebraic curve surface, and by the effect of the resistance and friction of the water against the shores, which cannot be considered as insensible.

The attention of Laplace does not appear to have been directed to the construction of Tide Tables for predicting the time and height of high water at any port; until very recently, this practical solution of the problem was attempted only in Great Britain.

---

moins en partie par l'inertie des eaux qui doivent être mises en mouvement et qui ne sauroient obéir assez promtement aux forces qui les sollicitent, pour leur faire suivre les loix que ces forces demanderoient."—p. 158. I use the word *retard* after Bernoulli.

Mathematical Theory of the Oscillations of the Surface of the Ocean.

If X, Y, Z denote the forces acting in the direction of the co-ordinate axes upon the fluid particle of which the rectangular co-ordinates are x, y, z, and if

$$u = \frac{dx}{dt}, \qquad v = \frac{dy}{dt}, \qquad w = \frac{dz}{dt},$$

$$u' = \frac{du}{dt} + u\frac{du}{dx} + v\frac{du}{dy} + w\frac{du}{dz}$$

$$v' = \frac{dv}{dt} + u\frac{dv}{dx} + v\frac{dv}{dy} + w\frac{dv}{dz}$$

$$w' = \frac{dw}{dt} + u\frac{dw}{dx} + v\frac{dw}{dy} + w\frac{dw}{dz},$$

then the differential equation to the surface of the fluid is

$$(X - u')\,dx + (Y - v')\,dy + (Z - w')\,dz = 0.$$

$X\,dx + Y\,dy + Z\,dz = 0$, would be the differential equation to the surface of the fluid if the system were at rest. See *Traité de Mécanique*, by M. Poisson, vol. ii. pp. 523 and 669.

If V is a certain function of x, y, z, the coordinates of the fluid molecule, and of x', y', z', the coordinates of the centre of the distant luminary,

$$dV = \frac{dV}{dx}dx + \frac{dV}{dx'}dx' + \frac{dV}{dy}dy + \frac{dV}{dy}dy' + \frac{dV}{dz}dz + \frac{dV}{dz'}dz'$$

$$= X\,dx + Y\,dy + Z\,dz + \frac{dV}{dx'}dx' + \frac{dV}{dy'}dy' + \frac{dV}{dz'}dz'.$$

The equation to the fluid surface is therefore

$$dV - \frac{dV}{dx'}dx' - u'\,dx - \frac{dV}{dy'}dy' - v'\,dy - \frac{dV}{dz'}dz' - w'\,dz = 0.$$

Bernoulli's theory of the tides rests upon the assumption that the equation to the fluid surface is

$$dV = 0, \qquad \text{or } V = \text{constant},$$

that is, it requires that the quantity

$$\frac{dV}{dx'}dx' + u'\,dx + \frac{dV}{dy'}dy' + v'\,dy + \frac{dV}{dz'}dz' + w'\,dz \;.\;.\;.\; \text{(A.)}$$

may be neglected.

Having *given* the general equation to the surface of the fluid, it remains to find when the distance from the centre of the earth is a maximum which is the time of *high water*.

The general equations of the motion of fluids referred to rectangular coordinates are given by M. Poisson, *Traité de Mécanique*, vol. ii. p. 669, and in other works.

$$\frac{1}{\varrho}\frac{dp}{dx} = X - \frac{du}{dt} - u\frac{du}{dx} - v\frac{du}{dy} - w\frac{du}{dz} \quad . \quad . \quad . \quad . \quad . \qquad \text{(A.)}$$

$$\frac{1}{\varrho}\frac{dp}{dy} = Y - \frac{dv}{dt} - u\frac{dv}{dx} - v\frac{dv}{dy} - w\frac{dv}{dz} \quad . \quad . \quad . \quad . \quad . \qquad \text{(B.)}$$

$$\frac{1}{\varrho}\frac{dp}{dz} = Z - \frac{dw}{dt} - u\frac{dw}{dx} - v\frac{dw}{dy} - w\frac{dw}{dz} \quad . \quad . \quad . \quad . \quad . \qquad \text{(C.)}$$

$$\frac{d\varrho}{dt} + \frac{d.\rho u}{dx} + \frac{d.\varrho v}{dy} + \frac{d.\varrho w}{dz} = 0 \quad . \quad . \quad . \quad . \quad . \quad . \quad . \quad . \qquad \text{(D.)}$$

Let $x = r\cos l\cos\mu \qquad y = r\cos l\sin\mu \qquad z = r\sin l$.

In the problem of the tides l may represent geographical latitude, and μ the sidereal time at the place. If

$$r' = \frac{dr}{dt} \qquad l' = \frac{dl}{dt} \qquad \mu' = \frac{d\mu}{dt}$$

$$X = \frac{dV}{dx} \qquad Y = \frac{dV}{dy} \qquad Z = \frac{dV}{dz}.$$

The general equations of motion referred to polar coordinates are

$$\frac{dp}{\varrho dr} = \frac{dV}{dr} - \frac{dr'}{dt} + \frac{r\,dl^2}{dt^2} + r\cos^2 l\frac{d\mu^2}{dt^2}$$

$$\frac{dp}{\varrho dl} = \frac{dV}{dl} - \frac{r^2 dl'}{dt} - 2r\frac{dr}{dt}\frac{dl}{dt} - r^2\sin l\cos l\frac{d\mu^2}{dt^2}$$

$$\frac{dp}{\varrho d\mu} = \frac{dV}{d\mu} - r^2\cos^2 l\frac{d\mu'}{dt} - 2r\cos^2 l\frac{dr}{dt}\frac{d\mu}{dt}$$
$$- 2r^2\sin l\cos l\frac{dl}{dt}\frac{d\mu}{dt}.$$

If $\mu = nt + \theta$,

$$\frac{d\mu}{dt} = n + \theta' = n + \frac{d\theta}{dt};$$

and if we neglect the quantities of the second order,

$$\frac{dl}{dt}\cdot\frac{d\theta}{dt}, \quad \frac{dl^2}{dt^2}, \text{ \&c.}$$

$$\frac{dp}{\varrho dr} = \frac{dV}{dr} - \frac{dr'}{dt} - n^2 r\cos^2 l - 2nr^2\cos^2 l\frac{d\theta}{dt}$$

$$\frac{\mathrm{d}p}{\rho\,\mathrm{d}l} = \frac{\mathrm{d}V}{\mathrm{d}l} - r^2\frac{\mathrm{d}l'}{\mathrm{d}t} - n^2 r^2 \sin l \cos l - 2nr^2 \sin l \cos l\frac{\mathrm{d}\theta}{\mathrm{d}t}$$

$$\frac{\mathrm{d}p}{\rho\,\mathrm{d}\mu} = \frac{\mathrm{d}V}{\mathrm{d}\mu} - r^2\cos^2 l\frac{\mathrm{d}\theta'}{\mathrm{d}t} - 2nr\cos^2 l\frac{\mathrm{d}r}{\mathrm{d}t}$$

$$- 2nr^2 \sin l \cos l\frac{\mathrm{d}l}{\mathrm{d}t},$$

and the equation to the surface will be

$$\left\{\frac{\mathrm{d}V}{\mathrm{d}r} - \frac{\mathrm{d}r'}{\mathrm{d}t} + n^2 r\cos^2 l - 2nr\cos^2 l\frac{\mathrm{d}\theta}{\mathrm{d}t}\right\}\mathrm{d}r$$

$$+\left\{\frac{\mathrm{d}V}{\mathrm{d}l} - r^2\frac{\mathrm{d}l'}{\mathrm{d}t} - n^2 r^2 \sin l \cos l - 2nr^2 \sin l \cos l\frac{\mathrm{d}\theta}{\mathrm{d}t}\right\}\mathrm{d}l$$

$$+\left\{\frac{\mathrm{d}V}{\mathrm{d}\mu} - r^2\cos^2 l\frac{\mathrm{d}\theta'}{\mathrm{d}t} - 2nr\cos^2 l\frac{\mathrm{d}r}{\mathrm{d}t}\right.$$

$$\left. - 2nr^2 \sin l \cos l\frac{\mathrm{d}l}{\mathrm{d}t}\right\}\mathrm{d}\mu = 0,$$

or if

$$\frac{\mathrm{d}V}{\mathrm{d}r}\mathrm{d}r + \frac{\mathrm{d}V}{\mathrm{d}l}\mathrm{d}l + \frac{\mathrm{d}V}{\mathrm{d}\mu}\mathrm{d}\mu = \mathrm{d}'V$$

$$\left\{\frac{\mathrm{d}r'}{\mathrm{d}t} + 2nr\cos^2 l\frac{\mathrm{d}\theta}{\mathrm{d}t}\right\}\mathrm{d}r$$

$$+\left\{r^2\frac{\mathrm{d}l'}{\mathrm{d}t} + 2nr^2\sin l\cos l\frac{\mathrm{d}\theta}{\mathrm{d}t}\right\}\mathrm{d}l$$

$$+\left\{r^2\cos^2 l\frac{\mathrm{d}\theta'}{\mathrm{d}t} + 2nr\cos^2 l\frac{\mathrm{d}r}{\mathrm{d}t} + 2nr^2\sin l\cos l\frac{\mathrm{d}l}{\mathrm{d}t}\right\}\mathrm{d}\mu$$

$$= \mathrm{d}\left\{\frac{n^2 r^2\cos^2 l}{2}\right\} + \mathrm{d}'V \quad . \quad . \quad . \quad . \quad . \quad . \quad . \quad . \quad (\mathrm{E.})$$

which is identical with Laplace's equation, *Méc. Cél.*, vol. i. p. 98.

When the system is in equilibrium and at rest, the equation to the surface is $\mathrm{d}'V = 0$. The solution of the problem of the tides has been obtained by neglecting the quantities on the left hand side of equation (E). With this simplification, and if the coordinates of the fluid are the only variables which enter into the composition of V, the equation to the surface is

$$\frac{n^2 r^2\cos^2 l}{2} + V = \text{constant.}$$

The remaining equations are to be deduced from the invariability of the mass of the element $\mathrm{d}\,m$.

The elementary parallelopiped

$$r^2\cos l\,\mathrm{d}r\,\mathrm{d}l\,\mathrm{d}\mu$$

is bounded by the sides

$$MA = dr, \qquad MB = r\,dl, \qquad MC = r\cos l\,d\mu,$$

the coordinates of the point M being r, l, μ,

———— A — $r + dr,\ l,\ \mu$

———— B — $r,\ l + dl,\ \mu$

———— C — $r,\ l,\ \mu + d\mu.$

Suppose that during the instant dt the points M, A, B, C, are transported to M′, A′, B′, C′, the coordinates r, l, μ, of M become

$$r + r'dt,\ l + l'dt, \text{ and } \mu + \mu'dt;$$

at the end of the instant dt these quantities are therefore the coordinates of the point M′, the coordinates A′, B′, C′ may be deduced from these by putting for r, l, μ the primitive coordinates of these points; thus the coordinates

of A′ will be $$r + dr + r'dt + \frac{dr'}{dr}dr\,dt$$

$$l + l'dt + \frac{dl'}{dr}dr\,dt$$

$$\mu + \mu'dt + \frac{d\mu'}{dr}dr\,dt$$

of B′ will be $$r + r'dt + \frac{dr'}{dl}dl\,dt$$

$$l + dl + l'dt + \frac{dl'}{dl}dl\,dt$$

$$\mu + \mu'dt + \frac{d\mu'}{dl}dl\,dt$$

and of C′ will be $$r + r'dt + \frac{dr'}{d\mu}d\mu\,dt$$

$$l + l'dt + \frac{dl'}{d\mu}d\mu\,dt$$

$$\mu + d\mu + \mu'dt + \frac{d\mu'}{d\mu}d\mu\,dt.$$

Let $r\cos l\cos\mu$, $\quad r\cos l\sin\mu$, $\quad r\sin l$,

$r'\cos l'\cos\mu'$, $\quad r'\cos l'\sin\mu'$, $\quad r'\sin l'$,

be the coordinates of any points P, P′, then

$$(PP')^2 = r^2 - 2rr'\{\cos l\cos l'\cos(\mu - \mu') + \sin l\sin l'\} + r'^2$$

$$= (r - r')^2 + 4rr'\sin^2\frac{(l - l')}{2} + 4rr'\sin l\sin l'\sin^2\frac{(\mu - \mu')}{2}$$

if $l - l'$ and $\mu - \mu'$ are small quantities,

$$(\mathrm{P\,P'})^2 = (r - r')^2 + r\,r'\,(l - l')^2 + r\,r' \cos l \cos l'\,(\mu - \mu')^2$$

therefore

$$(\mathrm{M'\,A'})^2 = \left(\mathrm{d}\,r + \frac{\mathrm{d}\,r'}{\mathrm{d}\,r}\,\mathrm{d}\,r\,\mathrm{d}\,t\right)^2 + r^2\left(\frac{\mathrm{d}\,l'}{\mathrm{d}\,r}\right)^2 \mathrm{d}\,r'\,\mathrm{d}\,t^2$$
$$+ r^2 \cos^2 l \left(\frac{\mathrm{d}\,\mu'}{\mathrm{d}\,t}\right)^2 \mathrm{d}\,r^2\,\mathrm{d}\,t^2$$

$$\mathrm{M'\,A'} = \mathrm{d}\,r + \frac{\mathrm{d}\,r'}{\mathrm{d}\,r}\,\mathrm{d}\,r\,\mathrm{d}\,t.$$

Similarly

$$\mathrm{M'\,B'} = r\,\mathrm{d}\,l \left\{1 + \frac{r'}{r}\,\mathrm{d}\,t + \frac{\mathrm{d}\,l'}{\mathrm{d}\,l}\,\mathrm{d}\,t\right\}$$

$$\mathrm{M\,C'} = r \cos l\,\mathrm{d}\,\mu \left\{1 + \frac{r'}{r}\,\mathrm{d}\,t - \frac{\sin l}{\cos l}\,l'\,\mathrm{d}\,t + \frac{\mathrm{d}\,\mu'}{\mathrm{d}\,\mu}\,\mathrm{d}\,t\right\}$$

the volume of the elementary parallelopiped becomes

$$r^2 \cos l\,\mathrm{d}\,r\,\mathrm{d}\,l\,\mathrm{d}\,\mu \left\{1 + \frac{\mathrm{d}\,r'}{\mathrm{d}\,r}\,\mathrm{d}\,t + \frac{\mathrm{d}\,l'}{\mathrm{d}\,l}\,\mathrm{d}\,t + \frac{\mathrm{d}\,l'}{\mathrm{d}\,l}\,\mathrm{d}\,t\right.$$
$$\left. + \frac{2\,r'}{r}\,\mathrm{d}\,t - \frac{\sin l}{\cos l}\,l'\,\mathrm{d}\,t\right\}$$

at the same time the density ρ becomes $\rho + \rho'\,\mathrm{d}\,t$,

$$\rho' = \frac{\mathrm{d}\,\rho}{\mathrm{d}\,t} + r'\,\frac{\mathrm{d}\,\rho}{\mathrm{d}\,r} + l'\,\frac{\mathrm{d}\,\rho}{\mathrm{d}\,l} + \mu'\,\frac{\mathrm{d}\,\rho}{\mathrm{d}\,\mu}.$$

Multiplying the volume by $\rho + \rho'\,\mathrm{d}\,t$, subtracting the primitive volume $r^2 \cos l\,\mathrm{d}\,r\,\mathrm{d}\,l\,\mathrm{d}\,\mu$, we get the variation of the mass during the instant $\mathrm{d}\,t$, and since this variation must equal zero, the following equation is obtained, which is equivalent to a transformation of equation (D):

$$\frac{\mathrm{d}\,\varrho}{\mathrm{d}\,t} + \frac{\mathrm{d}\,.\,\varrho\,r'}{\mathrm{d}\,r} + \frac{\mathrm{d}\,.\,\varrho\,l'}{\mathrm{d}\,l} + \frac{\mathrm{d}\,.\,\varrho\,\mu'}{\mathrm{d}\,\mu} + \frac{2\,\varrho\,r'}{r} - \rho\,\frac{\sin l}{\cos l}\,l' = 0,$$

or

$$\varrho' + \varrho \left\{\frac{\mathrm{d}\,r'}{\mathrm{d}\,r} + \frac{\mathrm{d}\,l'}{\mathrm{d}\,l} + \frac{\mathrm{d}\,\mu'}{\mathrm{d}\,\mu} + \frac{2\,r'}{r} - \frac{\sin l}{\cos l}\,l'\right\} = 0.$$

For incompressible fluids, when the effect of changes of temperature is neglected, $\varrho' = 0$ separately, and

$$\frac{\mathrm{d}\,r'}{\mathrm{d}\,r} + \frac{\mathrm{d}\,l'}{\mathrm{d}\,l} + \frac{\mathrm{d}\,\mu'}{\mathrm{d}\,\mu} + \frac{2\,r'}{r} - \frac{\sin l}{\cos l}\,l' = 0,$$

or $$\frac{\mathrm{d}\,(r^2\,r')}{\mathrm{d}\,r} + r^2 \left\{\frac{\mathrm{d}\,l'}{\mathrm{d}\,l} + \frac{\mathrm{d}\,\mu'}{\mathrm{d}\,\mu} - \frac{\sin l}{\cos l}\,l'\right\} = 0,$$

which equation is identical with that given by Laplace, *Méc. Cél.*, vol. i. p. 101.

If τ denote the temperature, Fourier has shown that

$$\frac{d\tau}{dt}+\frac{d.u\tau}{dx}+\frac{d.v\tau}{dy}+\frac{d.w\tau}{dz}=\frac{K}{C}\left\{\frac{d^2\tau}{dx^2}+\frac{d^2\tau}{dy^2}+\frac{d^2\tau}{dz^2}\right\}. \quad \text{(F.)}$$

and if e denote the temperature which corresponds to a given temperature b,

$$\varrho=e\{1+h(\tau-b)\} \quad . \quad . \quad . \quad . \quad . \quad . \quad \text{(G.)}$$

K, C, and h being constants. *Mémoires de l'Institut*, vol. xiii. p. 519. When the temperature varies, the two last equations supply the place of the equation $\rho'=0$.

The left hand side of the equation (F.) is of the same form as equation (D.), p. 28; hence by the help of a known transformation it is easy to transform equation (F.) to polar coordinates, and we obtain

$$\frac{d\tau}{dt}+\frac{d.\tau r'}{dr}+\frac{d.\tau l'}{dl}+\frac{d.\tau\mu'}{d\mu}+\frac{2\tau r'}{r}-\tau\frac{\sin l}{\cos l}l'$$

$$=\frac{K}{Cr}\left\{\frac{d^2.r\tau}{dr^2}+\frac{1}{r^2\cos^2 l}\left(\frac{d^2.r\tau}{d\mu^2}\right)\right.$$

$$\left.+\frac{1}{r^2\cos^2 l}\left(\frac{d.\cos l\frac{d.r\tau}{dl}}{dl}\right)\right\} \quad \text{(F.)}$$

If r be the distance of the sun's centre from that of the earth, ζ the sun's zenith distance, m the mass of the sun; if the same quantities accented refer to the moon; and if $\frac{M}{R^2}$ be the force of gravity, R being the distance of the fluid element d M from the earth's centre;

$$V=\frac{M}{R}-m\left\{\frac{R\cos\zeta}{r^2}-\frac{1}{(R^2-2rR\cos\zeta+r^2)^{\frac{1}{2}}}\right\}$$

$$-m'\left\{\frac{R\cos\zeta'}{r'^2}-\frac{1}{(R^2-2r'R\cos\zeta'+r'^2)^{\frac{1}{2}}}\right\}$$

because the fluid is disturbed by the difference of the attractions of m and m' upon each of its particles, and upon the centre of the earth. Expanding the radical

$$V=\frac{M}{R}+\frac{m}{r}+\frac{3mR^2}{2r^3}\left(\cos^2\zeta-\frac{1}{3}\right)+\frac{m'}{r'}+\frac{3m'R^2}{2r'^3}\left(\cos^2\zeta'-\frac{1}{3}\right)$$

nearly, $\frac{m}{r}$ is constant by hypothesis, and supposing the quantity

$\frac{m R^2}{2 r^3}$ to be also constant, the equation to the surface of the ocean is

$$\frac{1}{R} + \frac{3 m R^2 \cos \zeta^2}{2 M r^3} + \frac{3 m' R^2 \cos \zeta'^2}{2 M r'^3} = \frac{1}{c}$$

c being a constant. If only one luminary be considered,

$$\frac{1}{R} = \frac{1}{c} - \frac{3 m R^2 \cos^2 \zeta}{2 M r^3}$$

$$R = c + \frac{3 m c^2 R^2 \cos^2 \zeta}{2 M r^3}$$

which is the equation to a spheroid nearly, of which the semi-major axis $= c + \frac{3 m c^2 R^2}{2 M r^3}$, and the eccentricity $= \sqrt{\frac{3 m c R^2}{M r^3}}$

The solid content of this spheroid is equal to that of the sphere which the surface of the ocean would assume if the attraction of the luminary were to cease because the mass of water is the same, the solid content of the spheroid is

$$\frac{4 \pi c^3}{3} \left(1 + \frac{3 m c R^2}{2 M r^3}\right)^3 \left(1 - \frac{3 m c R^2}{M r^3}\right)$$

which (if the radius of the sphere be taken $= 1$), $= \frac{4 \pi}{3}$, π being the ratio of the diameter to the circumference, whence $c = 1 - \frac{m R^2}{2 M r^3}$ nearly, and

$$R = 1 + \frac{3 m R^2}{2 M r^3} \left(\cos \zeta^2 - \frac{1}{3}\right)$$

this equation is identical with that which Euler has obtained by a method which is the same in substance: *Principia*, vol. iii. p. 270, line 17, Glasgow edition.

The preceding theory amounts to supposing, that when the luminaries are in motion these spheroids have a motion round the earth in the same time as the luminary, and such that if the luminaries move in the equator, the major axis of the spheroid passes through the meridian of a given place a given time after the luminary passes through, and that the height of the water above the level which it would have if their attractions should cease, is the sum of the heights due to each spheroid separately. So that if, as in the figure, B F represent a section of the spheroid due to the

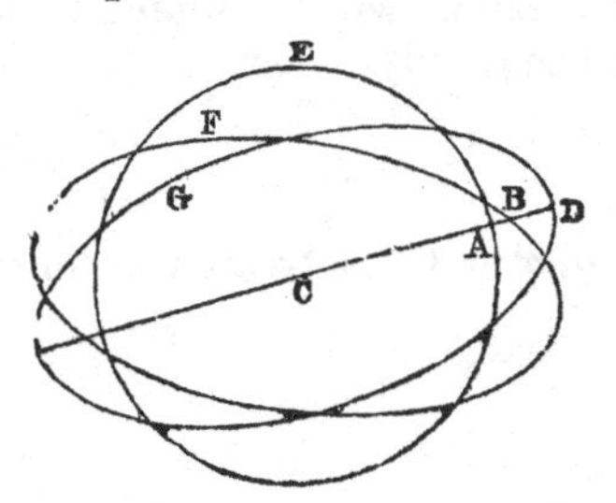

attraction of the moon by the equator, B G a section of that duc to the attraction of the sun, A E a section of the sphere which is such that C A is the mean height of the ocean, the excess of the height of the water above its mean height in any place will be (A B + A D) nearly.

If α denote right ascension, δ declination, l geographical latitude, and μ sidereal time,

$$\cos\zeta = \cos\delta\cos l\cos(\mu - \alpha) + \sin\delta\sin l.$$

If we neglect the quantity

$$u'\,dx + v'\,dy + w'\,dz,$$

the equation to the surface of the fluid is

$$\frac{dV}{dx}dx + \frac{dV}{dy}dy + \frac{dV}{dz}dz = 0,$$

which is the differential of V with respect only to the coordinates of the particle of fluid.

$$V = \frac{3}{2}\frac{mR^2}{r^3}\left\{\cos^2\zeta - \frac{1}{3}\right\} + \frac{3}{2}\frac{m'R^2}{r'^3}\left\{\cos^2\zeta' - \frac{1}{3}\right\} +, \&c.$$

$$= \frac{mR^2}{4r^3}\left\{\sin^2\delta - \frac{1}{2}\cos^2\delta\right\}\left\{1 - 3\cos 2l\right\} \qquad (1)$$

$$+ \frac{3mR^2}{r^3}\sin 2\delta\sin 2l\cos(\mu - \alpha) \qquad (2)$$

$$+ \frac{3}{4}\frac{mR^2}{r^3}\cos^2\delta\cos^2 l\cos(2\mu - 2\alpha) \qquad (3)$$

$$+ \frac{m'R^2}{4r'^3}\left\{\sin^2\delta' - \frac{1}{2}\cos^2\delta'\right\}\left\{1 - 3\cos 2l\right\} \qquad (1)$$

$$+ \frac{3m'R^2}{r'^3}\sin 2\delta'\sin 2l\cos(\mu - \alpha') \qquad (2)$$

$$+ \frac{3}{4}\frac{m'R^2}{r'^3}\cos^2\delta'\cos^2 l\cos(2\mu - 2\alpha') +, \&c. \qquad (3)$$

The coordinates R and l of the fluid molecule may be considered constant in comparison with μ, and hence the differential equation to the surface of the ocean will be found by differentiating this equation with regard to μ only. Let

$$\frac{d\mu}{dt} = c, \qquad \frac{d\alpha}{dt} = q, \qquad \frac{d\alpha'}{dt} = q',$$

c is constant, and if q, q' and δ be treated as constants, the equation to the surface after integration will be, considering only at present the terms of which the arguments are $2\mu - 2\alpha$ and $2\mu - 2\alpha'$)

$$h = D + E\left\{\frac{c}{(c-q)} A \cos(2\mu - 2\alpha) + \frac{c}{(c-q')} \cos(2\mu - 2\alpha')\right\} +, \&c.$$

where h is the height of the water, and D a constant depending only on the zero line from which the heights are reckoned.

$$A = \frac{m \cos^2 \delta\, P^3}{m' \cos^2 \delta'\, P'^3}, \qquad E = C m' \cos^2 \delta'\, P'^3,$$

P being the horizontal parallax, and C a constant depending upon geographical latitude. The quantities arising from $\frac{n^2 R^2 \cos^2 l}{2}$ (see p.9) and $\frac{M}{R}$ may be considered as included in the constant D.

If $\mu - \alpha' = \psi$, $\alpha - \alpha' = \phi$, ψ is the hour angle of the moon at the time of high water, and (in the perfect sphere) is an angle differing little from 0 or 180°.

$$h = D + E\left\{\frac{c}{(c-q)} A \cos(2\psi - 2\phi) + \frac{c}{(c-q')} \cos 2\psi\right\}$$

$$= D + E\left\{\left(1 + \frac{q}{c}\right) A \cos(2\psi - 2\phi) + \left(1 + \frac{q'}{c}\right) \cos 2\psi\right\}$$

The hourly motions of the moon and sun are respectively 32′ 56″·5 and 2′ 27″·8; the hourly variation of μ is 15° 2′ 28″, hence,

$$\frac{c}{c-q} = 1{\cdot}0027 \qquad \frac{c}{c-q'} = 1{\cdot}0379 \qquad \frac{c-q}{c-q'} = 1{\cdot}035.$$

The quantities $\frac{q}{c}$, $\frac{q'}{c}$, appear to correspond to the quantities denoted $m\,x$, $m'\,x$, by Laplace, and we see thus how they may arise from the differential equation to the surface. Laplace says, "il est *assez naturel* de supposer que ces constantes varient d'un astre à l'autre, proportionnellement à la différence des moyens mouvemens."

If the luminaries be supposed to move in the ecliptic,

$\tan \delta = \tan \omega \sin \alpha$, ω being the obliquity

$$\cos \zeta = \frac{\cos l \cos(\mu - \alpha) + \sin l \tan \omega \sin \alpha}{\sqrt{1 + \tan^2 \omega \sin^2 \alpha}}.$$

By expanding this quantity, a multitude of arguments would arise, and in order to treat the question rigorously it would then be necessary to substitute expressions for α and α', in terms of the time or *mean longitude*. I shall therefore employ the approximate equation

$$h = D + E\left\{\frac{c}{(c-q)} A \cos(2\psi - 2\phi) + \frac{c}{(c-q')}\cos 2\psi\right\}$$

Differentiating in order to find when this is a maximum,

$$A \sin(2\psi - 2\phi) + \sin 2\psi = 0$$

$$\tan 2\psi = \frac{A \sin 2\phi}{1 + A\cos 2\phi}.$$

When the average is taken of many observations made at different seasons of the year the declinations of the sun and moon tend to equal 15° in the results, and the parallax of the moon to equal 57′. I denote the value of A which obtains when the declinations of the luminaries are equal, and which corresponds to their mean parallaxes by the symbol (A), and the similar value of E by the symbol (E). Then the *semi-menstrual* inequality is given in the INTERVAL and in the HEIGHT by the expressions

$$\tan 2\psi = \frac{(A)\sin 2\phi}{1 + (A)\cos 2\phi}$$

$$h = D + (E)\left\{\left(1 + \frac{q}{c}\right)(A)\cos(2\psi - 2\phi) + \left(1 + \frac{q'}{c}\right)\cos 2\psi\right\}$$

if the greatest and least heights of high water be called h_1, h_2, (*Springs* and *Neaps*)

$$h_1 = D + (E)\left\{\frac{c}{(c-q)}(A) + \frac{c}{(c-q')}\right\}$$

$$h_2 = D + (E)\left\{-\frac{c}{(c-q)}(A) + \frac{c}{c-q')}\right\}$$

$$(E)\frac{c}{(c-q)} = \frac{h_1 - h_2}{2(A)}$$

$$h = D + (E)\frac{c}{(c-q)}\left\{(A)\cos(2\psi - 2\phi) + \frac{(c-q)}{(c-q')}\cos 2\psi\right\}$$

$$= D + \frac{h_1 - h_2}{2(A)}\left\{(A)\cos(2\psi - 2\phi) + \frac{(c-q)}{(c-q')}\cos 2\psi)\right\}.$$

If $\frac{q}{c}$ and $\frac{q'}{c}$ be neglected, the expression for the height may be put in the form

$$h = D + (E)(A)\frac{\sin 2\phi}{\sin 2\psi} = D + \left(\frac{h_1 - h_2}{2}\right)\frac{\sin 2\phi}{\sin 2\psi}.$$

Differentiating the expression

$$\tan 2\psi = \frac{A \sin 2\phi}{1 + A \cos 2\phi},$$

we find

$$\mathrm{d}\psi = \frac{\mathrm{d}A \sin 2\phi}{2\{1 + 2A\cos 2\phi + A^2\}}$$

$$= \frac{\mathrm{d}A \sin 2\phi}{2} \text{ nearly,}$$

$$\mathrm{d}A = 3A\frac{\mathrm{d}P}{P} - 2A\tan\delta\,\mathrm{d}\delta - 3A\frac{\mathrm{d}P'}{P'} + 2A\tan\delta'\,\mathrm{d}\delta'.$$

Hence the correction of the *semi-menstrual inequality* in the INTERVAL,

for the Sun's Parallax varies nearly as	$\sin 2\phi\,\mathrm{d}P$
—— Sun's Declination	$-\sin 2\phi\tan\delta\,\mathrm{d}\delta$
—— Moon's Parallax	$-\sin 2\phi\,\mathrm{d}P'$
—— Moon's Declination	$\sin 2\phi\tan\delta'\,\mathrm{d}\delta'$.

Differentiating the expression for the height of high water

$$h = D + AE\frac{\sin 2\phi}{\sin 2\psi}$$

$$\mathrm{d}h = \mathrm{d}(AE)\frac{\sin 2\phi}{\sin 2\psi} - \frac{2AE\sin 2\phi\cos 2\psi\,\mathrm{d}\psi}{\sin^2 2\psi}$$

$$= \mathrm{d}(AE)\frac{\sin 2\phi}{\sin 2\psi} - \frac{E}{A}\cos 2\psi\,\mathrm{d}A$$

$$\mathrm{d}(AE) = 3AE\frac{\mathrm{d}P}{P} - 2AE\tan\delta\,\mathrm{d}\delta.$$

Hence the correction of the *semi-menstrual inequality* in the HEIGHT,

for the Sun's Parallax varies nearly as	$(1 - \cos 2\psi)\,\mathrm{d}P$
—— Sun's Declination	$-(1 - \cos 2\psi)\tan\delta\,\mathrm{d}\delta$
—— Moon's Parallax	$\cos 2\psi\,\mathrm{d}P'$
—— Moon's Declination . . .	$-\cos 2\psi\tan\delta'\,\mathrm{d}\delta'$.

The correction for the sun's parallax is insensible.

Laplace, in the *Traité de Mécanique Céleste*, vol. ii. p. 232, gives the following expression for the height of the tide:

$$\alpha y = P\left\{\frac{L}{r^3}\cos v^2 \cos 2(nt + \pi - \psi - \lambda) + \frac{L'}{r^3}\cos v'^2\cos 2(nt + \pi - \psi' - \lambda)\right\}.$$

which coincides with the expression given above for the height.

r is the distance of the sun from the centre of the earth.
$n\,t + \pi - \psi$, the apparent time.
L, the mass of the sun.
v, the declination of the sun.
r', the distance of the moon from the centre of the earth.
$n\,t + \pi - \psi'$, the hour angle of the moon.
L', the mass of the moon.
v', the declination of the moon.
P and λ, constants which depend upon the local circumstances of the port, and which must be determined by observation.

The difference in height between the morning and evening tide depends upon the angles $\psi - \phi$ and ψ: if this difference be called d h,

$$\mathrm{d}\,h = B\,\{A \sin 2\,\delta \cos (\psi - \phi) + \sin 2\,\delta' \cos \psi\}$$

for the diurnal inequality of the interval,

$$\mathrm{d}\,\psi = \frac{F \cos 2\,\psi}{1 + A \cos 2\,\phi}\left\{\frac{A \sin 2\,\delta}{\cos^2 \delta'} \sin (\psi - \phi) + 2 \tan \delta' \sin \psi\right\};$$

B and F being constants which are not the same at different places.

$$\mathrm{d}\,h = C \sin 2\,\delta'$$

is an approximate form for the *diurnal inequality* in the height, C being a constant.

The inequalities of the heights at different places depending upon the angles $2\,\psi - 2\phi$ and $2\,\psi$ are proportional to the quantity (E), so that, if they have been obtained for any place P, they may be obtained for any other place P$'$ by multiplying the former by $\frac{(E')}{(E)}$, supposing the constant (A) to be the same for both places. Also, upon this supposition, all the inequalities in the height for different places are proportional to the difference between the greatest and least heights in the *semi-menstrual* inequality. By this assumption the Tables VII., VIII., IX., X., annexed may be rendered available for any place when the *semi-menstrual* inequality in the height at that place has been obtained.

When the moon's transit is at 2 o'clock P.M., α' is greater than α, ϕ is negative, tan $2\,\psi$ is negative, and ψ, the variable quantity to be *added* to the apparent solar time of the moon's transit, or the interval (in the perfect sphere), is negative.

The breadth of a channel in which a derived tide-wave flows, affects the amount of the rise. Suppose A B C D, E F G H to be

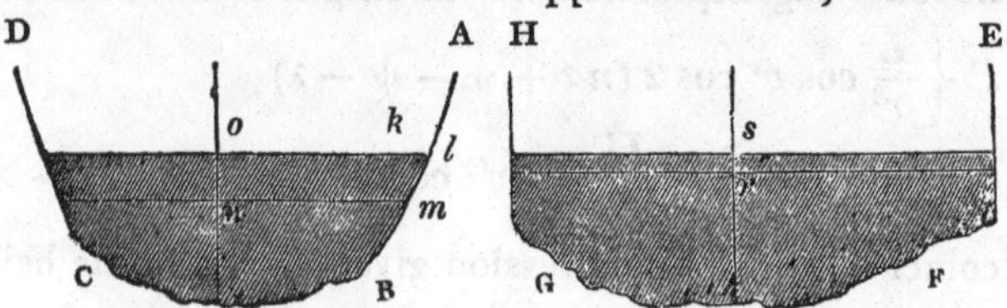

vertical sections so situate that the same derived tide flows through both, the first bounded by shores inclined to the horizon at any given angle, the latter bounded by perpendicular cliffs; $m\,n$, $q\,r$ the average height of high water, $l\,m\,n\,o$, $p\,q\,r\,s$ sections of fluid added by the tide at any epoch. We may suppose that the area $l\,m\,n\,o = p\,q\,r\,s$.

Let $n\,o = y \quad n\,m = x \quad r\,s = y' \quad r\,q = x' \quad k\,l = n\,y$. Since the areas $l\,m\,n\,o$, $p\,q\,r\,s$ are equal,

$$x\,y + \frac{n\,y^2}{2} = x'\,y' \qquad\qquad y + \frac{n\,y^2}{2\,x} = \frac{x'}{x}\,y'.$$

The term $\frac{n\,y^2}{2\,x}$ will generally be small, because the variations in the height of high water are always small in comparison with the breadth of the channel. Upon the preceding assumptions the variations in the height of high water will be inversely as the breadth of the channel, and if y' follows the law assigned to it by Bernoulli or any other, y will follow nearly the same law.

According to Mr. Russell of Edinburgh (see Seventh Report of the British Association, p. 426.), "The Tide Wave appears to be the only wave of the ocean which belongs to the first order, and appears to be identical with the great primary wave of translation; its velocity increases with the depth of the fluid, and appears to approximate closely to the velocity due to half the depth of the fluid in the rectangular channel *; and to a certain mean depth, which is that of the centre of gravity of the section of the channel. It is, however, difficult to determine the limits within which the tide wave retains its unity; where portions of the same channel differ much in depth at points remote from each other, the tide waves appear to separate.

"The tide appears to be a compound wave, one elementary wave bringing the first part of flood tide, another the high water, and so on; these move with different velocities according to the depth. On approaching shallow shores the anterior tide waves move more slowly in the shallow water, while the posterior waves moving more rapidly, diminish the distance between successive waves. The tide wave becomes thus dislocated, its anterior surface rising more rapidly, and its posterior surface descending more slowly than in deep water.

"A tidal bore is formed when the water is so shallow at low water that the first waves of flood tide move with a velocity so much less than that due to the succeeding part of the tidal wave, as to be overtaken by the subsequent waves, or wherever the tide rises so rapidly, and the water on the shore or in the river is so shallow that the height of the first wave of the tide is greater than the

* See the *Mécanique Anal.*, vol. ii. p. 335.

depth of the fluid at that place. Hence in deep water vessels are safe from the waves of rivers which injure those on the shore.

"The identity of the tide wave, and of the great wave of translation, show the nature of certain variations in the establishment of ports situated on tidal rivers. Any change in the depth of the rivers produces a corresponding change on the interval between the moon's transit and the high water immediately succeeding. It appears from the observations in this report, that the mean time of high water has been rendered 37 minutes earlier than formerly by deepening a portion of about 12 miles in the channel of a tidal river, so that a tide wave which formerly travelled at the rate of 10 miles an hour, now travels at the rate of nearly 15 miles an hour."

The semi-menstrual inequality of the times and heights may, I believe, be deduced, together with the constants (*A*), *D*, and (*E*), from all the observations of the tides at any place for one year, with accuracy. When more numerous observations are employed, and when they are all classed merely with reference to the moon's age for the purpose of obtaining the semi-menstrual inequality, the corresponding horizontal. parallax of the moon and the declinations of the luminaries will be nearly the same throughout, or a small and almost insensible correction will suffice as in my discussions of the Liverpool and London observations to render them so. But when few observations only are employed to determine the semi-menstrual inequality, the deviations will be greater and the correction required of greater consequence. The following example of the determination of the constants for Liverpool from the semi-menstrual inequality, deduced from the observations of 19 years, will serve to explain the method I employ.

SEMI-MENSTRUAL INEQUALITY AT LIVERPOOL.

Moon's Horizontal Parallax . . . . = 57′.
Moon's Declination . . . . = 15°.
Sun's Horizontal Parallax . . . . . . = 8″·8.
Sun's Declination . . . . = 15°.

☽'s Transit *A*.	Observed.		☽'s Transit *A*.	Observed.	
	Interval.	Height.		Interval.	Height.
h m	days h m	feet.	h m	days h m	feet.
0 30	2 0 12·5	17·67	6 30	2 0 38·2	12·46
1 30	1 23 57·5	16·96	7 30	2 1 1·7	13·82
2 30	1 23 44·3	15·85	8 30	2 1 5·6	15·29
3 30	1 23 37·2	14·44	9 30	2 0 57·8	16·57
4 30	1 23 40·2	12·99	10 30	2 0 44·1	17·40
5 30	2 0 2·7	12·10	11 30	2 0 28·8	17·85

If we take the average of a great many intervals between the time of the moon's transit and the time of high water, the time of the moon's transit which corresponds to this average interval may be considered as the *epoch* or zero point of the semi-menstrual inequality in the interval.

Each of the numbers in the second and third columns is the average of more than 1000 observations.

I find the average *interval* $2^d\ 0^h\ 20{\cdot}9^m$, which corresponds exactly with syzygy. When the moon passes the meridian at 12 o'clock, $\phi = 0$; when she passes the meridian at 3 o'clock, $\phi = -45°$; when she passes the meridian at 9 o'clock, $\phi = -135°$.

I find by interpolation the interval corresponding to the moon's transit at . . . . $3^h\ 0^m = 1^d\ 23^h\ 40^m{\cdot}7$

. . . . $9\ \ 0 = 2\ \ 1\ \ 1\ {\cdot}7$.

The difference $= 1^h\ 21^m$, which converted into space $= 20°\ 15'$,

$$\text{Log tan } 20°\ 15' = \log (A) = 9{\cdot}56693$$

$$(A) = {\cdot}3689 \qquad \frac{1}{(A)} = 2{\cdot}71.$$

If we take the difference between the greatest and least ***heights*** in the semi-menstrual inequality $= 5{\cdot}75$ feet, neglecting $\frac{q}{c}$ and $\frac{q'}{c}$,

$$(E) = \frac{5{\cdot}75}{2\,(A)} = 7{\cdot}79 \text{ for Liverpool.}$$

If we take the greatest height $= 17{\cdot}85$,

$$17{\cdot}85 = D + (E)\,\{1 + (A)\} = D + 10{\cdot}66$$

$D = 7{\cdot}19$ feet for Liverpool, from the datum on the east wall of the Canning Dock.

If m denote the mass of the sun, m' the mass of the moon, M the mass of the earth, P the mean horizontal parallax of the sun, P' that of the moon

$$\frac{m\,P^3}{(m' + M)\,P'^3} = ({\cdot}0748013)^2; \qquad (A) = \frac{m\,P^3}{m'\,P'^3}$$

$$\frac{(A)\,m'}{m' + M} = ({\cdot}0748013)^2 \qquad \frac{m'}{m' + M} = \frac{({\cdot}0748013)^2}{(A)}$$

${\cdot}0748013$ being the quantity generally denoted by m in the Lunar Theory. Combining the last equation with the value of (A) above, I find

$$\frac{m'}{(m' + M)} = \frac{1}{65{\cdot}93} \text{ and } \frac{m'}{M} = \frac{1}{64{\cdot}93}.$$

By the discussion of the London Dock tides, with reference to the transit B, this average interval has been found to correspond very nearly with the moon's transit at $12^h\ 25^m$.

From more than 24,000 observations of the tides at the London Docks, I found the interval corresponding to the moon's transit at . . . . $3^h\ 30^m = 2^h\ 26^m{\cdot}7$

. . . . $9\ \ 30 = 3\ \ 50\ {\cdot}4$.

See Phil. Trans. 1837, p. 131.

The difference $= 1^h\ 23{\cdot}7^m$, which converted into space $= 20°\ 55'$,

$$\text{Log tan } 20°\ 55' = \log (A) = 9{\cdot}58229$$

$$(A) = {\cdot}3822 \qquad\qquad \frac{1}{(A)} = 2{\cdot}616$$

If we take the difference between the greatest and least heights in the semi-menstrual inequality $= 3{\cdot}17$ feet, neglecting $\frac{q}{c}$ and $\frac{q'}{c}$,

$$(E) = \frac{3{\cdot}17}{2\,(A)} = 4{\cdot}17 \text{ for London.}$$

If we take the greatest height $= 22{\cdot}72$,

$$22{\cdot}72 = D + (E)\,\{1 + (A)\} = D + 5{\cdot}75,$$

$D = 16{\cdot}97$ feet for London, reckoned from the sill of the gates at the Wapping entrance of the London Docks.

In a previous discussion of 13,370 observations made at the London Docks, with reference to the moon's transit B, I found the interval corresponding to the moon's transit $3^h\ 30^m = 2^d\ 2^h\ 25^m{\cdot}1$

. . . . $9\ \ 30 = 2\ \ 3\ \ 48\ {\cdot}9$.

The difference $= 1^h\ 23^m{\cdot}8$, which converted into space $= 21°$ nearly,

$$\text{Log tan } 21° = \log (A) = 9{\cdot}58418$$

$$(A) = {\cdot}3838 \qquad\qquad \frac{1}{(A)} = 2{\cdot}605.$$

I found moreover $(E) = 4{\cdot}43$.

$D = 16{\cdot}69$ feet for London, reckoned from the sill of the gates at the Wapping entrance of the London Docks.

See Phil. Trans. 1836, p. 224, and the accompanying plate, in which the *semi-menstrual inequality* in the interval and in the height is laid down from theory and from observation in order that the agreement may be perceived.

If we take the difference in the intervals $= 1^h\ 24^m$ instead of $1^h\ 23{\cdot}7$ (and it is impossible to depend upon such a quantity as $\frac{3}{10}$ths of a minute), this difference converted into space is $21°$;

$$\text{Log } (A) = 9{\cdot}58418.$$

This value of (A) gives

$$\frac{m' + M}{m'} = 68{\cdot}61, \quad \frac{m'}{M} = \frac{1}{67{\cdot}61}.$$

In the preceding determinations $\frac{q}{c}$ and $\frac{q'}{c}$ have been neglected. If these quantities be taken into account, we have for Liverpool,

$$(A) = \cdot 3689 \text{ as before,}$$

$$\frac{(E)\,c}{c-q} = \frac{5 \cdot 75}{2\,(A)} = 7 \cdot 79$$

$$17 \cdot 85 = D + \left\{\frac{c-q}{c-q'} + (A)\right\} \frac{(E)\,c}{(c-q)} = D + 10 \cdot 94$$

$$D = 6 \cdot 91.$$

Hence in both cases we have for Liverpool,

$$\tan 2\,\psi = \frac{\cdot 3689 \sin 2\,\phi}{1 + \cdot 3689 \cos 2\,\phi}$$

but in the first case we have, in feet,

$$h = 7 \cdot 19 + 2 \cdot 87 \cos(2\,\psi - 2\,\phi) + 7 \cdot 79 \cos 2\,\psi, \qquad (a)$$

in the second case

$$h = 6 \cdot 91 + 2 \cdot 87 \cos(2\,\psi - 2\,\phi) + 8 \cdot 07 \cos 2\,\psi. \qquad (b)$$

the difference is so trifling, that it is of little consequence whether one expression or the other be employed. For London

$$(A) = \cdot 3822, \text{ as before,}$$

$$\frac{(E)\,c}{c-q} = \frac{3 \cdot 17}{2\,(A)} = 4 \cdot 17$$

$$22 \cdot 72 = D + \left\{\frac{c-q}{c-q'} + (A)\right\} \frac{(E)\,c}{(c-q)} = D + 5 \cdot 89$$

$$D = 16 \cdot 83.$$

Hence in both cases we have for the London Docks,

$$\tan 2\,\psi = \frac{\cdot 3822 \sin 2\,\phi}{1 + \cdot 3822 \cos 2\,\phi}$$

but in the first case we have, in feet,

$$h = 16 \cdot 97 + 1 \cdot 58 \cos(2\,\psi - 2\,\phi) + 4 \cdot 17 \cos 2\,\psi, \qquad (a)$$

in the second case,

$$h = 16 \cdot 83 + 1 \cdot 58 \cos(2\,\psi - 2\,\phi) + 4 \cdot 31 \cos 2\,\psi. \qquad (b)$$

The value of $\frac{m'}{M}$ which results from Delambre's coefficient of the lunar inequality of the sun's longitude is $\frac{1}{69 \cdot 22}$.—See M. Poisson's *Mémoire sur le Mouvement de la Lune autour de la Terre*, p. 118.

Dr. Brinkley's constant of nutation, lately confirmed by the researches of Dr. Robinson of Armagh, gives $\frac{m'}{M} = \frac{1}{69{\cdot}65}$, but M. de Lindenau's constant gives $\frac{m'}{M} = \frac{1}{88{\cdot}298}$.—See the Mémoire of M. Poisson, p. 126.

Laplace makes $\frac{1}{(A)} = 2{\cdot}35333$, *Méc. Cél.* vol. v. p. 206. from the Brest tide observations, which gives $\frac{m'}{M} = \frac{1}{75}$, but the considerations through which he arrived at this value do not seem free from obscurity.

We cannot account for the difference in the value of (A), as deduced from the London and Liverpool observations, it is therefore impossible to place much reliance upon the method which has been explained, of deducing the moon's mass from the quantity (A). Still less will any advantage ensue in practice from employing for tide-predictions a value of (A) obtained *à priori* from a supposed value of the moon's mass deduced from other considerations.

According to Mr. Mailly, for Ostend

$$\log (A) = 9{\cdot}00300 \qquad (A) = {\cdot}1007$$

While at London

$$\log (A) = 9{\cdot}58229 \qquad (A) = {\cdot}3822.$$

This value of (A) at Ostend is remarkable and difficult to account for.

In cases where the tide arrives by two different channels, the value of the constant (A) may be very different from its value for places not far distant where the tide is single. See *Méc. Cel.* vol. ii p. 225. But Ostend does not appear to be so situated that the value of (A) given above admits obviously of this kind of explanation.

I conceive that the best method of investigating alterations in the height of the land above the water in any given locality, where the water is influenced by the tides, will be to examine carefully whether any alteration has taken place in the values of the constants D and (E) for that place, the height of high water being of course always reckoned from some mark fixed in the land.

The practice which at present obtains of referring the heights of buildings and mountains to the *level of the sea* or to *high or low water mark*, seems objectionable. The heights of *spring* or *neap* tides, although not subject to so much uncertainty, are also quantities too vague to be used with propriety as standards of reference.

It has been usual to suppose that the time of high water at different places differs only by a constant quantity, or that the semi-menstrual inequality has everywhere the same form. This, however is not the case, and, if the figures in Tables I. and VI. be carefully * examined, they will be found to present differences which have not yet been accounted for. Up to the present time the semi-menstrual inequality in the interval has been carefully determined for but few places.

I shall now distinguish successive transits of the moon by the letters *A, B, C, D, E, F.* So that, if

A denotes the time of the moon's transit on Monday morning,
B may denote the time of the moon's transit on Monday afternoon;
C may denote the time of the moon's transit on Tuesday morning,
D may denote the time of the moon's transit on Tuesday afternoon;
E may denote the time of the moon's transit on Wednesday morning,
F may denote the time of the moon's transit on Wednesday afternoon.

I will also suppose that *F* denotes the time of the transit of the moon *immediately preceding* the time of high water at the London Docks. This being the case, if the progress of the tide round the eastern coast of England and Scotland be examined, some place will be found nearly on the same meridian at which high water takes place at the same instant as at London, produced by the succeeding tide-wave. Other arguments might be mentioned, but this is sufficient to show that some distinction should be introduced, and that the transit from which *the establishment* of any port is reckoned should not be left in ambiguity. All discussions of the tides made with a view of obtaining the semi-menstrual inequality, should have reference to one transit. Transit *B* seems to me the most convenient. I am aware that I have departed from this precept myself in the discussion of the Liverpool observations, but it was owing to an inadvertence, which was not discovered until it was too late to be rectified.

I call the time which elapses between the moon's transit and the time of high water at any given place, the INTERVAL; and the time of high water, the HEIGHT.

As the luminaries cease to have any sensible influence upon the tide in narrow seas, the tide has travelled unchanged a considerable distance before it arrives at the river Thames, and hence it is desirable to consider the phenomena with reference to a remote transit.

* The easiest way of examining them is by laying them down in curves on paper ruled in squares.

Let $F + i'$ denote the time of high water, that is, let i' be the *interval* with reference to the transit F, or the time to be added to F, in order to obtain the time of high water at London on any given day. Let i be the interval (for the same tide) with reference to the transit B, so that

$$B + i = F + i'$$
$$B - F = i' - i.$$

It is now evident that if the interval of time which intervenes between the transits B and F were always the same, that is, if $B - F$ were constant, $i' - i$ would be constant also. But $B - F$ varies considerably. The interval between successive transits may be considered constant with reference to the *age of the moon* or time of transit, and depending only upon her parallax and declination.

TABLE showing the interval between the moon's transit and the next succeeding, with a given moon's parallax and declination.

Moon's Parallax.

54′	55′	56′	57′	58′	59′	60′	61′
m 22·6	m 23·2	m 24·1	m 25·1	m 26·1	m 27·1	m 28·0	m 29·0

Moon's Declination.

0°	3°	6°	9°	12°	15°	18°	21°	24°	27°
m 23·2	m 23·3	m 23·5	m 23·8	m 24·3	m 24·9	m 25·6	m 26·3	m 27·1	m 27·9

Bernoulli's theory, upon which my tables are founded, amounts to supposing that if the earth were a perfect sphere, or spheroid, covered throughout by an ocean, the ocean would assume the same form at any given instant as it would do if the forces then acting upon it were invariable in magnitude and direction. In other words, Bernoulli supposes the form of the ocean, encompassing a perfect sphere or spheroid, to be that which it would assume at any moment* if it and the luminaries which produce the tides were at rest. If we suppose this state of things to obtain approximately in the great expanse of water in the southern hemisphere, and that the tide-wave is regularly propagated thence to our coasts with a velocity independent of the positions of the sun and moon, it is

* "Nous supposerons que le surface de la mer prend dans un instant sa juste figure, tout comme si l'eau n'avoit point d'inertie, ni résistances."—*Bernoulli*, p. 135.

remarkable how closely, in many circumstances, the phenomena accord with Bernoulli's theory. For having compared the results furnished by Bernoulli's theory with the results afforded by my discussions of the London and Liverpool observations, with reference to the transit B*, I have ascertained that tables of the parallax and declination corrections, founded upon Bernoulli's theory, may safely be employed, having for their argument that transit. Such tables have been employed in predicting the times and heights of high water in the British Almanac, in the tide tables published by the Hydrographic Office of the Admiralty, and for the times of high-water in the Nautical Almanac.

This agreement with theory is to be received with some limitations: thus, for example, the semi-menstrual inequality in the height † furnished by Bernoulli's theory, differs slightly in *position* from that which results from the observations at the London Docks (see Plate XVIII. Phil. Trans. 1836. Part 2.), and by a quantity greater than can possibly be attributed to any error in the method of discussion employed. The curves would agree better if the theory curve were moved a little to the left, which would amount to altering one of the constants (that which accompanies ϕ and which determines the epoch); but if this alteration were introduced, the semi-menstrual inequality in the *interval* would no longer coincide in position with that furnished by observation ‡. To introduce such an alteration in the inequality of the height and not in that of the interval, would be inconsistent with a rigid interpretation of Bernoulli's theory, with which the law of the inequality is in remarkable accordance. This curious circumstance may be roughly explained by saying that, while the semi-menstrual inequality in *the interval* would lead us to suppose, if we adopt Bernoulli's views, that the tide is due to the attractions of the luminaries at the time of the transit A, the inequality of *the height* seems to be due to the attractions of the luminaries at the time of the transit B. I have ascertained that the semi-menstrual inequality of the height at Liverpool also presents the same curious feature.

* The expense of these laborious calculations, which consist chiefly of adding up *intervals* and *heights* of tides happening under similar circumstances, and taking the averages, has been defrayed out of a grant of money which was placed at my disposal for that purpose by the British Association for the Advancement of Science.

† The *height* and *interval* are variable, but the inequality, which is far more considerable than the rest, has for its period half a lunation. The law of this inequality was discovered by Bernoulli; its existence was known before. Mr. Whewell assigned to it the appropriate designation of the *semi-menstrual* inequality, which has been generally adopted.

‡ According to Bernoulli's theory, the greatest and least heights of high water are coincident with the average *interval*, while according to observation they happen a tide sooner. I avoid the use of the terms *spring* and *neap* tides, as too vague to be admissible where precision is required.

This might be accounted for by supposing the wave of high water of spring tides to travel *more slowly* than the wave of high water of neap tides, whereas, according to Mr. Russell of Edinburgh, the contrary is the case. It is difficult to conceive how the coincidence between theory and observation in the semi-menstrual inequality of the *interval* can take place to so great an extent as is found to exist, unless the difference of velocities of the derived tide-wave at springs and neaps be very inconsiderable.

If I had discussed the observations with reference to the transit *A*, the *retard* would have come out zero, as the *average interval* would have corresponded exactly with the instant of syzygy. This is confirmed by my discussion of the Liverpool tides, and it is certain that, if we adopt Bernoulli's view of the semi-menstrual inequality of *the interval*, the tides on the shores of the Atlantic, which are afterwards propagated round Great Britain, must be considered as due to the attraction of the luminaries at the time of the transit *A*. So that, for example, the tide at the London Docks on Wednesday afternoon is produced (in the Pacific Ocean?) by the attraction of the luminaries on Monday morning. I conclude, therefore, that it is important to ascertain and to define throughout the earth the *interval* with reference either to transit *A* or to transit *B*,

When the moon's parallax = 57′
. . . . . declination = 15°
. . . sun's declination = 15°

and when the moon is in syzygy, that is, when the transit *A* or *B* happens at 12 o'clock *exactly**.

The influence of the change in the interval between the moon's transits upon the different corrections to be applied to a given transit, in order to obtain the interval, appears to have escaped the attention of Bernoulli, and of all writers who have since re-published his tables or described the phenomena. Bernoulli referred the phenomena to transit *A*, in consequence of remarking that the highest spring-tides take place on the coast of France about thirty-six† hours after syzygy; but he supposed that the transit *D* might

* The semi-menstrual inequality in the interval varies here rapidly, about sixteen minutes between $11^h\ 30^m$ and $12^h\ 30^m$, but the parallax and declination corrections are insensible. See the Tables annexed.

† "Nous avons encore fait voir, que sans le concours des causes secondes, les plus grandes marées devroient se faire dans les syzygies, et les plus petites dans les quadratures. Cependant on a observé que les unes et les autres se font un ou deux jours plus tard.... Quoiqu'il en soit, comme ce retardement a été observé le même à-peu près après les syzygies et après les quadratures, nous pouvons encore supposer qu'il est le même pendant toute la révolution de la lune, c'est-à-dire que les marées sont toujours telles qu'elles devroient être sans les dites causes, un ou deux jours auparavant."—*Bernoulli*, p. 158.

be employed with the same parallax correction, which is inconsistent with his own reasoning, quoted at foot. On the coast of France and in the British Channel, the highest spring tides are about two days after syzygy, but at London they are later, for the tides at London are caused by the wave which comes along the eastern coast of England, and which meets off the coast of Kent a tide produced twelve hours later.

If we examine the progress of the tide-wave, we find that the *establishment* of the several places, in the usual acceptation of that term, is as follows :—

	h	m		
Brest	3	48	reckoned from the transit	*D*
Plymouth Dock-Yard	5	33	———	*D*
Isle Brehat	5	52	———	*D*
Pembroke Dock-Yard	6	4	———	*D*
Bristol, Cumberland Gates	7	15	———	*D*
Howth Harbour	11	8	———	*D*
Liverpool Dock	11	25	———	*D*
Portsmouth Dock-Yard	11	40	———	*D*
Leith	2	0	———	*E*
London Docks	1	57	———	*F*

All these determinations were obtained for me by Mr. Dessiou. It is evident from the above table that the *establishments* of ports given in various works (often very inaccurately) are referred to different transits of the moon without distinction, thereby creating great confusion.

Hence also we find (see Table I.) that the tide arrives at

	days	h	m	
Brest	1	4	27	After the transit *B*, when the Moon is in syzygy.
Portsmouth	1	12	21	
Liverpool	1	12	2	
Leith	1	15	15	
London Docks	2	3	16	

So that the tide takes 23 hours 46 minutes in travelling from Brest to the London Docks, and the time of a given high water at Brest may be found nearly by subtracting 23 hours 46 minutes from the time given for the London Docks.

Many details concerning the tides on the coast of Great Britain may be found in the following works :—Directions for Navigating in the North Sea, by M. J. F. Dessiou. Directions for Navigating throughout the English Channel, by the same author. And, Sailing Directions for St. George's Channel, by Messrs. T. and A. Walker.

In the Philosophical Transactions for 1832, I published a map of the world, in which were inserted the times of high water at new and full moon (or the *establishments*), at many places on the globe, collected from various sources, works on navigation, sailing directions, voyages, &c., both *in the time at the place*, and *in the time at Greenwich.* The latter was inserted for the purpose of tracing successively the march of the tide-wave over the surface of the ocean. The crest of the wave travels over the open ocean with immense rapidity, and gives rise to a slow current of the particles of water, with which, however, it must not be confounded.

Mr. Whewell has since pursued this branch of the subject, in a paper published in the Philosophical Transactions for 1833, and, availing himself partly of *à priori* considerations, has drawn in my map the *cotidal lines**, or the lines on the surface of the ocean, throughout which *high water* takes place at the same instant of time. These lines are, however, almost entirely hypothetical, for we have few opportunities of determining the time of high water at a distance from the coast, although this is sometimes possible, by means of a solitary island, as St. Helena. It is evident, by a comparison of the *establishment* of St. Helena, and those of places on the neighbouring coasts of Africa and South America, that the *cotidal line* curves outwards, as represented in the map, although we must continue ignorant of its precise form. We are deficient of information with respect to the course of the tide-wave in the Pacific Ocean, and even on our own coast the number of places of which the *establishment* is accurately known is probably very small. As an example of the kind of information which the annexed map is intended to afford, it may be seen that the tide takes about twelve hours in proceeding from the Cape of Good Hope to Cape Blanco; from thence it reaches Brest in about four hours, its course being northwards up the Atlantic Ocean.

"A flood arriving at the mouth of a river, must act precisely as the great wave does. It must be propagated up the river in a certain time, and we shall have high water at all the different places in succession. This is distinctly seen in rivers. The most remarkable instance of this kind is the Maragnon or Amazon river in South America. It appears by the observations of Condamine and others, that between Para, at the mouth of the river, and the conflux of the Madera and Maragnon, there are seven coexistent high waters, with six low waters between them. Nothing can more evidently show that the tides in these places are nothing but the propagation of a wave."—*Encyclopædia Britt.*

The tides in the Pacific are very anomalous, and we know little about them.

Mr. Williams, of the London Missionary Society, says in his Narrative of Missionary Enterprises, p. 172,

* The cotidal lines and the establishments have been inserted in the six maps of the World published by the Society for the Diffusion of Useful Knowledge.

"It is to the Missionaries a well-known fact that the tides in Tahiti and the Society Islands are uniform throughout the year, both as to the time of the ebb and flow, and the height of the rise and fall, it being high-water invariably at noon and at midnight, and consequently the water is at its lowest point at six o'clock in the morning and evening. The rise is seldom more than eighteen inches or two feet above low-water mark. It must be observed that mostly once and frequently twice in the year a very heavy sea rolls over the reef, and bursts with great violence on the shore. But the most remarkable feature in the periodically high sea is that it invariably comes from the west and south-west, which is the opposite direction to that from which the trade wind blows. The eastern sides of the islands are, I believe, never injured by these periodical inundations. I have been thus particular in my observations, for the purpose, in the first place, of calling the attention of scientific men to this remarkable phenomenon, as I believe it is restricted to the Tahitian and Society groups in the South Pacific and the Sandwich Islands in the north. I cannot, however, speak positively respecting the tides at the islands eastward of Tahiti; but at all the islands I have visited in the same parallel of longitude to the southward, and in those to the westward in the same parallel of latitude, the same regularity is not observed, but the tides vary with the moon, both as to the time and the height of the rise and fall, which is the case at Raratonga. Another reason for which I have been thus minute is to correct the erroneous statements of some scientific visitors. One of these, Kotzebue, observes, 'Every noon, the whole year round, at the moment the sun touches the meridian, the water is the highest, and falls with the sinking sun till midnight.'

"Captain Beechy states, 'The tides in all harbours formed by coral reefs are very irregular and uncertain, and are almost wholly dependent upon the sea breezes. At Outrotauma it is usually low-water about six every morning, and high-water half an hour after noon. To make this deviation from the ordinary course of nature intelligible, it will be better to consider the harbour as a basin, over the margin of which, after the breeze springs up, the sea beats with considerable violence, and throws a larger supply into it than the narrow channels can carry off in the same time, and consequently during that period the tide rises. As the wind abates the water subsides, and the nights being generally calm, the water finds its lowest level by the morning.'

"This statement is certainly most incorrect, for not only have I observed for years the undeviating regularity of the tides, but this is so well understood by the natives, that the hours of the day and night are distinguished by the terms descriptive of its state. As, for example, instead of asking, 'What is the time?' they say 'Where is the tide?' Nor can the tides, as Capt. Beechy says, be 'wholly dependent on the sea breeze,' for there are many days during the year when it is perfectly calm, and yet the tide rises and falls with the same regularity as when the trade winds blow, and we frequently have higher tides in calms than during the prevalence of the trade wind. Besides which, the tides are equally regular on the westward or leeward side of the islands, which the trade wind does not reach, as on the eastward, from which point it blows. But the perfect fallacy of Capt. Beechy's theory will be still more apparent if it be recollected that the trade wind is most powerful from midday till about four or five o'clock, during which time the tide is actually ebbing so fast that the water finds its lowest level by six in the evening, and that in opposition to the strength of the sea breeze. Capt. Beechy adds, 'that the night being calm the water finds its lowest level by the morning'; whereas the fact is, that the water finds its highest point at midnight, when it is perfectly calm. How then can the tides be dependent on the sea-breeze?"

The continent of South America operates as a dam preventing the derived tide-wave from flowing freely into the Pacific. This may serve perhaps to account for these extraordinary anomalies. For further information respecting the march of the Tide-wave I refer to Mr. Whewell's Essay towards a first approximation to a Map of Cotidal Lines.—Phil. Trans. 1833. Part 1.

The *establishment* of London (*i. e.* the interval between moon's transit *F* at $0^h\ 0^m$ and the time of high water) appears now to be very different from what it was in the time of Flamsteed. It seems to have fluctuated more than ten minutes even since the commencement of this century. At the London Docks, in 1807, it was $2^h\ 0{\cdot}9^m$; in 1818, $1^h\ 57{\cdot}0^m$; in 1835, $2^h\ 4{\cdot}4^m$. See Phil. Trans. 1837, p. 136. and the accompanying plate. These numbers were obtained by Mr. E. Russell, and the corrections requisite in order that they might refer to the same parallax and declinations were carefully applied.

This perplexing fluctuation presents an insuperable obstacle to extreme accuracy in tide predictions, until it can be explained. It is probably owing to changes in the bed of the river, the drainage of the banks, &c., which it is impossible to embrace in any mathematical formula. Perhaps the manner of taking the observations may have varied slightly.

I am indebted to Mr. Yates for notice of a very ancient tide table which exists in a MS. in the British Museum. It is in the Codex Cottonianus, Julius DVII., which appears to have been written in the 13th century, and to have belonged to St. Alban's Abbey. It contains calendar and other astronomical or geographical matters, some of which are the productions of John Wallingford, who died Abbot of St. Albans, A.D. 1213. At p. 45 b. is a table on one leaf, showing the time of high water at London Bridge, "flod at london brigge", thus:

Ætas Lunæ.	h	m
1	3	48
2	4	36
3	5	24
4	6	12
......	......	......
......	......	......
28	1	24
29	2	12
30	3	0

N.B. The numbers increase by a constant difference of forty-eight minutes. The first column gives the moon's age in days.

Hence it would appear that high water at London on full and change was at that epoch $3^h\ 48^m$, or more than an hour later than at present. The time of high water at London on full and change is given in Mr. Riddle's *Navigation* and in other works $2^h\ 45^m$; Flamsteed made it 3^h.

Comparison of Theory with Observation.

The expression for V, p. 14, contains three terms :*

(1.) A term independent of μ, this appears to give rise to no sensible inequality on our coasts.

(2.) This term gives rise to the *diurnal* or *menstrual inequality*.

(3.) This term gives rise to the *semi-diurnal* or *semi-menstrual inequality*, which is by far the most considerable on the coasts of Great Britain.

In order to obtain a comparison of theory with observation, I separated the observations into various classes or categories, having reference to the particular inequalities I sought to determime. This plan is essentially the same as that pursued by Laplace in the *Mécanique Céleste.* But the results given in the Phil. Trans. are not only more extensive and founded upon a greater number of observations, but Laplace has omitted in taking the average of the results of the observations to take also the averages of the corresponding transits of the moon, so that the argument or age of the moon to which they refer is not determined with precision.

The following example, copied from the computation books containing the discussion of the London Dock observations, will serve to explain the method I employed. These books are deposited in the Library of the Royal Society. Those containing the discussion of the observations made at Liverpool by Mr. Hutchinson are deposited in the Athenæum at that place.

* "Ces oscillations sont de trois espèces. Celles de la première espèce sont indépendantes du mouvement de rotation de la terre, et leur détermination offre peu de difficultés. Les oscillations dépendantes de la rotation de la terre, et dont le période est d environ un jour, forment la seconde espèce; enfin la troisième espèce est composée des oscillations dont le période est à peu près d'un demi-jour. Elles surpassent considérablement les autres, dans nos ports."—*Exposition du Système du Monde*, p. 279.

January, 0 to 1 hour.						
Upper Transit; A.M.						
Date.	☽'s Transit *B.*	Time of H. W.	Height of H. W.	☽'s Dec.	☽'s H. P.	☉'s Dec.
	h m	h m	ft. in.	°	′	°
1808, Jan. 14	0 16	3 25	22 0	N. 16	59	S. 21
1809, Jan. 2	0 3	3 15	22 8	N. 18	55	S. 23
1825, Jan. 6	0 58	3 58	23 6	N. 18	60	S. 23
1826, Jan. 25	0 48	3 47	22 5	N. 11	59	S. 19
No. of Obs. 23	699	757	59 7	433	1314	485
Mean	0 30·4	3 32·9 0 30·4	22 7·1	N. 18·8	57·1	S. 21·1
		3 2·5				

Hence I conclude that with reference to the moon's transit *B* at 30^{m}·4 A.M. apparent solar time,

The moon's declination being . . . . N. 18°·8
The moon's parallax being . . . . . 57′·1
The sun's declination being . . . . . S. 21·1;

the INTERVAL between the moon's transit *B* and the corresponding high water at the London Docks is 2 days 3 hours and 2·5 minutes, and that the HEIGHT of the high-water is 22 feet and 7·1 inches.

The moon's transit in the above example is given in apparent solar time taken from the Nautical Almanac. The time of high water is also given in apparent solar time by adding the equation of time to the time of high water in the books containing the London Dock observations, presuming the observations to be recorded in mean solar time.

The moon's parallax and declination are given for the time of the moon's transit, and were obtained by interpolation from those quantities as given for noon in the Nautical Almanac.

The results so obtained were reduced by interpolation to the even half hour, and a slight correction was introduced where it became necessary in order to refer them all to parallax 57′. The semi-menstrual inequality was found by taking the average of all the calendar months.

The number of observations employed was so considerable, that the moon's parallax corresponding to each of these categories belonging to the calendar months never deviated much from 57′.

When the semi-menstrual inequality is sought to be determined from a few observations this correction is sensible, and strictly should not be omitted. The labour is not much increased by taking out the H. P. corresponding to the moon's transit.

The following Table shows a comparison between the semi-menstrual inequality at the London Docks in the INTERVAL and in the HEIGHT, as deduced from theory and from results of observation.

SEMI-MENSTRUAL INEQUALITY AT LONDON.

Moon's Horizontal Parallax . . . . = 57′.
Moon's Declination . . . . = 15°.
Sun's Horizontal Parallax . . . . . = 8″·8.
Sun's Declination . . . . = 15°.

Apparent Solar Time of Moon's Transit *B*.	Interval $\psi + 2^d\ 3^h\ 8^m{\cdot}4$ Theory.	Interval Observation.	Height. *h*. Theory.	Height. Observation.	$-\phi$
h m	d h m	d h m	feet.	feet.	h m
0 30	2 3 7·0	2 3 7·1	22·72	22·72	0 5
1 30	2 2 50·7	2 2 50·9	22·54	22·44	1 5
2 30	2 2 36·3	2 2 36·5	22·07	21·92	2 5
3 30	2 2 26·0	2 2 26·7	21·36	21·14	3 5
4 30	2 2 24·2	2 2 24·0	20·55	20·23	4 5
5 30	2 2 38·5	2 2 37·5	19·83	19·57	5 5
6 30	2 3 11·5	2 3 10·8	19·56	19·55	6 5
7 30	2 3 42·1	2 3 41·5	19·93	20·26	7 5
8 30	2 3 53·2	2 3 53·4	20·69	21·15	8 5
9 30	2 3 49·6	2 3 50·4	21·49	21·89	9 5
10 30	2 3 38·3	2 3 39·0	22·16	22·42	10 5
11 30	2 3 23·4	2 3 23·6	22·59	22·70	11 5

The mean interval is $2^d\ 3^h\ 8^m{\cdot}4$, which corresponds to the Moon's transit *B* at $0^h\ 25^m{\cdot}2$ or $0^h\ 25^m$ nearly. This shows that if the observations had been discussed with reference to transit *A*, the epoch would have come out $0^h\ 0^m$.

The columns headed theory have been calculated from the expressions

$$\tan 2\psi = \frac{\cdot 3822 \sin 2\phi}{1 + \cdot 3822 \cos 2\phi},$$

$$h = 16{\cdot}97 + 1{\cdot}58 \cos(2\psi - 2\phi) + 4{\cdot}17 \cos 2\psi. \qquad (a)$$

The agreement between theory and observation in the inequality in the INTERVAL is very remarkable.

The inequality in the HEIGHT furnished by theory agrees in *form* with that given by the observations, but it appears to deviate in *position*.

In order to diminish the irregularities in the moon's parallax inequality, as deduced from observation, and to employ the concourse of all the observations, I have employed the following method. Let δ P be the difference of parallax, or

The ☽'s horizontal parallax − 57′.

I suppose the parallax correction to be proportional to δ P; hence the correction for parallax 54′ = three times the correction for parallax 56′, and the total of the absolute corrections for parallaxes 54′, 55′, 56′, 58′, 59′, 60′, 61′ = $\frac{16}{3}\times$ by the correction for parallax 54′. Whatever be the law of the parallax correction, it may certainly be considered as proceeding according to powers of δ P; and the preceding hypothesis amounts to neglecting all the powers except the first. I next employ only the total of the corrections deduced from the discussions, and I multiply it by $\frac{3}{16}$, or the equivalent multiplier, in order to have the correction for 54′.

This has been executed by Mr. Russell, and the results are exhibited in the following Table.

TABLE showing a comparison between the Moon's Parallax Inequality resulting from Theory, and that resulting from Observations at the London Docks.

H. P. 54′.

Moon's Transit B.	Inequality of the Interval.		Inequality of the Height.	
	Theory.	Observation.	Theory.	Observation.
h m	m	m	ft.	ft.
0 30	0·0	− 1·5	− ·66	− ·55
1 30	− 2·0	− 2·3	·66	·62
2 30	− 4·2	− 5·8	·64	·62
3 30	− 6·5	− 10·9	·62	·75
4 30	− 8·7	− 13·2	·61	·77
5 30	− 8·4	− 12·2	·64	·96
6 30	0·0	− 4·7	·66	·90
7 30	+ 8·4	+ 1·9	·64	·78
8 30	+ 8·7	+ 3·9	·61	·64
9 30	+ 6·5	+ 4·1	·62	·59
10 30	+ 4·2	+ 2·6	·64	·58
11 30	+ 2·0	+ 1·0	− ·66	− ·53

Diagram showing the Moon's Parallax Inequality.

H.P. 54′.

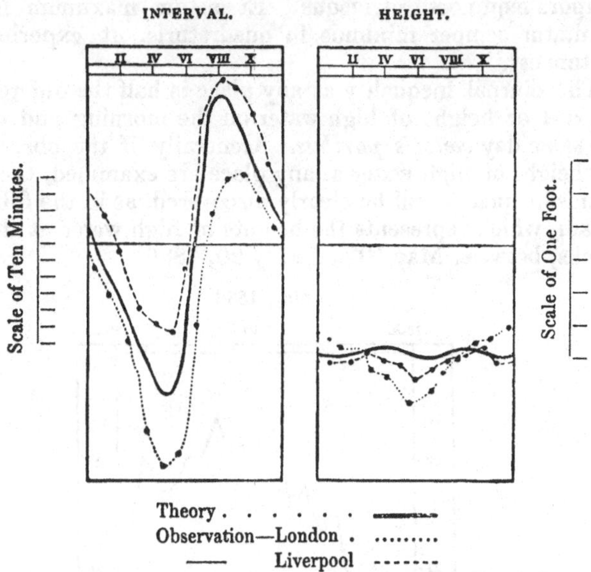

In these curves the Abscissa represents the Moon's Transit *B*.

Each of the Liverpool *dots* in the preceding diagrams may be considered as representing the mean of more than 1000 observations, and each of the London *dots* as representing the mean of more than 2000 observations.

The London *interval* curve, although agreeing in form with the Liverpool interval curve, differs from it throughout by several minutes. This difference seems to me remarkable. The height curves agree closely, showing that the height inequality varies as the quantity (E), as I have supposed.

Laplace says "Elles [les marées] augmentent et diminuent avec le diamètre et le parallaxe lunaire, *mais dans un plus grand rapport*"; but this is true only at *neap* tides.

The calendar month inequality results implicitly from the inequalities due to changes in the declinations of the luminaries and in the sun's parallax, and agrees generally with the equilibrium theory. (See the Report of the Seventh Meeting of the British Association, p. 108.) The spring equinoctial tides are greater than the neap equinoctial tides, and the neap solstitial tides are greater than the spring solstitial tides, as is stated by Laplace in the *Exposition du Système du Monde*, 5e ed., p. 83, and by Newton: "In quadraturis autem solstitialibus majores ciebunt æstus quam in quadraturis æquinoctialibus, eo quod Lunæ jam in æquatore constitutæ effectus maxime superat effectum Solis. Incidunt igitur æstus maximi in syzygias et minimi in quadraturas luminarium, circa tempora æquinoctii utriusque. Et æstum maximum in syzygiis comitatur semper minimus in quadraturis, ut experientia compertum est."

The diurnal inequality at any place is half the difference in the interval or height of high water on the morning and evening of the same day *cæteris paribus*. Generally, if the observations of the height of high water at any place are examined, the existence of this inequality will be clearly discovered, as in the following diagram, which represents the heights of high water at the London Docks, between May 10 and May 20, 1836.

May, 1836.

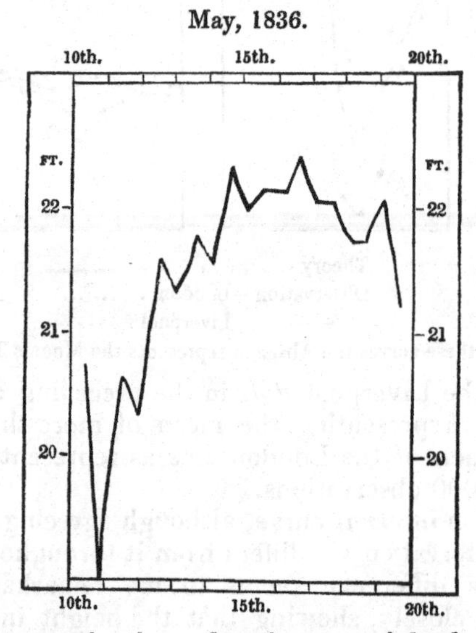

On a cursory examination, the zigzags might be attributed to the wind, the carelessness of the observers, or to other accidental causes. Mr. Whewell was the first to notice these irregularities, and to refer them to their true cause, as belonging chiefly to the *diurnal inequality*.

Mr. Whewell also remarked that the *diurnal inequality* in the height may be represented by the expression

$$d\,h = C \sin 2\,\delta',$$

which is, in fact, an approximate form of the expression

$$d\,h = B\,\{A \sin 2\,\delta \cos(\psi - \phi) + \sin 2\,\delta' \cos\psi\},$$

since the angle ψ is nearly constant at a given place for the time of high water. See p. 18. C being constant for any given place.

Probably the amount depends also upon the moon's parallax, and then the expression for d h will be

$$C \times \frac{P'^3}{(57')^3} \sin 2\,\delta' \text{ (for a given transit A.M. or P.M.).}$$

But this expression will not afford results agreeing with those which I have obtained from the observations, if the declination of the moon be employed belonging to the time of the transit B, and it is necessary to employ the moon's declination at some time previous. This is not at variance with what is stated in the *Exposition*, except that, although Laplace considers the two waves separately *, he has not, I think, referred distinctly to the change in the epoch for different places, or to the difference between the epoch of the original diurnal and semidiurnal waves, which produce the derived tides observed on our coasts. If, however, the diurnal inequality-wave travels more slowly than the semidiurnal inequality-wave, the epoch also will be different, and thus it may depend upon the moon's declination several days earlier.

If this view be correct, the diurnal inequality of high water has a maximum (geographically) at those places on the coast at which the diurnal inequality-wave and the semidiurnal inequality-wave arrive simultaneously, and there will be places intermediate at which the diurnal inequality of high water is imperceptible, but where the diurnal inequality of low water is a maximum. This theory agrees with observation in giving no difference in the diurnal inequality for upper or lower transits.

The diurnal inequality in the interval is inappreciable on our coasts.

By examining the results which were afforded by the London and Liverpool discussions, I discovered that the diurnal inequality, in passing from Liverpool to London, becomes reversed, that is to say, if a and b denote two successive heights of high water at Liverpool, and a' and b' successive heights at London, *caused by the same tides*:

$$\text{if } a > b, \text{ then, generally, } a' < b'.$$

—See Plate IV. Phil. Trans. 1837. Part I. It has sometimes been supposed that the tide flows on unchanged in character, but the march of the diurnal wave differs from that of the semidiurnal inequalities, and must be examined separately. I think it will be found, that at certain places on the coast (as at Leith?) the diurnal inequality in the height of high water will be found imperceptible, while precisely at those places the diurnal inequality in the height of low water will be found considerable. Mr. Whewell however

* "J'ai déterminé la grandeur de ce flux et l'heure de son *maximum* dans le port de Brest. J'ai trouvé un cinquième de mètre [7·4 inches] à fort peu près pour sa grandeur ; et un dixième de jour environ, pour le temps dont il précède à Brest, l'heure du *maximum* de la marée semidiurne."—*Exposition du Système du Monde*, 5e ed., p. 286.)

is of opinion that the differences of the diurnal inequality at different places are governed by local circumstances, and that the motion of the diurnal wave is extremely irregular. See the Phil. Trans. 1837, p. 231. Observations are wanted to settle these points, which are not unimportant with reference to navigation. I believe that much of the apparent irregularity manifested by observations of the tides which has been attributed to carelessness on the part of the observers, has really been due to the diurnal inequality.

Method of predicting the Time and Height of High Water.

As the *semi-menstrual inequality* in the INTERVAL deduced from observation is, *when the proper constants are employed*, identical with that given by Bernoulli's expression, it is immaterial whether in the predictions of the interval, a table be used deduced solely from observation or from Bernoulli's expression. The same may be said of the inequality due to changes in the moon's distance from the earth (*moon's parallax inequality*), the transit *A* or *B* being employed for the argument, which is indispensable.

The inequality due to changes in the declination of the luminaries (*calendar month inequality*) is more minute in amount, but it accords generally with Bernoulli's theory.

The form or law of the *semi-menstrual inequality* in the HEIGHT is also identical in theory with that deduced from the observations, but the epoch appears to be different from that given by the interval. Hence, in predictions of the height of high water, either the semi-menstrual inequality deduced empirically may be employed, or the epoch of the inequality as deduced from the interval must be changed. See Plate XVIII. Phil. Trans. 1836. Part I.

The inequalities in the height of high water due to changes in the moon s distance from the earth (*moon's parallax inequality*) and to changes in the declinations of the luminaries (*calendar month inequality*) accord generally with Bernoulli's theory. See Plates I. and XI. Phil. Trans. 1837.

The following tables will serve to calculate the time and height of high water generally, without taking into account the diurnal inequality which appears to vary at different places, and has not yet been ascertained with sufficient precision except at a few ports. In order to determine it, the height of high water may be calculated from the following tables, omitting that correction, and an examination of the errors will then serve to show its amount and afford the means of estimating the constant *C*.

The *diurnal inequality* in the time of high water on our coasts is too minute to be detached from the inevitable errors of observation.

Tables to be used in Predicting the Time of High Water.

TABLE I.—Showing the Semi-menstrual Inequality + a constant in the Interval between the Moon's Transit and the time of High Water, with reference to the apparent Solar time of the Moon's Transit B, the Moon's Parallax being 57′, her Declination 15°, the Sun's Parallax 8″·8, and Declination 15°.

Moon's Transit *B.*	Brest Harbour.	Plymouth Dock-yard.	Portsmouth Dock-yard.	Sheerness Dock-yard.	London Docks.	Pembroke Dock-yard.	Bristol Cumberland Gates.	Liverpool Docks.	Howth Harbour.	Leith Docks.	Moon's Transit *B.*
h m	d h m	d h m	d h m	d h m	d h m	d h m	d h m	d h m	d h m	d h m	h m
0 0	1 4 27	1 6 12	1 12 21	2 1 48	2 3 16	1 6 42	1 7 53	1 12 2	1 11 43	1 15 15	0 0
0 30	1 4 18	1 6 5	1 12 13	2 1 40	2 3 7	1 6 35	1 7 45	1 11 54	1 11 32	1 15 8	0 30
1 0	1 4 11	1 5 56	1 12 6	2 1 32	2 2 59	1 6 27	1 7 36	1 11 46	1 11 22	1 15 0	1 0
1 30	1 4 3	1 5 47	1 11 58	2 1 24	2 2 51	1 6 20	1 7 29	1 11 40	1 11 13	1 14 53	1 30
2 0	1 3 56	1 5 38	1 11 50	2 1 16	2 2 43	1 6 12	1 7 19	1 11 33	1 11 5	1 14 47	2 0
2 30	1 3 48	1 5 29	1 11 43	2 1 10	2 2 36	1 6 6	1 7 12	1 11 27	1 11 0	1 14 40	2 30
3 0	1 3 43	1 5 20	1 11 37	2 1 4	2 2 30	1 6 1	1 7 4	1 11 22	1 10 58	1 14 36	3 0
3 30	1 3 40	1 5 12	1 11 33	2 1 0	2 2 26	1 5 56	1 6 56	1 11 17	1 10 58	1 14 34	3 30
4 0	1 3 39	1 5 6	1 11 30	2 0 58	2 2 24	1 5 52	1 6 48	1 11 14	1 11 0	1 14 34	4 0
4 30	1 3 40	1 5 3	1 11 30	2 0 57	2 2 24	1 5 50	1 6 44	1 11 14	1 11 2	1 14 37	4 30
5 0	1 3 33	1 5 1	1 11 33	2 1 3	2 2 29	1 5 49	1 6 43	1 11 17	1 11 7	1 14 45	5 0
5 30	1 3 51	1 5 4	1 11 40	2 1 11	2 2 38	1 5 54	1 6 47	1 11 27	1 11 16	1 14 56	5 30
6 0	1 4 3	1 5 14	1 11 55	2 1 26	2 2 56	1 6 5	1 6 55	1 11 43	1 11 25	1 15 13	6 0
6 30	1 4 18	1 5 31	1 12 12	2 1 44	2 3 11	1 6 18	1 7 10	1 12 0	1 11 40	1 15 30	6 30
7 0	1 4 32	1 5 52	1 12 25	2 1 59	2 3 27	1 6 37	1 7 35	1 12 15	1 11 55	1 15 42	7 0
7 30	1 4 43	1 6 9	1 12 37	2 2 16	2 3 42	1 6 54	1 7 57	1 12 28	1 12 10	1 15 53	7 30
8 0	1 4 52	1 6 22	1 12 47	2 2 22	2 3 50	1 7 2	1 8 9	1 12 36	1 12 19	1 15 58	8 0
8 30	1 4 58	1 6 31	1 12 51	2 2 27	2 3 53	1 7 9	1 8 14	1 12 40	1 12 22	1 15 57	8 30
9 0	1 4 59	1 6 36	1 12 53	2 2 25	2 3 52	1 7 12	1 8 16	1 12 40	1 12 24	1 15 55	9 0
9 30	1 4 57	1 6 36	1 12 50	2 2 23	2 3 50	1 7 11	1 8 16	1 12 38	1 12 22	1 15 52	9 30
10 0	1 4 53	1 6 34	1 12 46	2 2 18	2 3 45	1 7 10	1 8 14	1 12 33	1 12 18	1 15 47	10 0
10 30	1 4 48	1 6 30	1 12 40	2 2 12	2 3 38	1 7 4	1 8 11	1 12 26	1 12 12	1 15 40	10 30
11 0	1 4 40	1 6 24	1 12 33	2 2 4	2 3 31	1 6 57	1 8 6	1 12 20	1 12 3	1 15 33	11 0
11 30	1 4 33	1 6 19	1 12 27	2 1 57	2 3 23	1 6 48	1 8 0	1 12 10	1 11 53	1 15 24	11 30

The semi-menstrual inequality for each place was found for me by Mr. Dessiou, with reference to the transit immediately preceding :

For Brest	from	532	observations
Plymouth Dock-yard . .	„	538	„
Pembroke Dock-yard. .	„	697	„
Bristol Cumberland Gates	„	1395	„
Howth Pier	„	254	„
Portsmouth Dock-yard .	„	689	„
Leith Harbour	„	705	„

These are I believe the only places for which the semi-menstrual inequality has yet been carefully ascertained, with the exception (very recently) of Antwerp, Nieuport, Ostend and Blankenberg, on the coast of Belgium by Mr. E. Mailly.

The numbers given for London and Liverpool are accurate, but those for the other places can only be considered as approximations, especially as the precaution of reducing to the same parallax and declination (see p. 20) was omitted. They present differences which seem beyond the limits of error arising from the method of discussion, and which deserve investigation.

TABLE II.—Showing the Correction for the Moon's Parallax.

Moon's Transit *B.*	H. P. 54′.	H. P. 55′.	H. P. 56′.	H. P. 57′.	H. P. 58′.	H. P. 59′.	H. P. 60′.	H. P. 61′.	Moon's Transit *B.*
h	m	m	m	m	m	m	m	m	h
0	+1	+1	+1	0	0	−1	−1	−1	0
1	−1	−1	−1	0	0	+1	+1	+1	1
2	−3	−2	−1	0	+1	+2	+3	+4	2
3	−5	−3	−1	0	+1	+3	+5	+7	3
4	−7	−5	−2	0	+2	+4	+6	+8	4
5	−9	−6	−3	0	+2	+5	+7	+9	5
6	−4	−2	−1	0	+1	+2	+3	+4	6
7	+4	+2	+1	0	−1	−2	−3	−4	7
8	+9	+6	+3	0	−2	−5	−7	−9	8
9	+7	+5	+2	0	−2	−4	−6	−8	9
10	+5	+3	+1	0	−1	−3	−5	−7	10
11	+3	+2	+1	0	−1	−2	−3	−4	11

TABLE III.—Showing the Correction for the Moon's Declination.

Moon's Transit *B.*	0° Dec.	3° Dec.	6° Dec.	9° Dec.	12° Dec.	15° Dec.	18° Dec.	21° Dec.	24° Dec.	27° Dec.	30° Dec.	Moon's Transit *B.*
h	m	m	m	m	m	m	m	m	m	m	m	h
0	−1	−1	0	0	0	0	+1	+1	+1	+1	+ 2	0
1	+1	+1	0	0	0	0	−1	−1	−1	−1	− 2	1
2	+2	+2	+1	+1	+1	0	−1	−1	−2	−3	− 4	2
3	+3	+3	+3	+2	+1	0	−1	−2	−3	−5	− 7	3
4	+3	+3	+3	+2	+1	0	−1	−3	−5	−7	−10	4
5	+3	+3	+3	+2	+1	0	−2	−4	−6	−9	−12	5
6	+2	+2	+2	+1	+1	0	−1	−2	−4	−4	− 5	6
7	−2	−2	−2	−1	−1	0	+1	+2	+4	+4	+ 5	7
8	−3	−3	−3	−2	−1	0	+2	+4	+6	+9	+12	8
9	−3	−3	−3	−2	−1	0	+1	+3	+5	+7	+10	9
10	−3	−3	−3	−2	−1	0	+1	+2	+3	+5	+ 7	10
11	−2	−2	−1	−1	−1	0	+1	+1	+2	+3	+ 4	11

TABLE IV.—Showing the Correction for the Sun's Declination.

Moon's Transit *B.*	0° Dec.	3° Dec.	6° Dec.	9° Dec.	12° Dec.	15° Dec.	18° Dec.	21° Dec.	24° Dec.	Moon's Transit *B.*
h	m	m	m	m	m	m	m	m	m	h
0	0	0	0	0	0	0	0	0	0	0
1	0	0	0	0	0	0	0	0	+1	1
2	−1	−1	−1	−1	0	0	0	+1	+2	2
3	−2	−1	−1	−1	0	0	+1	+2	+4	3
4	−3	−2	−1	−2	0	0	+2	+3	+5	4
5	−3	−2	−2	−1	−1	0	+2	+3	+5	5
6	−1	−1	−1	−1	0	0	+1	+2	+2	6
7	+1	+1	+1	+1	0	0	−1	−2	−2	7
8	+3	+2	+2	+1	+1	0	−2	−3	−5	8
9	+3	+2	+1	+2	0	0	−2	−3	−5	9
10	+2	+1	+1	+1	0	0	−1	−2	−4	10
11	+1	+1	+1	+1	0	0	0	−1	−2	11

TABLE V.—Showing the Correction for the Sun's Parallax.

Moon's Transit *B.*	Jan. Dec.	Feb. Nov.	March, Oct.	April, Sep.	May, August.	June, July.	Moon's Transit *B.*
	H. P. 8″·94	H. P. 8″·90.	H. P. 8″·84.	H. P. 8″·76.	H. P. 8″·70.	H. P. 8″·66.	
h	m	m	m	m	m	m	h
0	0	0	0	0	0	0	0
1	0	0	0	0	0	0	1
2	−1	−1	0	0	0	0	2
3	−1	−1	0	0	+1	+1	3
4	−2	−2	0	0	+2	+2	4
5	−3	−2	−1	+1	+3	+3	5
6	−2	−1	−1	+1	+2	+2	6
7	+2	+1	+1	−1	−2	−2	7
8	+3	+2	+1	−1	−3	−3	8
9	+2	+2	0	0	−2	−2	9
10	+1	+1	0	0	−1	−1	10
11	+1	+1	0	0	0	0	11

Tables to be used in predicting the Height of High Water.

TABLE VI.—Showing the Semi-menstrual Inequality + a constant in the Height of High Water with reference to the apparent Solar Time of the Moon's Transit B, the Moon's Parallax being 57′, her Declination 15°, the Sun's Parallax 8″·8, and Declination 15°. (*Empirical.*)

Moon's Transit *B.*		Plymouth Dock-yard.	Portsmouth Dock-yard.	Sheerness Dock-yard.	London Docks.	Pembroke Dock-yard.	Bristol Cumberland Gates.	Liverpool Docks.	Leith Docks.	Moon's Transit *B.*	
h	m	Feet.	Feet.	Feet.	Feet.	Feet.	Feet.	Feet.	Feet.	h	m
0	0	17·87	19·09	25·63	22·71	22·77	32·23	17·57	16·29	0	0
0	30	17·85	19·05	25·56	22·72	22·73	32·25	17·62	16·18	0	30
1	0	17·72	18·97	25·50	22·58	22·55	32·16	17·50	16·00	1	0
1	30	17·58	18·82	25·33	22·44	22·25	31·77	17·17	15·78	1	30
2	0	17·38	18·62	25·15	22·18	21·86	31·08	16·75	15·54	2	0
2	30	17·10	18·46	24·92	21·92	21·37	30·25	16·30	15·25	2	30
3	0	16·75	18·25	24·65	21·53	20·84	29·38	15·73	14·88	3	0
3	30	16·37	18·00	24·25	21·14	20·21	28·20	15·10	14·42	3	30
4	0	15·99	17·71	23·86	20·68	19·54	26·82	14·37	13·92	4	0
4	30	15·50	17·45	23·50	20·23	18·88	25·27	13·65	13·46	4	30
5	0	15·04	17·12	23·20	19·90	18·11	23·91	13·05	13·05	5	0
5	30	14·70	16·86	23·00	19·57	17·67	22·53	12·65	12·73	5	30
6	0	14·52	16·69	22·96	19·56	17·40	21·72	12·40	12·58	6	0
6	30	14·46	16·65	23·08	19·55	17·29	21·73	12·32	12·67	6	30
7	0	14·63	16·83	23·31	19·84	17·46	22·39	12·55	12·87	7	0
7	30	15·00	17·16	23·64	20·26	17·92	22·88	13·10	13·21	7	30
8	0	15·50	17·49	23·94	20·71	18·71	25·25	13·86	13·60	8	0
8	30	16·00	17·79	24·27	21·15	19·50	26·60	14·55	14·08	8	30
9	0	16·42	18·11	24·58	21·52	20·12	27·90	15·20	14·61	9	0
9	30	16·82	18·37	24·92	21·89	20·75	29·00	15·80	15·14	9	30
10	0	17·12	18·58	25·20	22·15	21·29	30·00	16·35	15·59	10	0
10	30	17·42	18·75	25·42	22·42	21·83	30·77	16·85	15·92	10	30
11	0	17·67	18·90	25·58	22·56	22·27	31·50	17·20	16·18	11	0
11	30	17·82	19·03	25·63	22·70	22·60	32·05	17·45	16·30	11	30

The *height* in this Table is reckoned as follows :—

At Plymouth, from a mark which is 2 feet above the sill of the North New Dock Gates.

At Portsmouth, from the sill of the North Dock Gates.

At Sheerness, from the entrance of the Basin, 31 feet below Lloyd's standard mark (+XXXI.) on the Quay.

At the London Docks, from the sill of the Gates at the Wapping entrance.

At Pembroke, from the marks cut in the stone at the entrance of the Dock.

At Bristol, from the marks at the Cumberland Gates.

At Liverpool, from the datum on the East Wall of the Canning Dock.

At Leith, from the sill at the entrance of the Docks.

Table VII.—Showing the Correction for the Moon's Parallax.

Moon's Transit *B.*	H.P. 54′	H.P. 55′	H.P. 56′	H.P. 57′	H.P. 58′	H.P. 59′	H.P. 60′	H.P. 61′	Moon's Transit *B.*
h.	feet.	feet.	feet.	feet.	feet.	feet.	feet.	feet.	h.
0	−·66	−·45	−·23	0	+·24	+·49	+·74	+1·00	0
1	−·66	−·45	−·23	0	+·24	+·49	+·74	+1·00	1
2	−·65	−·44	−·23	0	+·23	+·47	+·72	+ ·98	2
3	−·63	−·43	−·22	0	+·22	+·46	+·71	+ ·96	3
4	−·61	−·42	−·21	0	+·22	+·45	+·69	+ ·94	4
5	−·63	−·43	−·22	0	+·23	+·46	+·70	+ ·96	5
6	−·65	−·45	−·23	0	+·24	+·48	+·73	+ ·99	6
7	−·65	−·45	−·23	0	+·24	+·48	+·73	+ ·99	7
8	−·63	−·43	−·22	0	+·23	+·46	+·70	+ ·96	8
9	−·61	−·42	−·21	0	+·22	+·45	+·69	+ ·94	9
10	−·63	−·43	−·22	0	+·22	+·46	+·71	+ ·96	10
11	−·65	−·44	−·23	0	+·23	+·47	+·72	+ ·98	11

Table VIII.—Showing the Correction for the Moon's Declination.

Moon's Transit *B.*	0° Dec.	3° Dec.	6° Dec.	9° Dec.	12° Dec.	15° Dec.	18° Dec.	21° Dec.	24° Dec.	27° Dec.	30° Dec.	Moon's Transit *B.*
h.	feet.	feet.	feet.	feet.	feet.	feet.	feet.	feet.	feet.	feet.	feet.	h.
0	+·32	+·31	+·27	+·21	+·12	0	−·13	−·29	−·47	−·66	−·87	0
1	+·32	+·31	+·27	+·21	+·12	0	−·13	−·29	−·47	−·66	−·87	1
2	+·31	+·30	+·26	+·20	+·11	0	−·13	−·28	−·46	−·65	−·86	2
3	+·30	+·30	+·26	+·20	+·10	0	−·13	−·28	−·45	−·63	−·83	3
4	+·30	+·29	+·25	+·19	+·11	0	−·13	−·27	−·44	−·61	−·80	4
5	+·30	+·29	+·25	+·19	+·11	0	−·13	−·27	−·44	−·61	−·79	5
6	+·31	+·30	+·26	+·20	+·12	0	−·13	−·28	−·46	−·63	−·83	6
7	+·31	+·30	+·26	+·20	+·12	0	−·13	−·28	−·46	−·63	−·83	7
8	+·30	+·29	+·25	+·19	+·11	0	−·13	−·27	−·44	−·61	−·79	8
9	+·30	+·29	+·25	+·19	+·11	0	−·13	−·27	−·44	−·61	−·80	9
10	+·30	+·30	+·26	+·20	+·10	0	−·13	−·28	−·45	−·63	−·83	10
11	+·31	+·30	+·26	+·20	+·11	0	−·13	−·28	−·46	−·65	−·86	11

Table IX.—Showing the Correction for the Sun's Declination.

Moon's Transit *B.*	0° Dec.	3° Dec.	6° Dec.	9° Dec.	12° Dec.	15° Dec.	18° Dec.	21° Dec.	24° Dec.	Moon's Transit *B.*
h.	feet.	feet.	feet.	feet.	feet.	feet.	feet.	feet.	feet.	h.
0	+·12	+·11	+·10	+·08	+·04	00	−·05	−·11	−·18	0
1	+·12	+·11	+·10	+·08	+·04	00	−·05	−·11	−·18	1
2	+·10	+·09	+·08	+·06	+·04	00	−·05	−·10	−·15	2
3	+·07	+·06	+·05	+·04	+·02	00	−·03	−·06	−·10	3
4	+·01	+·01	+·01	+·01	00	00	−·01	−·01	−·02	4
5	−·06	−·06	−·05	−·04	−·02	00	+·03	+·05	+·08	5
6	−·11	−·10	−·10	−·07	−·04	00	+·05	+·10	+·15	6
7	−·11	−·10	−·10	−·07	−·04	00	+·05	+·10	+·15	7
8	−·06	−·06	−·05	−·04	−·02	00	+·03	+·05	+·08	8
9	+·01	+·01	+·01	+·01	00	00	−·01	−·01	−·02	9
10	+·07	+·06	+·05	+·04	+·02	00	−·03	−·05	−·10	10
11	+·10	+·09	+·08	+·06	+·04	00	−·05	−·10	−·15	11

TABLE X.—Showing the Correction for the Sun's Parallax.

Moon's Transit *B.*	Jan. Dec. H.P. 8″·94	Feb. Nov. H.P. 8″·90	March, October. H.P. 8″·84	April, Sept. H.P. 8″·76	May, August. H.P. 8″·70	June, July. H.P. 8″·66	Moon's Transit *B.*
h.	feet.	feet.	feet.	feet.	feet.	feet.	h.
0	+·9	+·06	+·03	−·03	−·05	−·08	0
1	+·9	+·06	+·03	−·03	−·05	−·08	1
2	+·07	+·04	+·02	−·02	−·04	−·07	2
3	+·04	+·03	+·01	−·01	−·03	−·04	3
4	+·01	+·01	+·01	−·01	−·01	−·01	4
5	−·04	−·03	−·01	+·01	+·02	+·03	5
6	−·08	−·05	−·03	+·03	+·05	+·07	6
7	−·08	−·05	−·03	+·03	+·05	+·07	7
8	−·04	−·03	−·01	+·01	+·02	+·03	8
9	+·01	+·01	+·01	−·01	−·01	−·01	9
10	+·04	+·03	+·01	−·01	−·03	−·04	10
11	+·07	+·04	+·02	−·02	−·04	−·07	11

TABLE XI.—Showing the Diurnal Inequality in the Height, $C = \cdot 5$.

Moon's Declination.	Moon's Horizontal Parallax. 54′	55′	57′	59′	61′
°	feet.	feet.	feet.	feet.	feet.
0	·00	·00	·00	·00	·00
3	·04	·05	·05	·06	·06
6	·09	·09	·10	·11	·12
9	·13	·14	·15	·17	·18
12	·17	·18	·20	·22	·24
15	·21	·22	·25	·27	·30
18	·25	·26	·29	·32	·35
21	·28	·30	·33	·37	·40
24	·31	·33	·37	·40	·44
27	·34	·36	·40	·44	·48
30	·37	·39	·43	·47	·51

The above Tables have been calculated from theory with the value of (E) which obtains at the London Docks. For any other place the inequalities of the height which belong to the semi-diurnal wave may be considered as proportional to the value of (E) for that place.

Thus in employing Tables VII., VIII., IX., and X., for other places than London, the quantities given in these Tables must be multiplied by a certain constant, which for

Plymouth is	1·0	Bristol	3·2
Portsmouth	0·7	Liverpool	1·6
Sheerness	0·7	Leith	1·0
Pembroke	1·6		

The Declination to be used in calculating the *Diurnal Inequal-*

ity is that which corresponds to the Transit, the *seventh* preceding the Moon's Transit *B*.

Suppose the preceding transits of the moon be denoted by the letters *f*, *e*, *d*, *c*, *b*, *a*, *A*, *B*, *C*, *D*, *E*, *F*, *F* being the transit immediately preceding the high water at London, of which the height is required, the declination which corresponds to the transit *f* is to be employed, in using the preceding table of the diurnal inequality; the *epoch* of the diurnal inequality being different from that of the semidiurnal inequalities. This arises, probably, from the circumstance that the waves which constitute these inequalities do not travel with the same velocity.

At London.—For the Moon's *upper* Transit the *diurnal Inequality* has the same sign as the Declination; for the *lower* Transit it has a contrary sign, $C = {\cdot}5$.

At Liverpool.—For the Moon's *lower* Transit, the *diurnal Inequality* has the same sign as the Declination; for the *upper* Transit it has a contrary sign, $C = {\cdot}8$.

At Plymouth.—The same rule as at Liverpool, $C = {\cdot}9$.

Example of the Use of the Tables.

		d	h	m
The Nautical Almanac gives the apparent Solar Time of the Moon's Transit (*B*) P.M. 12th January, 1837		0	5	5
On the same day, at the Time of Transit, the Moon's Parallax is 57′·8, Moon's Decl. 2°, Sun's Decl. 22°, and Sun's Par. 8″·94.				
Table I. gives		2	2	29
— II. —	+ 2m			
— III. —	+ 3			
— IV. —	+ 3	+0	0	14
— V. —	− 3			
Equation of Time	+ 9			
		2	7	48

which gives 7^h 48^m for the mean Solar Time of High Water on the 14th January, P.M., at the London Docks. Ten minutes are afterwards added to give the time of High Water at London Bridge.

		feet.
Table VI. gives		19·71
— VII. —	+ ·18	
— VIII. —	+ ·29	+ ·49
— IX. —	+ ·06	
— X. —	− ·04	
The Moon's Declination and Parallax corresponding to transit *f*, being 20.S. and 60′·8, the transit being lower, Table XI. gives		+ ·37
		20·57

which gives 20·57 feet for the height of high water on the 14th January, 1837, P.M. at the London Docks.

In consequence of the many irregularities to which the times and heights of high water are subject from accidental circumstances and the comparative magnitude of the semi-menstrual inequality; the superior accuracy of predictions founded upon the tables here given, over other predictions based upon the correct *establishment* and semi-menstrual inequality, but with different corrections for the moon's parallax and declination, may not be apparent. The fluctuation of the establishment introduces a difficulty which the best tables cannot surmount, and in order to render the predictions as accurate as possible, this fluctuation must be carefully watched from year to year. With regard to the accuracy of predictions of the time of high water, they will sometimes, but rarely, be out an hour, owing to accidental causes; generally, however, they may be depended upon to within ten minutes; and it should be recollected that it is difficult to observe the instant of high water to within five minutes, as the fluctuations near the maximum height are very minute, and the water is said to *hang*.

The heights are liable to still greater irregularities from accidental causes; and this is the more to be regretted, as disastrous effects often arise from unexpected and unusually high tides.

M. Daussy has ascertained that at Brest the height of high water varies inversely as the height of the barometer, and that the British Channel there rises more than eight inches for a fall of about half an inch in the barometer. I have found that at Liverpool a fall of one-tenth of an inch in the barometer corresponds to a rise in the river Mersey of about an inch, and that at the London Docks a fall of one-tenth of an inch in the barometer corresponds to a rise in the river Thames of about seven-tenths of an inch. So that with a *low* barometer the tides may be expected to be *high*, and *vice versâ, cæteris paribus.*

In violent storms it is obvious that the progress of the tide-wave is liable to be disturbed. Thus during a violent hurricane (Jan. 8, 1839) I am informed that there was no tide at Gainsborough, which is twenty-five miles up the Trent, a circumstance unknown before. At Saltmarsh, only five miles up the Ouse from the Humber, the tide went on ebbing and never flowed till the river was dry in places, while at Ostend, towards which the wind was blowing, contrary effects were observed.

It has been remarked, that in consequence of the sheltered situation of the port of London, the great undulations produced by the winds will be less sensible there than on the coasts of France, as at Brest for example. But it should be recollected, that if the tide at London is to be considered as a *derived* tide, transmitted from the Atlantic, the irregularities which are felt at Brest will equally tend to affect it at all those places which it reaches subsequently.

"During strong north-westerly gales, the tide marks high water earlier in the river Thames than otherwise, and does not give so much water, whilst the ebb-tide runs out later, and marks lower; but, upon the gales abating and the weather moderating, the tides put in and rise much higher, whilst they also run longer before high water is marked, and with more velocity of current, nor do they run out so long or so low." For this information with respect to the influence of the wind on the tides in the river Thames, I am indebted to Sir J. Hall.

It has been found, that since the construction of the new London Bridge and the removal of the old foundations, there is less water at the St. Katherine Docks at low water by about 18 inches than formerly, but as respects the depth of high water it is the same; in other words, the flood-tide at the entrance of the St. Katherine Docks lifts about 18 inches more within the time of flood than formerly. I am indebted to Sir J. Hall for this information. I do not however attribute the fluctuations of the *establishment* to the removal of the old bridge. They began long before the foundations were touched. See the Plate annexed.

My tables are employed for the purpose of calculating the times and heights of high water in the British Almanac, in the tide tables published by the Hydrographic Office of the Admiralty, and the times in the Nautical Almanac. By comparing such calculations with observations and by examining the errors, improvements may hereafter suggest themselves; but considering, on the one hand, how well the semi-menstrual inequality, deduced from theory, agrees with observation, and on the other, how minute are the other corrections, especially when the transit B is used as the argument, it will be difficult to effect any material improvements in the prediction of the phenomena.

Observations of the Tides.

Observations of the tides should record particularly,

The time and height of high water.
The time and height of low water.
The direction and force of the wind.

The following example may serve as a specimen of the manner in which the observations should be recorded, and is extracted from the *Observations of the Tides taken at Her Majesty's Dockyards, and printed by order of the Lords of the Admiralty.*

Register of the Tides at Portsmouth Dock-yard.

1834. January.		High Water.		Low Water.		Range of Tide.	Wind.	
		Time.	Height.	Time.	Height.		Direction..	Force.
Day.		h m	ft. in.	h m	ft. in.	ft. in.		
1	A. M.	3 15	19 3·5	8 20	8 11	10 4·5	N.N.W.	6
	P. M.	3 30	18 4·7	8 45	8 0·7	10 4	N.W.	4
2	A. M.	4 0	18 6	9 15	8 3	10 3	W.W. by N.	5
	P. M.	4 15	17 5·5	9 25	8 7	8 10·5	N. by W.	2

The *Time* is *Mean Time,* obtained through the Royal Naval College; and the Dock-yard clock, which is used for ascertaining the *Times* of High and Low Water, is regulated thereby, by Mr. Smithers, Clock-maker, Portsea.

The Height of the Tide is ascertained by Lloyd's Tide Gauge.

The Line from which the Heights are measured is the Sill of the North Dock gates.

The force of the wind is indicated by figures, according to the following scale, contrived by Captain Beaufort:

0 *Calm.*

1 *Light Air* .... Or just sufficient to give steerage way.

2 *Light Breeze*	Or that in which a well conditioned man-of-war with all sail set, and clean full, would go in smooth water from	1 to 2 knots.
3 *Gentle Breeze* ...		3 to 4 knots.
4 *Moderate Breeze* .		5 to 6 knots.
5 *Fresh Breeze*	Or that to which such a ship could just carry in chase, full and by	Royals, &c.
6 *Strong Breeze* ...		Single-reefed topsails and top-gallant sails.
7 *Moderate Gale* ...		Double-reefed topsails, jib, &c.
8 *Fresh Gale*		Triple-reefed topsails, &c.
9 *Strong Gale*		Close-reefed topsails, and courses.

10 *Whole Gale* .... Or that with which she could scarcely bear close-reefed main topsail and reefed foresail.

11 *Storm* .... Or that which reduces her to storm stay-sails.

12 *Hurricane* .... Or that which no canvass could withstand.

The circumstances of high water are more interesting, and admit generally of more accurate observation, than those of low water.

The height of the water must be given from some fixed mark or line*, which should be described accurately, so that it may be easily recovered. It should also be carefully stated whether the time in which the observations are given is mean or apparent solar time, and how obtained.

The name of the observer, or his initials, should be attached to each observation. The simplest method of observation appears to be by means of a staff, carefully graduated, connected with a float, and working through a collar where the height is read off. The staff must be kept in a vertical position by means of friction rollers; the float should be in a chamber to which the water has access by a small opening, in order that the ripple may be as much diminished as possible. It is convenient to have a clock close to the tide gauge; and if made to strike minutes, so much the better. The observer should note the height of the water at the end of every minute, for half an hour before the expected time of high water, and until there can be no doubt that the time of high water is past. The minute at which the water stood the highest, or the time of high water, is then easily seen. This process is tedious, and it might be imagined that it would suffice to note the time when the water reaches a certain height shortly before high water, and the time when it reaches the same line in its descent; but the water rises and falls by jerks, and much too irregularly for this plan to be adopted with safety, at least in our river.

Mr. Palmer has described, in the Philosophical Transactions, a self-registering machine, which is intended to give the time and height of high water. Such a machine is in operation in Her Majesty's Dock-yard at Sheerness, and is described in the Nautical Magazine for October 1832.

The principle consists in a style, or pencil, which is moved horizontally by the tide along the summit of a cylinder which is turned round slowly and uniformly; the pencil describes a curve upon paper wound round the cylinder, which curve indicates the fluctuations of the water. The motion of the tide being originally vertical, is changed by a simple mechanical contrivance.

When it is intended to make a long series of observations, it is of course desirable to adopt every precaution to ensure accuracy; but many persons have it in their power to make observations, which may be useful in determining the *establishment* of a port, or the mean interval between the moon's transit and the time of high water, without any expensive apparatus.

For this purpose the observations during one lunation, or even less, may suffice, where, as in the river Thames, the rise is considerable and the tides little subject to irregularities. In the open

* I consider this of particular importance, and I allude to it because it is frequently omitted.

ocean, where the rise on the contrary is small, the tide often hangs half an hour at high water, and the phenomena take place very irregularly. At St. Helena the rise in springs, according to Dr. Maskelyne, is 39 inches, and in neaps 20 inches ; and I apprehend that less information could be elicited from a year's observations there, than from a month's observations at the London Docks. When a few observations only are made with a view of determining the establishment, they should not be used to determine that quantity absolutely, but they should be compared with observations at some place of which the establishment is accurately known, or where observations are continually carried on.

The tides in this port continue to be observed at the London and St. Katherine Docks. These Docks are contiguous, so that the places at which the observations are made are not distant from each other more than 900 yards. We might therefore conclude, that whatever cause affects the tide at one place will equally affect it at the other ; and hence, if we find the difference in the registers of the times and heights of high water much greater than the average difference, suspicion arises that the observation at one or the other place must be erroneous.—See Tables A. and B. Phil. Trans. 1836. p. 229. The observations at the London Docks are made by a person who notes the time when the water has begun to fall, that is, *has made its mark*. Those at the St. Katherine Docks are made by noting upon a slate (ruled for the purpose) the height of the water every minute for a short time before high water is expected, all which is afterwards copied into a book ruled in the same manner, and the time of high water, with the height, is easily inferred. The height is ascertained by means of a rod or tide-gauge, connected with a float, which is placed in a chamber, into which the water enters through a culvert, so that the ripple or agitation of the water in the river is avoided as much as possible. A clock, carefully regulated, stands close at hand. This plan has been adopted at my suggestion, and, if the observer and transcriber of the observation do their duty, it does not seem to me to be susceptible of any improvement.

I find by examining the registers of the observations at the London and the St. Katherine Docks, that the tide is on the average about five minutes later there than at the former place, and the difference in height between the lines or zero points, from which the rise is measured, is about five feet. But the differences are extremely irregular. I believe that the late Mr. Peirce devoted much attention to the observations of the tides at the London Docks, and I believe that Sir J. Hall has done all in his power at the St. Katherine Docks to cause the observations there to be made accurately. I am therefore unwilling to attribute these discrepancies to the carelessness of those to whom the observations are entrusted. If they do not proceed upon a uniform system this will operate disadvantageously, for the observations of one person will not be comparable with those of another. Errors sometimes occur no doubt in transcription, and the continual swell created by the steamers in the river may also interfere with the accuracy of the observations.

I think it possible, however, that owing to causes which have not yet been sufficiently investigated, in the progress of the tide-wave through a narrow channel of varying dimensions the height of high water (reckoned from the same zero-line,) may vary somewhat at two contiguous places, and that the intervals between the times of high water may also be subject to irregular fluctuations. It will then be impossible, even if the observations be carefully made, to employ single observations in certain localities as a test of the accuracy of the astronomical corrections which are employed in calculating the phenomena. I believe that the predicted times and heights of high water calculated for the river Thames now agree nearly as well with the phenomena as the observations at the contiguous docks agree with each other. M. Chazallon has lately published the observations of the time of high water at Brest in June 1835, and hence I have been able to ascertain that the times of high water for that place in June 1835, predicted in our *Tide Tables for the English and Irish Channels*, agree better with the observations at Brest than the times of high-water given for the London Docks in the same work agree with those observed at the latter place, although the constants employed for Brest were only roughly obtained from a year's observations, and the parallax and declination corrections employed were the same in both cases.

Although the tide at London is principally due to the wave which descends along the eastern coast of Great Britain, it may be somewhat modified by the tide which comes up the British Channel. If such be the case, this influence is probably irregular and varying with the weather; hence another difficulty which may interfere with the accuracy of our tide predictions for London, and for all places where the tide is not *single*.

LOCAL CONSTANTS

corresponding to { Moon's Parallax ... . 57′. ; Moon's Declination .... 15°. ; Sun's Parallax .... 8″·8 ; Sun's Declination........... 15°. } for Transit B.

	a			*b*		*c*	*d*
	d	h	m	h	m	feet.	feet.
Brest Harbour	1	4	18	1	17	?	?
Plymouth Dock-yard	1	6	5	1	24	17·85	3·39
Portsmouth Dock-yard......	1	12	13	1	17	19·05	2·40
Sheerness Dock-yard	2	1	40	1	23	25·56	2·48
London Docks	2	3	7	1	24	22·72	3·17
Pembroke Dock-yard	1	6	35	1	15	22·73	5·44
Bristol, Cumberland Docks	1	7	45	1	20	32·25	10·52
Liverpool Dock	1	11	54	1	21	17·62	5·30
Howth Harbour	1	11	32	1	24	?	?
Leith Dock	1	15	8	1	18	16·18	3·51

a = *interval* between the moon's transit B at $0^h\ 30^m$ and the time of high water.

b = difference between *interval* corresponding to moon's transit B at $3^h\ 30^m$ and that corresponding to moon's transit B at $9^h\ 30^m$.

c = *height* of high water corresponding to moon's transit B at $0^h\ 30^m$ (*Spring*).

d = difference between the *height* of high water corresponding to moon's transit B at $0^h\ 30^m$ and that corresponding to moon's transit B at $6^h\ 30^m$.

The above table is merely offered as a specimen of the kind of table which seems to be required of *Local Constants*. At present the data do not exist for the construction of such a Table.

From the above, the quantities (A), D and (E), can be immediately determined. A column may hereafter be added with advantage, affording the maximum of the diurnal inequality. It will be necessary also to specify the date for which the constants are supposed to obtain, and the zero-line from which the heights in column c is measured. I think that the *average interval* will be found always to correspond to transit B at $0^h\ 25^m$, but if not, another column will be required, containing the constant which determines the *epoch* of the tide at different ports.

THE END.

Printed by Richard and John E. Taylor, Red Lion Court, Fleet Street.

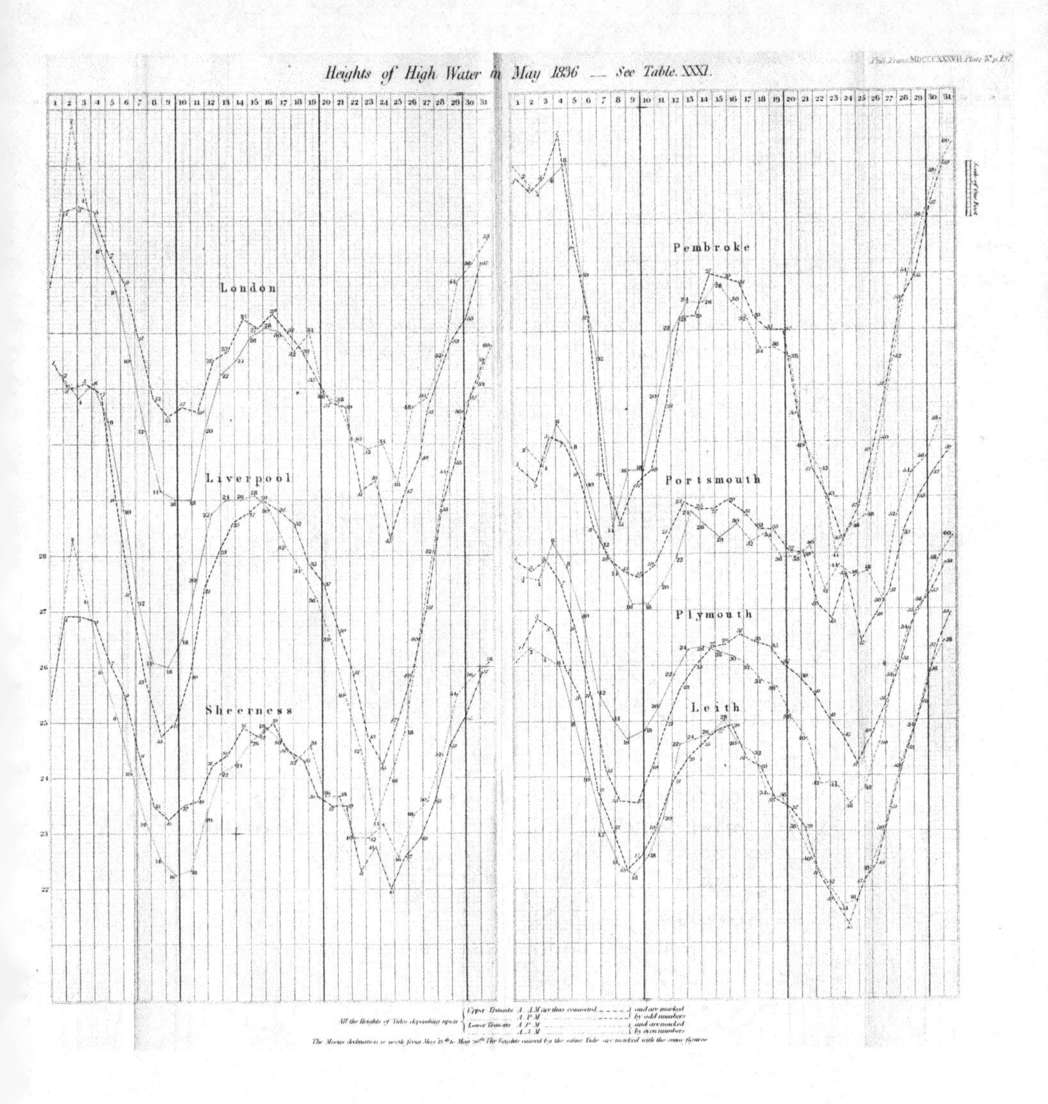

The material originally positioned here is too large for reproduction in this reissue. A PDF can be downloaded from the web address given on page iv of this book, by clicking on 'Resources Available'.

Diagram showing the Establishment of the Port of Liverpool. — See Table XIV. p. 121.

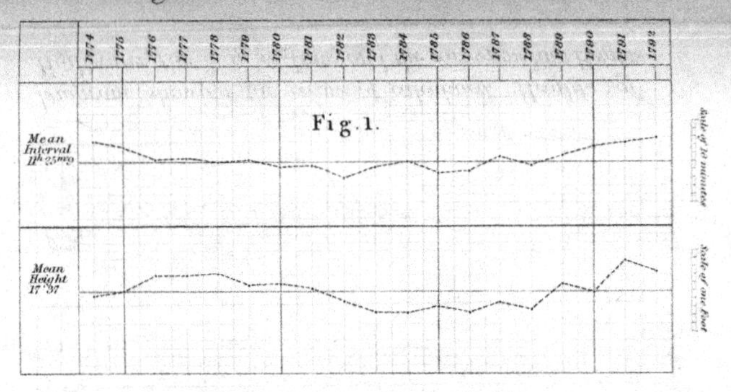

Diagram showing the Establishment of the Port of London. — See Table XXX. p. 136.

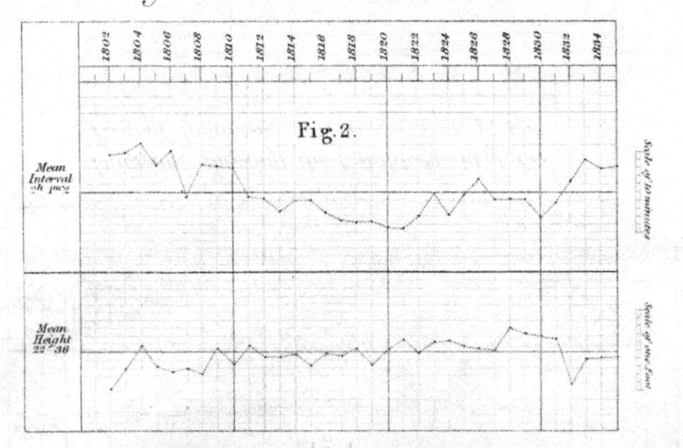

Diagram showing the errors of calculated Heights of Highwater for May & June and the corresponding Heights of the Barometer at Liverpool & London. — See. p. 104.

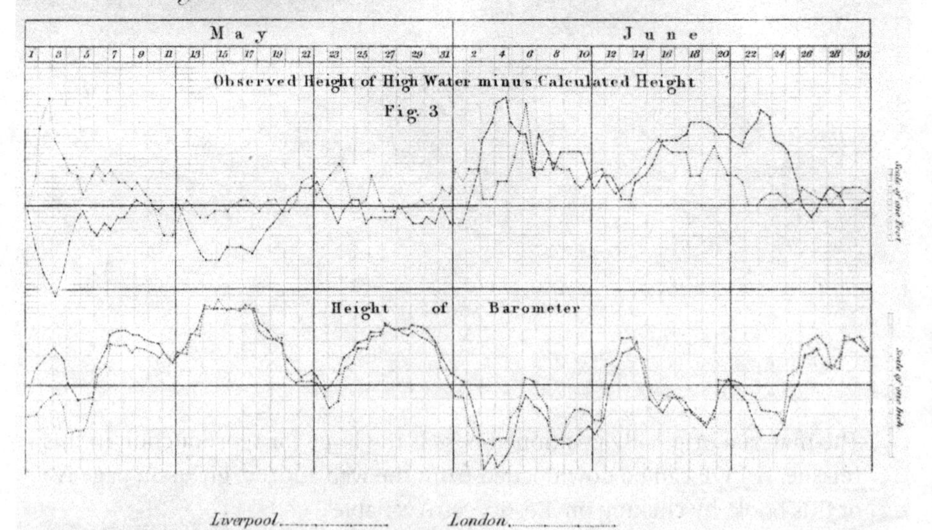

Liverpool .... *London* ....

The material originally positioned here is too large for reproduction in this reissue. A PDF can be downloaded from the web address given on page iv of this book, by clicking on 'Resources Available'.

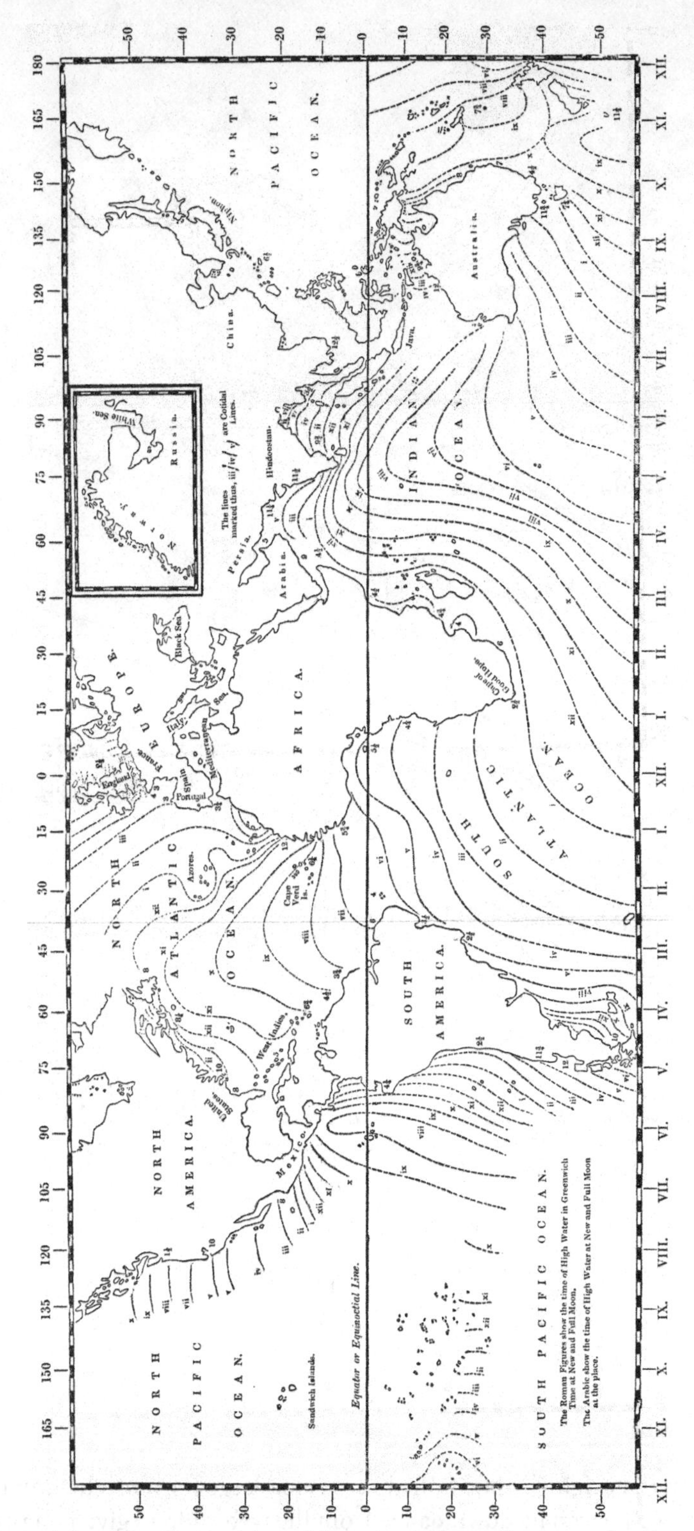

The material originally positioned here is too large for reproduction in this reissue. A PDF can be downloaded from the web address given on page iv of this book, by clicking on 'Resources Available'.

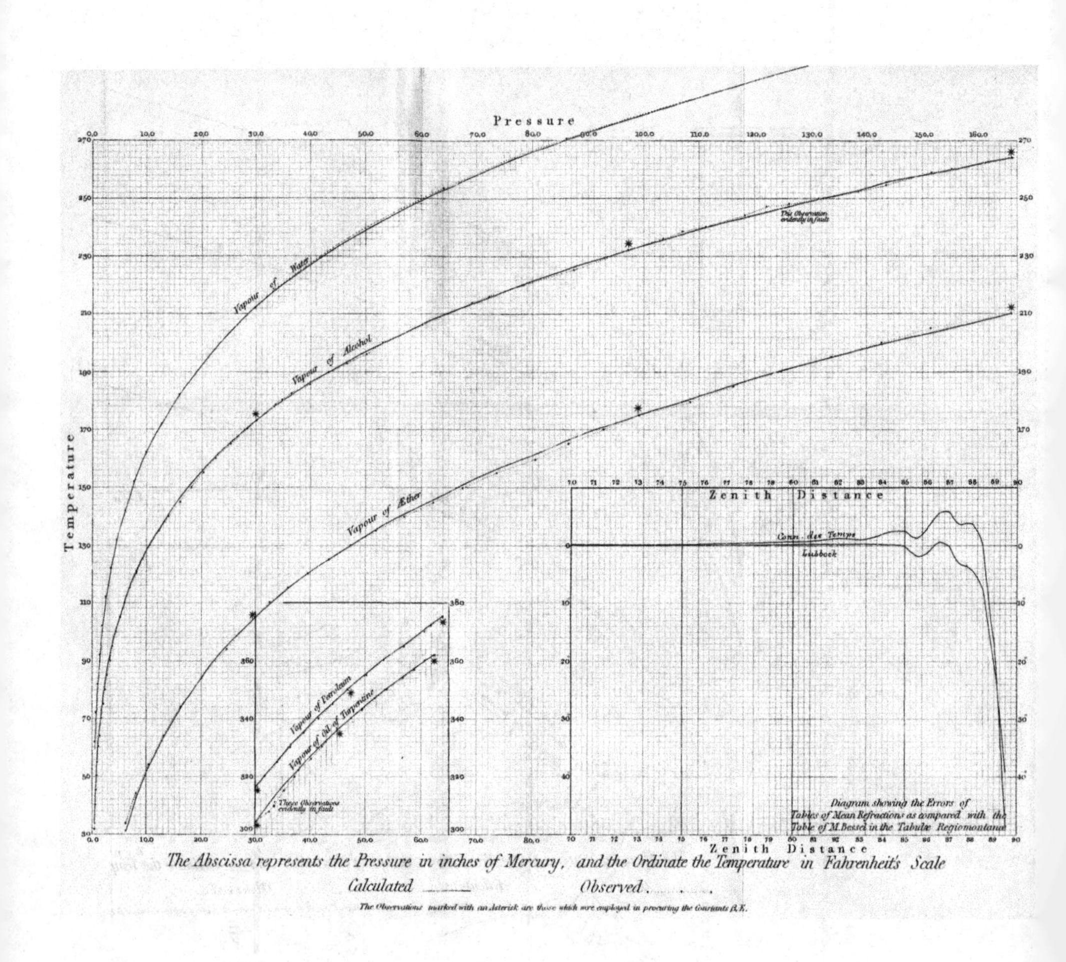

The material originally positioned here is too large for reproduction in this reissue. A PDF can be downloaded from the web address given on page iv of this book, by clicking on 'Resources Available'.

ON

THE HEAT OF VAPOURS

AND ON

ASTRONOMICAL REFRACTIONS.

BY

JOHN WILLIAM LUBBOCK, Esq., Treas. R.S.,
F.R.A.S. & F.L.S.,
VICE-CHANCELLOR OF THE UNIVERSITY OF LONDON,
MEMBER OF THE AMERICAN ACADEMY OF ARTS AND SCIENCES,
AND OF THE ACADEMY OF PALERMO.

LONDON:
CHARLES KNIGHT AND CO., LUDGATE HILL.
1840.

PREFACE.

THE connexion between the temperature and the pressure (or elasticity) of elastic vapours is a desideratum in Physics. A knowledge of it is indispensable to an exact theory of the Steam Engine, to an exact theory of Astronomical Refractions, and to an accurate solution of other important problems. The want of it has hitherto been supplied by unsatisfactory approximations; but these questions cannot be completely investigated without a more careful attention to the premises than has hitherto been possible, owing to a want of the proper key to these researches, which consists in a knowledge of the mathematical law which connects the temperature and the pressure in elastic fluids, and which is required in addition to the law of Mariotte and Gay Lussac to complete their theory.

If V represent the absolute heat or *caloric*, i the *latent* heat, c the *sensible* heat or that which affects the *thermometer*,

$$V = i + c.$$

If θ be the *temperature* as indicated by a thermometer, there can be little doubt that V is capable of being expressed in a series proceeding according to positive powers of θ, so that

$$V = a + b\theta + c\theta^2 + \&c.$$

a, b, c, &c., have a certain signification in Taylor's theorem, but without being able to determine their values, *à priori*, or to obtain any relations between them, they may be treated as constants. If the *latent* heat be constant, which is probable, and if the effect indicated by the thermometer is proportional to the sensible heat,

$$c = b\,\theta, \quad V = a + b\,\theta.$$

It must, however, be left to experiment to decide how many terms are to be taken into account for any given substance, within any given range of the thermometric scale, and in order to satisfy the results of observation within any given quantity. The other suppositions upon which my theory is founded are those of Laplace, viz. that the quantity called γ by M. Poisson is constant for the same substance at different temperatures, and that the equation

$$V = A + B\frac{p^{\frac{1}{\gamma}}}{\varrho}$$

is the solution of a certain differential equation. See *Méc. Cél.*, vol. v. p. 108. Poisson, *Méc.*, vol. ii. p. 640.

The theorems which are given by M. Poisson in the second volume of the *Mécanique*, and which are also to be found in the works of Pouillet and Navier, rest upon the condition that the absolute heat is constant, while the sensible heat varies. This is the most restricted hypothesis which can be made upon the nature of heat, and it does not satisfy the observations. In this Treatise I have gone a step further, by supposing the absolute heat to vary with the sensible heat, or to be represented by an expression of the form $a + b\,\theta$, (or what is the same, $V = C + D\,(1 + \alpha\,\theta)$. See p. 2.) θ being the temperature reckoned from some fixed point, a and b constants. This includes implicitly the other hypothesis, which if true, in determining a and b by means of observations, the constant b should come out zero. This in the case of steam is certainly not the case, nor is it so in any case which I have examined.

The experiments of Dulong and Arago upon steam at high temperatures, those of Southern and Dalton, and those of Dr. Ure, furnish data by which the supposition I have adopted and the formulæ which flow from it can be scrutinized; and if the expressions which result from it fail to represent those observations, we have at least arrived at this conclusion, that the condition of the invariability of the quantity called γ by M. Poisson does not obtain in nature, or that the absolute heat cannot be represented by so simple a function of the temperature or sensible heat. Recourse must then be had to more complicated expressions. If, on the contrary, my formula represents the observations of the temperature of vapours with accuracy, its origin in a simple theoretical notion of the quantity of absolute heat, and its simplicity, are great additional recommendations in its favour. The formula which I have obtained does, I believe, represent the observations better than any hitherto devised; at low temperatures and pressures it deviates a little, but a very slight error in the observed pressures may account for this discrepancy. Dalton says that it is next to impossible to free any liquid entirely from air; of course if any air enter, it unites its force to that of the vapour. Moreover, when the pressures are small, the variation of temperature becomes great for a small variation of pressure; so that the agreement of theory with observation may be considered as complete, even if the absolute amount of the error of the calculated temperature is then more considerable.

My formula has also been compared with the observations of Dr. Ure, on the vapour of alcohol, æther, petroleum, and oil of turpentine, recorded in the Philosophical Transactions for 1818.

I think that the comparisons contained in this treatise afford sufficient evidence that my formula is established, and that the deviations of the calculated results from those of observation are within the limits of the errors of the latter; but this point I leave to be decided by those more conversant with the nature of the experiments. It would not militate against my views if it were found necessary to take in an additional term and to make

$$V = C + D(1 + \alpha\theta) + E(1 + \alpha\theta)^2 + \&c.$$

but the expressions for the temperature and density in terms of the pressure would not be quite so simple, although more pliable.

As the same principles must be applicable to the constitution of the atmosphere, I have examined the observations made by M. Gay Lussac in his aëronautic ascent from Paris, and which are published by M. Biot in the *Conn. des Temps.* My calculated temperatures may be considered as identical with the *températures regularisées* of M. Biot, which are given by that distinguished philosopher as representing the condition of the atmosphere divested of the irregularities and errors incidental to observations made under circumstances so difficult and so disadvantageous. But the altitude to which man can ascend is so limited, that observations of the temperature made in aëronautic ascents will never furnish so complete a test of the accuracy of any formula professing to give the relation between the pressure and the temperature in elastic fluids, as observations of the temperature of the vapour of water and other substances, which can be carried through a greater range of the thermometric scale, and above all through the low pressures where the character of the curve is more decided.

M. Biot has dwelt with reason upon the importance of introducing into the theory of Astronomical Refractions a greater conformity with the conditions of the problem than has hitherto been attempted; and he has also noticed the imperfection in principle of the present mode of calculating heights by observations of the barometer, a method which must of course be abandoned (at least in any accurate exposition of this theory) whenever the discovery of the true connexion between the temperature and the pressure of the higher regions of the atmosphere renders it possible to adopt a more rigorous mode of eliminating the density from the differential equation which connects d p and d z. The correct expression which connects the difference of altitude with the pressures at the upper and lower stations ought to be the foundation of the theory of Refractions. Considering on the one hand the notions upon which my formula is ultimately founded, its identity with the results offered by the observations of steam and other vapours, and moreover the agreement afforded by the direct comparison with the observations of M. Gay Lussac, there can be no doubt that it represents the density of the atmosphere at different altitudes with greater fidelity than any hypothesis which has up to the present time been made the basis of the theory of Astronomical Refractions.

I think that my table of mean refractions represents the observed quantities within the limits of their probable errors, and I have obtained this result without any arbitrary alterations of the constants.

In the higher regions of the atmosphere the cold is intense*, depriving the air of its elasticity and converting it into a liquid or solid substance. My formula of course is only applicable as long as the air continues in the state of an elastic vapour; and if at any altitude it ceases to maintain that condition, the density must be represented by a discontinuous function. But the density of this frozen air must be extremely small, and it probably has little effect upon the amount of Refraction.

I am indebted to Mr. Russell for his kind assistance in the numerical calculations which accompany this treatise.

* See Poisson, *Théorie de la Chaleur*, p. 460.

29, *Eaton Place*,
March 2, 1840.

CONTENTS.

ON

THE HEAT OF VAPOURS

AND ON

ASTRONOMICAL REFRACTIONS.

GENERAL EXPRESSIONS.

LET V be the quantity of absolute heat, considered as a function of the sensible heat or temperature θ,

$$\frac{dV}{d\theta} = \frac{dV\,d\varrho}{d\varrho\,d\theta} + \frac{dV}{dp}\frac{dp}{d\theta}$$

$$p = k\varrho\,(1 + \alpha\theta).$$

ϱ being the density, p the pressure, k and α constants,

$$\frac{d\varrho}{d\theta} = -\frac{\alpha\varrho}{1 + \alpha\theta}$$

$$\frac{dp}{d\theta} = \frac{\alpha p}{1 + \alpha\theta}$$

if

$$-\frac{dV}{d\varrho}\,\frac{\alpha\varrho}{(1 + \alpha\theta)} = \gamma\,\frac{dV}{dp}\,\frac{\alpha p}{(1 + \alpha\theta)}$$

$$\varrho\,\frac{dV}{d\varrho} + \gamma p\,\frac{dV}{dp} = 0.$$

If γ be considered as a constant quantity the integral of this partial differential equation is

$$\frac{p^{\frac{1}{\gamma}}}{\varrho} = \text{funct}^{\text{n}}.\ V*.$$

The simplest form which can be assigned to this function of V is such that

$$V = A + B\,\frac{p^{\frac{1}{\gamma}}}{\varrho}$$

A and B being constants.

---

* So far the reasoning is identical with that contained in the *Mécanique* of M. Poisson, but M. Poisson proceeds further upon the limited supposition of V being constant.

Laplace arrived at this equation (*Méc. Cél.* vol. v. p. 128.).

See Poisson, *Annales de Chimie*, tom. xxiii. p. 342; *Mécanique*, vol. ii. p. 648; Navier, *Leçons données à l'Ecole des Ponts et Chaussées*, tom. ii.

$$V = A + B \frac{p}{\varrho}^{\frac{1}{\gamma}} \qquad V' = A + B \frac{p'}{\varrho'}^{\frac{1}{\gamma}}$$

$$p = k \varrho (1 + \alpha \theta) \qquad p' = k \varrho' (1 + \alpha \theta')$$

k and α being constants.

I will now introduce the additional condition that the heat is *proportional* to the temperature, in which case

$$V = C + D(1 + \alpha \theta)$$
$$V' = C + D(1 + \alpha \theta')$$

C and D being constants. These equations include implicitly the hypothesis attributed to Watt and also that of Southern, respecting the vapour of water: on the former $D = 0$. Hence

$$V = C + D(1 + \alpha \theta) = A + B \frac{p}{\rho}^{\frac{1}{\gamma}}$$

$$V' = C + D(1 + \alpha \theta') = A + B \frac{p'}{\rho'}^{\frac{1}{\gamma}}$$

$$1 + \alpha \theta' = (1 + \alpha \theta) \frac{p'^{\frac{\gamma-1}{\gamma}}}{p^{\frac{\gamma-1}{\gamma}}} \left\{ \frac{1 - \frac{D}{k B} p^{\frac{\gamma-1}{\gamma}}}{1 - \frac{D}{k B} p'^{\frac{\gamma-1}{\gamma}}} \right\}$$

If $\frac{D}{k B} = E$ and if θ correspond to the boiling point, $\theta = 180°$ in Fahrenheit's scale, if the pressure be measured in atmospheres $p = 1$, but generally

$$1 + \alpha \theta' = (1 + \alpha \theta) \frac{(p^{\frac{1-\gamma}{\gamma}} - E)^*}{(p'^{\frac{1-\gamma}{\gamma}} - E)} \qquad [1]$$

* This equation must not be confounded with another equation which may be deduced from it by making $E = 0$, and which is not reconcileable with phenomena, as was long since noticed by M. Poisson in the case of steam. An equation of this kind is given by M. Pouillet in the form

$$x = \cdot 2669 \left(\frac{760}{p}\right)^{1 - \frac{1}{1 \cdot 375}}$$

as $$p' = k\,\varrho'\,(1 + \alpha\,\theta')$$

$$\frac{\varrho'}{\varrho} = \frac{p'\,(p'^{\frac{1-\gamma}{\gamma}} - E)}{p\,(p^{\frac{1-\gamma}{\gamma}} - E)}, \qquad [2]$$

$$= \left(\frac{p'}{p}\right)^{\frac{1}{\gamma}} \left\{ \frac{1 - E\,p'^{\frac{\gamma-1}{\gamma}}}{1 - E\,p^{\frac{\gamma-1}{\gamma}}} \right\}$$

if $$\frac{E\,p^{\frac{\gamma-1}{\gamma}}}{1 - E\,p^{\frac{\gamma-1}{\gamma}}} = -H$$

$$\left(\frac{p'}{p}\right)^{\frac{\gamma-1}{\gamma}} = 1 - q \qquad \frac{1 - E\,p'^{\frac{\gamma-1}{\gamma}}}{1 - E\,p^{\frac{\gamma-1}{\gamma}}} = 1 - H\,q$$

$$\frac{\varrho'}{\varrho} = (1 - q)^{\frac{1}{\gamma-1}}\,(1 - H\,q),$$

if $$\log\,(1 - H q) = -\,u \qquad c^{-u} = 1 - H\,q$$

c being the number of which the hyperbolic logarithm equals unity

$$\frac{\varrho'}{\varrho} = H^{\frac{1}{1-\gamma}}\;c^{-u}\left\{c^{-u} - 1 + H\right\}^{\frac{1}{\gamma-1}}.$$

if $\dfrac{\varrho'}{\varrho} = 1 - \omega$

$$\omega = 1 - H^{\frac{1}{1-\gamma}}\,c^{-u}\left\{c^{-u} - 1 + H\right\}^{\frac{1}{\gamma-1}}$$

Since $C + D\,(1 + \alpha\,\theta) = A + B\,\dfrac{p^{\frac{1}{\gamma}}}{\varrho}$

---

for atmospheric air. *Elémens de Physique*, vol. i. p. 400, and by Navier, *Leçons données à l'Ecole des Ponts et Chaussées*, vol. ii. p. 310, in the form

$$v' = \frac{(1 + \alpha\,v)}{\alpha}\left(\frac{\varpi'}{\varpi}\right)^{\cdot 3748} - \frac{1}{\alpha}$$

If $\frac{1}{\varrho} = v$

$$\alpha D(\theta' - \theta) = B p^{\frac{1}{\gamma}} \{v' - v\} = V' - V,$$

supposing the heat and the volume to vary, the pressure remaining constant.

According to Dulong the following laws obtain, which however, are not admitted by Dr. Apjohn (see Phil. Mag. 1838, p. 339):

"1°. Des volumes égaux de tous les fluides élastiques pris à une même température et sous une même pression, étant comprimés ou dilatés subitement d'une même fraction de leur volume, dégagent ou absorbent la même quantité absolue de chaleur. 2°. Les variations de température qui en résultent sont en raison inverse de leur chaleur spécifique à volume constant."—*Mém. de l'Institut*, tom. x. p. 188.

According to the first of these laws the quantity B must be the same for different vapours; of the second I am unable to offer any satisfactory interpretation.

In what follows I propose to ascertain how far the equations [1] and [2] satisfy the best observations on record. The general relation gives

$$1 + \alpha \theta'' = (1 + \alpha \theta) \frac{(p^{\frac{1-\gamma}{\gamma}} - E)}{(p''^{\frac{1-\gamma}{\gamma}} - E)}.$$

Eliminating E between this equation and that which connects θ' and p',

$$(\theta'' - \theta)(1 + \alpha \theta')(p'^{\frac{1-\gamma}{\gamma}} - p^{\frac{1-\gamma}{\gamma}}) = (\theta' - \theta)(1 + \alpha \theta'')(p''^{\frac{1-\gamma}{\gamma}} - p^{\frac{1-\gamma}{\gamma}})$$

If $\frac{1-\gamma}{\gamma} = \beta$

$$\frac{\left(\frac{p''}{p}\right)^{\beta} - 1}{\left(\frac{p'}{p}\right)^{\beta} - 1} = \frac{(\theta'' - \theta)\left(\frac{1}{\alpha} + \theta'\right)}{(\theta' - \theta)\left(\frac{1}{\alpha} + \theta''\right)}.$$

From this equation, knowing θ'', θ', θ, p'', p', p; β may be determined for any gas or vapour. Knowing β, E may be found from the equation

$$E = \frac{p'^{\beta}\left(\frac{1}{\alpha} + \theta'\right) - p^{\beta}\left(\frac{1}{\alpha} + \theta\right)}{\theta' - \theta}.$$

ON THE PRESSURE OF STEAM.

The most accurate and extensive experiments by which the accuracy of these relations can be tested are those which have been made upon the conditions of steam. The following are the experiments of Arago and Dulong, as recorded in tom. x. of the *Mémoires de l'Institut*, p. 231; together with the temperatures calculated by the best empirical formulæ.

Nos. des observations.	Elasticité en mètres de mercure à 0°.	Elasticité en atmosph. de 0m·76.	Température observée.	Température calculée par la formule de Tredgold.	Température calculée par la formule de Roche cöeff. moyen.	Température calculée par la formule de Coriolis.	Température calculée par la formule adoptée.
			Cent.	Cent.	Cent.	Cent.	Cent.
1	1·62916	2·14	123°·7	123°·54	123°·58	123°·45	122°·97
3	2·1816	2·8705	133·3	133·54	133·43	133·34	132·9
5	3·4759	4·5735	149·7	150·39	150·23	150·3	149·77
8	4·9383	6·4977	163·4	164·06	163·9	164·1	163·47
9	5·6054	7·3755	168·5	169·07	169·09	169·3	168·7
15	8·840	11·632	188·5	188·44	188·63	189·02	188·6
21	13·061	17·185	206·8	206·15	207·04	207·43	207·2
22	13·137	17·285	207·4	206·3	206·94	207·68	207·5
25	14·0634	18·504	210·5	209·55	210·3	211·06	210·8
28	16·3816	21·555	218·4	216·29	218·01	218·66	218·5
30	18·1894	23·934	224·15	222·09	233·4	224·0	224·02

There are reasons which make it probable that in inquiries of this nature the scale of temperature as indicated by the expansion of air is to be preferred, although the difference between the indications of a mercury thermometer with that of air is not considerable.

The following table is given by M. Pouillet (*Elémens de Physique*, vol. i. p. 259) for the centigrade scale:

Températures indiquées par le therm. à mercure, à enveloppe de verre.	Températures indiquées par un therm. à air, et corrigées de la dilatation du verre.	Volumes correspondans d'une même masse d'air.
– 36°	– 36°	0·8650
0	0	1·0000
100	100	1·3750
150	148·70	1·5576
200	197·05	1·7389
250	245·05	1·9189
300	292·70	2·0976
Ebull. du merc. 360	350·00	2·3125

From the above I have deduced the following Table for Fahrenheit's scale:

Merc. therm.	Air therm.	Merc. therm.	Air therm.
212°	212°	482°	478°·1
302	299·7	572	558·9
392	386·7	680	662·0

I now proceed to determine for steam the constants γ and E by means of the observations of Dulong and Arago which I have quoted in p. 5.

For the air thermometer on Fahrenheit's scale the experiments of Dulong and Arago (*Mém. de l'Institut*, vol. x.) give, θ being reckoned in Fahrenheit's scale and from the freezing point of water,

$$p = 1 \qquad \theta = 180 \qquad \frac{1}{\alpha} = 480^\circ$$

$$p' = 11{\cdot}632 \qquad \theta' = 334{\cdot}7 \qquad \frac{1}{\alpha} + \theta' = 814{\cdot}7$$

$$p'' = 23{\cdot}934 \qquad \theta'' = 396{\cdot}4 \qquad \frac{1}{\alpha} + \theta'' = 876{\cdot}4.$$

I find from these observations

$$\frac{p''^{\beta} - 1}{p'^{\beta} - 1} = [0{\cdot}1140623],$$

the quantity within brackets being the logarithm of the corresponding number; and hence I find

$$\beta = {\cdot}0134^* \qquad \gamma = {\cdot}98677 \qquad \frac{1}{\gamma} = 1{\cdot}0134$$

$$E = 1{\cdot}17602 \qquad \log E = 0704184 \qquad H = 6{\cdot}6809.$$

The pressure at the boiling point of water (212°) being unity,

$$\frac{1}{\alpha} + \theta = -\frac{[2{\cdot}0651059]}{p^{\cdot 0134} - 1{\cdot}17602};$$

so that if τ is the number of degrees on Fahrenheit's scale of the air thermometer, and the pressure p be reckoned in atmospheres,

---

* This value of β appears to me to be the only one which will satisfy the equation.

$$\tau = -\frac{[2{\cdot}0651059]}{p^{{\cdot}0134} - 1{\cdot}17602} - 448^{\circ},$$

and if ϱ be the density of steam, the relative volume

$$\frac{\varrho}{\varrho'} = \frac{p\,\{p^{{\cdot}0134} - 1{\cdot}17602\}}{p'\,\{p'^{{\cdot}0134} - 1{\cdot}17602\}}.$$

In order to ascertain how far the new expression here given for τ represents the totality of the observations, I have calculated the temperatures corresponding to all the observed pressures in the observations of Arago and Dulong, and the results are exhibited in the following table.

Pressure in atmospheres.	Temperature.				Error of temperature calculated by Lubbock. Fahr.
	Observed.			Calculated.	
	Merc. therm. Cent.	Merc. therm. Fahr.	Air therm. Fahr.	Air therm. Fahr.	
2·1400	123°·7	254°·66	253°·6	252°·8	−°·8
2·8705	133·3	271·94	270·4	270·1	−·3
4·5735	149·7	301·46	299·2	299·4	+·2
6·4977	163·4	326·12	323·0	323·2	+·2
7·3755	168·5	335·30	331·9	332·3	+·4
11·6320	188·5	371·30	366·7*	366·7	0
17·1850	206·8	404·24	398·6	398·9	+·3
17·2850	207·4	405·32	399·5	399·4	−·1
18·5040	210·5	410·90	404·9	405·3	+·4
21·5550	218·4	425·12	418·5	418·8	+·3
23·9340	224·15	435·47	428·4*	428·5	+·1

The observations marked with an asterisk were employed in determining the constants.

The error of the temperature calculated by the formula adopted by Arago and Dulong corresponding to the first observation is − ·73 cent. or −1°·3 of Fahr. I have no doubt that the observed temperature is in excess, and the agreement with the rest of the observations is so complete that within this range of temperature the formula may, I think, be considered as exactly representing the phenomena. The errors of the temperatures, calculated by the various empirical expressions which have been hitherto proposed, are much greater, as may be seen in the table of Dulong and Arago. The following observations are those of Southern, given in p. 172, vol. ii., of Dr. Robison's Mechanical Philosophy.

Pressure.	Temperature.		Error of calculated temp.	Pressure.	Temperature.		Error of calculated temp.
	Observed.	Calculated.			Observed.	Calculated.	
Inch.		°	°	Inch.			°
·52	62	59·5	−2·5	4·68	132	131·4	−·6
·73	72	69·3	−2·7	6·06	142	141·3	−·7
1·02	82	79·3	−2·7	7·85	152	151·6	−·4
1·42	92	89·9	−2·1	9·99	162	161·7	−·3
1·95	102	100·2	−1·8	12·64	172	171·8	−·2
2·65	112	110·7	−1·3	15·91	182	182·0	0
3·57	122	121·3	− ·7	29·80	212	212·	

The formula deviates slightly from the observations at very low pressures. Dalton says that it is next to impossible to free any liquid entirely from air; of course if any air enter, it unites its force to that of the vapour.—*Manchester Memoirs*, vol. v. p. 570. It must be recollected that according to theory the constants γ and E are the same only as long as the chemical constitution of the vapour remains the same, and they vary for different substances.

With regard to the nature of the accurate expression which connects the pressure with the temperature, opinions have hitherto been various. According to Dr. Robison, Mr. Watt found that water would distil *in vacuo* when of the temperature of 70°, and that in this case the latent heat of the steam appeared to be about 100°; and some other experiments made him suppose that the sum of the sensible and latent heats is a constant quantity. This, Dr. Robison says, is a curious and not improbable circumstance. Southern, on the contrary, concluded from experiments on the latent heat of steam at high temperatures that the *latent heat* is a constant quantity, instead of the latent heat + sensible heat being so. M. de Pambour, in speaking of Southern's view, says, "Cette opinion a paru plus rationelle à quelques auteurs, mais le première nous semble mise hors de doute par les observations que nous allons rapporter." It appears to me by no means clear that Watt entertained the opinion here attributed to him, for in a note in the Appendix to Sir David Brewster's edition of Robison's Mechanical Philosophy, vol. ii. p. 167, he professes to agree in the opinion there delivered by Southern. In p. 166 Southern records three experiments, from which he obtained 1171°, 1212°, and 1245°, for the sums of the latent + sensible heat corresponding to the temperatures or sensible heat 229°, 270°, 295°. If we take the two extreme observations, we find a difference in the sum of the latent + sensible heat of 74 degrees, corresponding to a difference in the sensible heat of 66 degrees.

If the conditions under which Laplace obtained the equation

$$V = A + B\frac{p}{\varrho}^{\frac{1}{\gamma}}$$

are admitted, the value of E different from zero shows that the absolute heat is not constant; but the preceding theory does not appear to me to furnish the means of determining the value of D, and hence of deciding with certainty whether the latent heat is constant, and whether in augmentations of heat the sensible heat only varies. I think there can be little doubt that the conditions assumed by Laplace actually obtain, and that the hypothesis attributed to Watt* must be abandoned. The experiments recorded by Mr. Parkes in the 3rd volume of the Transactions of the Society of Civil Engineers, p. 71, which show that the quantity of fuel required to evaporate a given weight of water is nearly the same whatever be the pressure of the steam, do not seem to me to authorize a different conclusion. For this is precisely what would take place if the *latent* heat be constant, and if the quantity of fuel required to generate the *latent* greatly exceed that required to generate the concomitant *sensible* heat.

The quantity γ has never before been determined for steam† or for the vapour of any liquid, properly so called, as far as I am aware. It may excite surprise that the value of γ should come out less than unity. Both Poisson and Dulong assert that it is evident that γ must surpass unity, but the reason which they assign appears to me inconclusive.

ON THE STEAM-ENGINE.

The law which connects the pressure and the temperature of steam having been unknown, various empirical rules have been given. As, however, the expressions which arise are not in a convenient form for the calculations which are required in order to ascertain the *duty* which steam-engines are capable of performing, or to solve other problems of the same nature, M. de Pambour‡, in his work on that subject, has employed another expression, viz.

$$\mu = \frac{1}{\varrho} = \frac{1}{n + qp},$$

in which ϱ is the density of steam, p the pressure, and n and q constants. According to my expression

* Mr. Sharpe has maintained the same opinion in the 2nd vol. of the Manchester Memoirs. See Dr. Thomson's Outline of Heat and Elasticity, p. 198.

† "Quant à la valeur de γ, elle nous est jusqu'à present tout-à-fait inconnue."—Poisson, *Méc.*, tom. ii. p. 652.

‡ *Théorie de la Machine à Vapeur*, p. 111.

$$\frac{1}{\varrho} = \frac{(1 - E)}{p^{\frac{1}{\gamma}} - E p}.$$

The pressure being reckoned in atmospheres, and the density of steam corresponding to the pressure of one atmosphere (or 14·706 lbs. per square inch) being unity. If we take the density of water for unity, then as the volume of steam at the pressure of one atmosphere is 1700 times greater than that of the same weight of water,

$$\mu = -\frac{K}{p^{\frac{1}{\gamma}} - E p} \qquad K = 1700\,(E - 1)$$

$$\log K = 2{\cdot}4765041 \qquad \frac{1}{\gamma} = 1{\cdot}0134 \qquad E = 1{\cdot}17602.$$

If we suppose that a certain volume of water represented by S be transformed into vapour at the pressure p, and that M is the absolute volume of vapour which results, we shall have

$$\frac{M}{S} = \mu = \frac{[0{\cdot}4109002]}{p}\left(\frac{1}{\alpha} + \theta\right).$$

If afterwards the same volume of water is transformed into vapour at the pressure p', and that the absolute volume which the resulting vapour occupies be called M', we shall have

$$\frac{M'}{S} = \mu'$$

$$\frac{M}{M'} = \frac{p'^{\frac{1}{\gamma}} - E p}{p^{\frac{1}{\gamma}} - E p}$$

" Soit* P la pression *totale* de la vapeur dans la chaudière, et p' la pression qu'aura cette vapeur à son arrivée dans le cylindre, pression qui sera toujours moindre que P, excepté dans un cas particulier que nous traiterons plus loin. La vapeur pénétrera donc dans le cylindre à la pression p', et elle continuera d'affluer avec cette pression et de produire un effet correspondant, jusqu'à ce que la communication entre la chaudière et le cylindre soit interceptée. Alors il cessera d'arriver de la vapeur nouvelle dans le cylindre, mais celle qui y est déjà parvenue, commencera à se dilater pendant le reste de la course du piston, en produisant par sa détente une certaine quantité de travail, qui s'ajoutera à celle déjà produite pendant la période d'admission de la vapeur.

---

* The reasoning here is taken from M. de Pambour's work.

"P étant la pression de la vapeur dans la chaudière, et p' la pression qu'elle prendra à son arrivée dans le cylindre avant la détente, soit π la pression de cette vapeur en un point quelconque de la détente. Soit en même temps l la longueur totale de la course du piston, l' la portion parcourue au moment où a commencé la détente, et λ celle qui correspond au point où la vapeur a acquis la pression π. Enfin, soit encore a l'aire du piston, et c la liberté du cylindre, c'est-à-dire l'espace libre qui existe à chaque bout du cylindre, au-delà de la portion parcourue par le piston, et qui se remplit nécessairement de vapeur à chaque course; cet espace, y compris les passages aboutissants, étant représenté par une longueur équivalente du cylindre.

" Si l'on prend le piston au moment où la longueur de course parcourue est λ, et la pression π, on verra que si le piston parcourt, en outre, un espace élémentaire $d\,\lambda$, le travail élémentaire produit dans ce mouvement sera $\pi\, a\, \mathrm{d}\,\lambda$. Mais en même temps, le volume $a\,(l' + c)$ occupé par le vapeur avant la détente sera devenu $a\,(\lambda + c)$." Hence,

$$\frac{M}{M'} = \frac{p'^{\frac{1}{\gamma}} - E\,p'}{\pi^{\frac{1}{\gamma}} - E\,\pi} = \frac{\lambda + c}{l' + c}.$$

$$\lambda + c = (l' + c)\,\frac{(p'^{\frac{1}{\gamma}} - E\,p')}{(\pi^{\frac{1}{\gamma}} - E\,\pi)}$$

The elementary work produced $= \pi\, a\, \mathrm{d}\,\lambda$.

$$\int \pi\, a\, \mathrm{d}\,\lambda = \pi\, a\,(\lambda + c) - \int a\,(\lambda + c)\,\mathrm{d}\,\pi$$

$$= \pi\, a\,(\lambda + c) - \int a\,(l' + c)\,(p'^{\frac{1}{\gamma}} - E\,p')\,\frac{\mathrm{d}\,\varpi}{(\pi^{\frac{1}{\gamma}} - E\,\pi)}$$

$$= \pi\, a\,(\lambda + c) - a\,(l' + c)\,(p'^{\frac{1}{\gamma}} - E\,p')\int \frac{\mathrm{d}\,\varpi}{\pi^{\frac{1}{\gamma}} - E\,\pi}$$

$$= \pi\, a\,(\lambda + c)$$

$$+ a\,(l' + c)\,\frac{(p'^{\frac{1}{\gamma}} - E\,p')\,\gamma}{E\,(\gamma - 1)}\,\log\left(1 - E\,\pi^{\frac{\gamma - 1}{\gamma}}\right) + \text{const.}$$

This integral is to be taken from $\lambda = l'$ to $\lambda' = l$, let $\pi = p$ when $\lambda = l$, when $\lambda = l'$, $\pi = p'$.

$$\int \pi a \, d\lambda = p a (l + c) - p' a (l' + c)$$

$$+ a (l' + c) \frac{(p'^{\frac{1}{\gamma}} - E p') \gamma}{E(\gamma - 1)} \log \left\{ \frac{1 - E p^{\frac{\gamma-1}{\gamma}}}{1 - E p'^{\frac{\gamma-1}{\gamma}}} \right\}$$

for the values of the constants E, γ. See p. 6.

To this must be added the work effected during the course of the piston through l', which is $p' a l$, and if R is the total pressure exerted upon unity of surface of the piston

$$p a (l + c) - p' a c$$

$$+ a (l' + c) \; (p'^{\frac{1}{\gamma}} - E p') \; \frac{\gamma}{E (\gamma - 1)} \log \left\{ \frac{1 - E p^{\frac{\gamma-1}{\gamma}}}{1 - E p'^{\frac{\gamma-1}{\gamma}}} \right\}$$

$$= a R l \qquad \text{(A)*}$$

$R = (1 + \delta) r + p'' + f$, f is the friction of the machine not loaded, δ the increase of this friction due to unity of the charge r, p'' the pressure on the surface of the piston, representing the atmospheric pressure when the machine works without condensation, and otherwise the pressure of condensation in the cylinder.

If S denote the volume of water converted into vapour by the boiler in unity of time, this volume in the cylinder becomes

$$- \frac{S K}{p'^{\frac{1}{\gamma}} - E p'}$$

K being the same constant as in p. 10.

It is evident, according to the reasoning of M. de Pambour, in p. 125. of his work, that if v denote the velocity of the piston

$$\frac{S K}{p'^{\frac{1}{\gamma}} - E p'} = - v a \frac{l' + c}{l} \qquad \text{(B.)}$$

---

* This equation is equivalent to the equation (A) of M. de Pambour, p. 123, which may be put into the form

$$p a (l + c) - p' a c + \frac{l S}{q v} \text{ Nap. log } \left\{ \frac{l + c}{l' + c} \right\} = a R l.$$

or

$$p a (l + c) - p' a c + [6{\cdot}9505960] \frac{l S}{v} \log \left\{ \frac{l + c}{l' + c} \right\} = a R l,$$

the pressure being reckoned in lbs. per square inch.

$$p'^{\frac{1}{\gamma}} - E p' = - \frac{l S K}{a v (l' + c)} = \frac{K}{\mu'},$$

Similarly,

$$p^{\frac{1}{\gamma}} - E p = - \frac{l S K}{a v (l + c)} = \frac{K}{\mu}$$

$$\frac{1}{\mu} = \frac{l S}{a v (l + c)} \qquad \frac{1}{\mu'} = \frac{l S}{a v (l' + c)}$$

$$p a (l + c) - p' a c$$

$$- \frac{l S K \gamma}{v E (1 - \gamma)} \text{ Nap.log} \left\{ \frac{1 - E p'^{\frac{\gamma - 1}{\gamma}}}{1 - E p^{\frac{\gamma - 1}{\gamma}}} \right\} = a R l$$

$$p a (l + c) - p' a c - (p'' + f) a l$$

$$- [4{\cdot}6411966] \frac{l S}{v} \log{}^{*} \left\{ \frac{1 - E p'^{\frac{\gamma - 1}{\gamma}}}{1 - E p^{\frac{\gamma - 1}{\gamma}}} \right\} = a l (1 + \delta) r$$

$$\log \left\{ \frac{1 - E p'^{\frac{\gamma - 1}{\gamma}}}{1 - E p^{\frac{\gamma - 1}{\gamma}}} \right\} = \log \left\{ \frac{l + c}{l' + c} \right\} - \frac{1}{\gamma} \log \left(\frac{p'}{p} \right).$$

If the machine work without expansion,

$$p = p', \qquad \log \left\{ \frac{1 - E p'^{\frac{\gamma - 1}{\gamma}}}{1 - E p^{\frac{\gamma - 1}{\gamma}}} \right\} = 0,$$

$$r = \frac{p - (p'' + f)}{1 + \delta}$$

The data upon questions relating to the steam-engine are the quantities a, l, l', S, and v, and it is evident that from these quantities the quantities μ and μ' may at once be found by an easy arithmetical operation; from these the following table will give the corresponding pressures p and p', and these pressures being introduced into equation A, the value of $a\,r$ may be easily found.

---

* Log. of Briggs, the pressure being reckoned in atmospheres, the log. of the constant is [7·9670537], the pressure being reckoned in lbs. per square foot.

Table showing the volume (compared with that of water at 212°), and the temperature of steam.

Pressure in lbs. per square inch. 14·706 × p.	Temperature. Fahrenheit. τ		Volume. μ†	Pressure in lbs. per square inch. 14·706 × p.	Temperature. Fahrenheit. τ		Volume. μ
	Air*. Therm.	Merc. Therm.			Air. Therm.	Merc. Therm.	
1	101°·5	101°·5	20816	56	287°·6	289°·5	498
2	126·0	126·0	10871	57	288·7	290·7	490
3	141·4	141·4	7442	58	289·8	291·8	482
4	153·0	153·0	5691	59	290·9	293·0	474
5	162·2	162·2	4622	60	292·0	294·1	467
6	170·1	170·1	3902	61	293·0	295·1	460
7	176·8	176·8	3381	62	294·0	296·2	453
8	182·8	182·8	2986	63	295·0	297·2	447
9	188·3	188·3	2678	64	296·0	298·2	440
10	193·2	193·2	2429	65	297·0	299·2	434
11	197·8	197·8	2223	66	298·0	300·3	428
12	202·0	202·0	2051	67	299·0	301·3	422
13	205·9	205·9	1905	68	300·0	302·3	417
14	209·5	209·5	1778	69	301·0	303·3	411
14·706	212·0	212·0	1700	70	301·9	304·2	406
15	212·9	212·9	1669	71	302·8	305·2	401
16	216·3	216·4	1572	72	303·7	306·1	396
17	219·3	219·5	1487	73	304·6	307·0	391
18	222·3	222·6	1410	74	305·5	308·0	386
19	225·1	225·5	1342	75	306·4	308·9	381
20	227·9	228·3	1280	76	307·3	309·8	377
21	230·4	230·9	1224	77	308·2	310·7	372
22	232·9	233·5	1172	78	309·1	311·6	368
23	235·2	235·8	1125	79	310·0	312·6	364
24	237·6	238·3	1082	80	310·9	313·5	359
25	239·8	240·6	1042	81	311·8	314·5	355
26	242·0	242·8	1005	82	312·7	315·4	351
27	244·0	244·9	971	83	313·5	316·3	348
28	246·1	247·0	939	84	314·3	317·1	344
29	248·0	249·0	909	85	315·1	317·9	340
30	250·0	251·0	881	86	315·9	318·7	337
31	251·8	252·8	855	87	316·7	319·6	333
32	253·7	254·8	831	88	317·5	320·4	330
33	255·4	256·5	808	89	318·3	321·2	326
34	257·2	258·3	786	90	319·1	322·0	323
35	258·9	260·1	765	91	319·9	322·9	320
36	260·6	261·8	746	92	320·7	323·7	317
37	262·2	263·5	727	93	321·5	324·5	313
38	263·8	265·1	709	94	322·3	325·3	310
39	265·3	266·6	693	95	323·0	326·0	307
40	266·8	268·2	677	96	323·7	326·8	305
41	268·2	269·6	662	97	324·4	327·5	302
42	269·7	271·2	647	98	325·1	328·3	299
43	271·1	272·6	633	99	325·8	329·0	296
44	272·5	274·1	620	100	326·5	329·7	293
45	273·8	275·4	608	105	330·0	333·3	281
46	275·2	276·8	596	120	339·7	343·4	249
47	276·5	278·2	584	135	348·4	352·4	223
48	277·8	279·5	573	150	356·5	360·8	203
49	279·1	280·8	562	165	363·9	368·4	186
50	280·4	282·1	552	180	370·7	375·5	172
51	281·7	283·5	542	195	377·0	382·0	160
52	282·9	284·7	532	210	383·3	388·4	150
53	284·1	286·0	523	225	389·0	394·4	141
54	285·2	287·1	514	240	394·5	400·1	133
55	286·4	288·3	506				

* $\tau = -\frac{[2\cdot0651059]}{p^{\cdot0134} - 1\cdot17602} - 448^\circ$ † $\mu = \frac{[0\cdot4109002]\left(\frac{1}{\alpha}+\theta\right)}{p}$.

The following example will serve to show in what manner the table was calculated.

Ex.—Calculation of the temperature and volume of steam for the pressure of 180 lbs. per square inch.

$$
\begin{array}{l}
\log 180 \quad = 2{\cdot}2552725 \\
\log 14{\cdot}706 = 1{\cdot}1674946 \\
\hline
\log p = 1{\cdot}0877779 \times {\cdot}0134 = {\cdot}01457622386 = \log 1{\cdot}03413
\end{array}
$$

$$
\begin{array}{rr}
1{\cdot}17602 & \\
1{\cdot}03413 & \\
\hline
 & 2{\cdot}0651059 \\
\log {\cdot}14189 = & 9{\cdot}1519518 \\
\hline
 & 2{\cdot}9131541 = \log 818{\cdot}7 \\
 & 448{\cdot}0 \\
\hline
\text{Temp. Fahr. Air Therm.} & = 370{\cdot}7 \\
 & 4{\cdot}8 \\
\hline
\text{Temp. Fahr. Merc. Therm.} & = 375{\cdot}5
\end{array}
$$

$$
\begin{array}{rl}
 & 0{\cdot}4109002 \\
\log \frac{1}{p} = & 8{\cdot}9122221 \\
 & 2{\cdot}9131541 \\
\hline
 & 2{\cdot}2362764 = \log 172 \qquad \mu = 172
\end{array}
$$

The following data are taken from M. de Pambour's work on the Steam Engine, p. 238.

$l' = {\cdot}25\, l \qquad v = 250 \qquad f = {\cdot}5$ (lb. per square inch) $\qquad \delta = {\cdot}14$

$l = 10$ ft. $\qquad a = 12{\cdot}566$ sq. ft. $\qquad S = {\cdot}927$ cub. ft.

$p'' = 4$ lbs. per square inch. $\qquad c = {\cdot}05\, l$

$$
\begin{array}{ll}
\log v = 2{\cdot}3979400 & \log v = 2{\cdot}3979400 \\
\log a = 1{\cdot}0991971 & \log a = 1{\cdot}0991971 \\
\log {\cdot}30 = 9{\cdot}4771213 & \log 1{\cdot}05 = 0{\cdot}0211893 \\
\hline
2{\cdot}9742584 & 3{\cdot}5183264 \\
\log S = 9{\cdot}9670797 & \log S = 9{\cdot}9670797 \\
\hline
3{\cdot}0071787 = \log 1016{\cdot}6 & 3{\cdot}5512467 = \log 3558{\cdot}3 \\
\mu' = 1016{\cdot}6 & \mu = 3558{\cdot}3
\end{array}
$$

Hence by the table $p' = 25{\cdot}686$, $p = 6{\cdot}627$ reckoned in lbs. per square inch.

$p' = 1{\cdot}7466$, $p = {\cdot}4506$, $p'' + f = {\cdot}3060$ *in atmospheres.*

$\log 1{\cdot}7466 = {\cdot}2422019$, $\log {\cdot}4506 = 9{\cdot}6538224$

$${\cdot}2422019 \times {\cdot}0134 = {\cdot}00324550546$$

$$9{\cdot}9967545$$
$$\log E = 0{\cdot}0704184$$
$$0{\cdot}0671729 = \log 1{\cdot}16727$$

$$0{\cdot}3461776 \times {\cdot}0134 = {\cdot}00463877984$$
$${\cdot}0704184$$
$${\cdot}0750571 = \log 1{\cdot}18865$$

$$\log {\cdot}18865 = 9{\cdot}2756568$$
$$\log {\cdot}16727 = 9{\cdot}2234181$$
$$0{\cdot}0522387$$

$\log p = 9{\cdot}6538224$	$\log p' = 0{\cdot}2422019$	$\log (p'' + f) = 9{\cdot}4857179$
$\log a = 1{\cdot}0991971$	$\log a = 1{\cdot}0991971$	$\log a = 1{\cdot}0991971$
$\log (l + c) = 1{\cdot}0211893$	$\log c = 9{\cdot}6989700$	$\log l = 1{\cdot}0000000$
$1{\cdot}7742088$	$1{\cdot}0403690$	$1{\cdot}5849150$
$59{\cdot}457$	$10{\cdot}974$	$38{\cdot}451$
	$38{\cdot}451$	
	$49{\cdot}425$	

$4{\cdot}6411966$	
$\log l = 1{\cdot}0000000$	$\log 94{\cdot}834 = 1{\cdot}9769641$
$\log S = 9{\cdot}9671554$	$\log 14{\cdot}706 = 1{\cdot}1674946$
$8{\cdot}7179923$	$\log 144 = 2{\cdot}1583625$
$4{\cdot}3263443$	$5{\cdot}3028212$
$\log v = 2{\cdot}3979400$	$\log l\,(1 + \delta) = 1{\cdot}0569049$
$1{\cdot}9284043$	$4{\cdot}2459163$
	$a\,r = 17616$
$84{\cdot}802$	
$59{\cdot}457$	
$144{\cdot}259$	
$49{\cdot}425$	
$94{\cdot}834$	

$a\,r = 17616$ expressed in lbs., M. de Pambour finds $a\,r = 17337$.

ON THE VAPOURS OF ÆTHER, ALCOHOL, PETROLEUM, AND OIL OF TURPENTINE.

The following Table is extracted from a valuable paper by Dr. Ure in the Phil. Trans. for 1818.

Table of the pressure of the vapours of æther, alcohol, petroleum or naphtha, and oil of turpentine.

Æther.	
Temp. τ	Pressure.
Fahr. °	Inch.
34	6·20
44	8·10
54	10·30
64	13·00
74	16·10
84	20·00
94	24·70
104	30·00
2nd.	Æther.
105°	30·00
110	32·54
115	35·90
120	39·47
125	43·24
130	47·14
135	51·90
140	56·90
145	62·10
150	67·60
155	73·60
160	80·30
165	86·40
170	92·80
175	99·10
180	108·30
185	116·10
190	124·80
195	133·70
200	142·80
205	151·30
210	166·00

Alcohol sp. gr. ·0·813.	
Temp. τ	Pressure.
Fahr. °	Inch.
32	0·40
40	0·56
45	0·70
50	0·86
55	1·00
60	1·23
65	1·49
70	1·76
75	2·10
80	2·45
85	2·93
90	3·40
95	3·90
100	4·50
105	5·20
110	6·00
115	7·10
120	8·10
125	9·25
130	10·60
135	12·15
140	13·90
145	15·95
150	18·00
155	20·30
160	22·60
165	25·40
170	28·30
173	30·00
178·3	33·50
180	34·73
182·3	36·40
185·3	39·90
190	43·20

Alcohol sp. gr. 0·813.	
Temp. τ	Pressure.
Fahr. °	Inch.
193·3	46·60
196·3	50·10
200	53·00
206	60·10
210	65·00
214	69·30
216	72·20
220	78·50
225	87·50
230	94·10
232	97·10
236	103·60
238	106·90
240	111·24
244	118·20
247	122·10
248	126·10
249·7	131·40
250	132·30
252	138·60
254·3	143·70
258·6	151·60
260	155·20
262	161·40
264	166·10

Petroleum.	
Temp. τ	Pressure.
Fahr.	Inch.
316	30·00
320	31·70
325	34·00
330	36·40
335	38·90
340	41·60
345	44·10
350	46·86
355	50·20
360	53·30
365	56·90
370	60·70
372	61·90
375	64·00

Oil of Turpentine.	
Temp.	Pressure.
304°	30·00
307·6	32·60
310	33·50
315	35·20
320	37·06
322	37·80
326	40·20
330	42·10
336	45·00
340	47·30
343	49·40
347	51·70
350	53·80
354	56·60
357	58·70
360	60·80
362	62·40

From these observations I take the following data for æther.

$$p = 1 \qquad \theta = 73^\circ$$

$$p' = \frac{99{\cdot}10}{30} \qquad \theta' = 143^\circ$$

$$p'' = \frac{166{\cdot}00}{30} \qquad \theta'' = 178^\circ$$

$$\frac{p''^\beta - 1}{p'^\beta - 1} = [0{\cdot}1523534].$$

Hence for æther

$$\beta = -{\cdot}03153 \qquad \gamma = 1{\cdot}0325 \qquad E = {\cdot}67086$$

For alcohol

$$p = 1 \qquad \theta = 141{\cdot}0^\circ$$

$$p' = \frac{97{\cdot}10}{30} \qquad \theta' = 200{\cdot}0^\circ$$

$$p'' = \frac{166{\cdot}10}{30} \qquad \theta'' = 232{\cdot}0^\circ$$

$$\frac{p''^\beta - 1}{p'^\beta - 1} = [0{\cdot}1830354].$$

$$\beta = {\cdot}04025 \qquad \gamma = {\cdot}96131 \qquad E = 1{\cdot}55796$$

For petroleum

$$p = 1 \qquad \theta = 284^\circ$$

$$p' = \frac{46{\cdot}86}{30} \qquad \theta' = 318^\circ$$

$$p'' = \frac{64}{30} \qquad \theta'' = 343^\circ$$

$$\frac{p''^\beta - 1}{p'^\beta - 1} = [0{\cdot}2259762]$$

$$\beta = -0{\cdot}6268 \qquad \gamma = 1{\cdot}0668 \qquad E = {\cdot}35294.$$

For oil of turpentine

$$p = 1 \qquad \theta = 272^\circ$$

$$p' = \frac{45}{30} \qquad \theta' = 304^\circ$$

$$p'' = \frac{62{\cdot}40}{30} \qquad \theta'' = 330^\circ$$

$$\frac{p''^{\beta} - 1}{p'^{\beta} - 1} = [0{\cdot}2441091].$$

$$\beta = -{\cdot}1816 \qquad \gamma = 1{\cdot}2219 \qquad E = -{\cdot}73937$$

And hence for the vapour of æther

$$\tau = \frac{[2{\cdot}2601058]}{p^{-{\cdot}03153} - {\cdot}67086} - 448^\circ.$$

For the vapour of alcohol

$$\tau = \frac{[2{\cdot}5396942]}{p^{{\cdot}04025} - 1{\cdot}57796} - 448^\circ.$$

For the vapour of petroleum

$$\tau = \frac{[2{\cdot}6940380]}{p^{-{\cdot}06268} - {\cdot}35294} - 448^\circ.$$

For the vapour of oil of turpentine

$$\tau = \frac{[3{\cdot}1166099]}{p^{-{\cdot}1816} + {\cdot}73937} - 448^\circ.$$

The temperature being reckoned in Fahrenheit's scale and the pressure in atmospheres.

Mr. E. Russell has calculated for me the following table, showing how far the above formulæ represent the observations of Dr. Ure. The results are exhibited in the plate annexed, and it will be seen that the discrepancies between the theory here suggested and the results of observations are chiefly owing to the irregularities of the latter, which arise doubtless from the great difficulties incidental to such experiments. When the pressures are small, the variation of temperature becomes great for a small variation of pressure, so that the agreement of theory with observation may be considered as complete, even if the absolute amount of the error of the calculated temperature is then more considerable.

Æther.		
Pressure.	Temp. calc. τ	Error.
Inch.	°	°
6·20	30·8	−3·2
8·10	42 2	−1·8
10·30	52·9	−1·1
13·00	63·5	−0·5
16·10	73·6	−0·4
20·00	84·2	+0·2
24·70	94·9	+0.9
30·00*	105·0	0·0
32·54	109·3	−0·7
35·90	114·7	−0·3
39·47	119·9	−0·1
43·24	125·0	0·0
47·14	129·8	−0·2
51·90	135·4	+0·4
56·90	140·8	+0·8
62·10	145·9	+0·9
67·60	151·0	+1·0
73·60	156·2	+1·2
80·30	161·6	+1·6
86·40	166·2	+1·2
·92·80*	170·8	+0·8
99·10	175·0	0·0
108·30	180·8	+0·8
116·10	185·4	+0·4
124·80	190·2	+0·2
133·70	194·9	−0·1
142·80	199·5	−0·5
151·30	203·5	−1·5
166·00*	210·0	0·0

Alcohol. Sp. gr. 0·813.		
Pressure.	Temp. calc. τ	Error.
Inch.	°	°
0·40	33·6	+1·6
0·56	42·8	+2·8
0·70	48·1	+3·1
0·86	53·3	+3·3
1·00	57·2	+2·2
1·23	62·6	+2·6
1·49	67·8	+2·8
1·76	72·4	+2·4
2·10	77·4	+2·4
2·45	81·9	+1·9
2·93	87·3	+2·3
3·40	91·8	+1·8
3·90	96·1	+1·1
4·50	100·7	+0·7
5·20	105·4	+0·4
6·00	110·2	+0·2
7·10	116·0	+1·0
8·10	120·7	+0·7
9·25	125·5	+0·5
10·60	130·5	+0·5
12·15	135·6	+0·6
13·90	140·8	+0·8
15·95	146·3	+1·3
18·00	151·1	+1·1
20·30	156·1	+1·1
22·60	160·7	+0·7
25·40	165·7	+0·7
28·30	170·4	+0·4
30·00*	173·0	0·0
33·50	178·0	−0·3
34·73	179·6	−0·4
36·40	181·8	−0·5
39·90	186·1	+0·8
43·20	189·9	−0·1

Alcohol. Sp. gr. 0·813 (continued).		
Pressure.	Temp. calc. τ	Error.
Inch.	°	°
46·60	193·6	+0·3
50·10	197·1	+0·8
53·00	199·9	−0·1
60·10	206·3	+0·3
65·00	210·3	+0·3
69·30	213·7	−0·3
72·20	215·8	−0·2
78·50	220·3	+0·3
87·50	226·2	+1·2
94·10	230·2	+0·2
97·10*	232·0	0·0
103·60	235·6	−0·4
106·90	237·4	−0·6
111·24	239·8	−0·2
118·20	243·3	−0·7
122·10	245·2	−1·8
126·10	247·1	−0·9
131·40	249·6	−0·1
132·30	250·0	0·0
138·60	252·8	+0·8
143·70	255·9	+1·6
151·60	258·3	−0·3
155·20	259·8	−0·2
161·40	262·2	+0·2
166·10*	264·0	0·0

Petroleum.		
Pressure.	Temp. calc. τ	Error.
Inch.	°	°
30·00*	316·0	0·0
31·70	320·1	+0·1
34·00	325·4	+0·4
36·40	330·7	+0·7
38·90	335·5	+0·5
41·60	340·7	+0·7
44·10	345·3	+0·3
46·86*	350·0	0·0
50·20	355·4	+0·4
53·30	360·2	+0·2
56·90	365·4	+0·4
60·70	370·7	+0·7
61·90	372·3	+0·3
64·00*	375·0	0·0

Oil of Turpentine.		
Pressure.	Temp. calc. τ	Error.
Inch.	°	°
30·00*	304·0	0·0
32·60	310·5	+2·9
33·50	312·7	+2·7
35·20	316·6	+1·6
37·06	320·6	+0·6
37·80	322·2	+0·2
40·20	327·1	+1·1
42·10	330·8	+0·8
45·00*	336·0	0·0
47·30	340·0	0·0
49·40	343·4	+0·4
51·70	347·0	0·0
53·80	350·2	+0·2
56·60	354·3	+0·3
58·70	357·1	+0·1
60·80	359·9	−0·1
62·40*	362·0	0·0

The observations marked with an asterisk are those which were employed in procuring the constants β, E.

The æther which was observed by Dr. Ure below the pressure of 30 inch. appears to have been slightly different in quality from the other; in estimating the comparison this circumstance should be born in mind.

ON THE CONDITIONS OF THE ATMOSPHERE, AND ON THE CALCULATION OF HEIGHTS BY THE BAROMETER.

The same principles are applicable to the constitution of the atmosphere; but we are far from possessing such extensive and satisfactory data for testing the accuracy of the formulæ. The best observations for this purpose are those of M. Gay Lussac, recorded by M. Biot in the *Connaissance des Temps* for 1841, in the following table.

Table des observations par ordre de hauteurs barométriques.

Numéro des observations par ordre de pression.	Températures en degrés du thermomètre contésimal.	Moyenne des indications des deux hygromètres.	Hauteur moyenne du baromètre dans l'atmosphère, ramenée à celle d'un baromètre à niveau constant.	Hauteurs correspondantes au-dessus de l'Observatoire de Paris, calculées par la formule barométrique de M. Laplace.
	°	°	m	m
1	+30·75	57·5	0·76568	0·00
2	12·50	62·0	0·5381	3032·01
3	11·00	50·0	0·5143	3412·11
4	8·50	37·3	0·4968	3691·45
5	10·50	33·0	0·4905	3816·79
6	12·00	30·9	0·4666	4264·65
7	11·00	29·9	0·4626	4327·86
8			0·4528	4511·61
9	8·75	29·4	0·4528	4511·61
10	8·25	27·6	0·4404	4725·90
11	6·50	27·5	0·4353	4808·74
12	5·25	30·1	0·4229	5001·85
13	1·00	33·0	0·4141	5175·06
14	4·25	27·5	0·4114	5267·73
15	2·50	32·7	0·3985	5519·16
16	0·00	35·1	0·3918	5631·65
17	+ 0·50	30·2	0·3901	5674·85
18	− 3·00	32·4	0·3717	6040·70
19	− 1·50	32·1	0·3696	6107·19
20	− 3·25	33·9	0·3670	6143·79
21	− 7·00	34·5	0·3339	6884·14
22	− 9·50		0·3288	6977·47

I shall employ the 1st, 5th, and 21st observations for the determination of the constants, and I propose then to calculate with these constants the temperatures corresponding to the intermediate observations. As the pressures are proportional to the heights of the barometer, if the variation of gravity be neglected we may take the heights of the barometer to represent them, and we have

$p = \cdot 76568$ $\theta = 30^\circ \cdot 75$ $\frac{1}{\alpha} = 266 \cdot 67$

$p' = \cdot 4905$ $\theta' = 10^\circ \cdot 50$ $\frac{1}{\alpha} + \theta' = 277 \cdot 17$

$p'' = \cdot 3339$ $\theta'' = -7^\circ \cdot 00$ $\frac{1}{\alpha} + \theta'' = 259 \cdot 67$

I find

$$\frac{\left(\frac{p''}{p}\right)^\beta - 1}{\left(\frac{p'}{p}\right)^\beta - 1} = \frac{(\theta'' - \theta)\left(\frac{1}{\alpha} + \theta'\right)}{(\theta' - \theta)\left(\frac{1}{\alpha} + \theta''\right)}$$

$$= [0 \cdot 2988164].$$

The quantity between brackets being the logarithm of the corresponding number

$$\beta = -\cdot 32931 \qquad \gamma = 1 \cdot 4910$$

$$E = \frac{\left(\frac{p''}{p}\right)^\beta\left(\frac{1}{\alpha} + \theta''\right) - \left(\frac{1}{\alpha} + \theta\right)}{\theta'' - \theta}$$

$$= -1 \cdot 1618 \qquad H = \cdot 53772$$

$$\tau = \frac{[2 \cdot 8081857]}{p^{-\cdot 32931} + 1 \cdot 1618} - 266^\circ \cdot 67$$

in the centigrade scale, the pressure corresponding to $\cdot 76568^{\mathrm{m}}$ of mercury in the barometer being unity. In Fahrenheit's scale,

$$\tau = \frac{[3 \cdot 0634582]}{p^{-\cdot 32931} + 1 \cdot 1618} - 448^\circ,$$

the pressure corresponding to 30·14 inches of mercury being unity.

If we take $\gamma = 1 \cdot 5$, assuming the 21st observation of M. Gay Lussac, $E = -1 \cdot 1920$.

The difference in the results obtained with these constants from those obtained with the other system of constants $\gamma = 1 \cdot 4910$ and $E = -1 \cdot 1618$, is quite insignificant, only changing the density slightly in the fifth place of decimals. By taking $\gamma = 1 \cdot 5$

$$\varrho' = \varrho\,(1 - q)^2\,(1 - H\,q),$$

so that the expression for the density becomes more simple, consisting of only three terms, c^{-3u}, c^{-2u}, c^{-u}, (as will be seen hereafter), which is advantageous in the theory of astronomical refractions.

Mr. Russell has calculated for me the following table, by means of my expressions, and with the constants

$$\gamma = 1{\cdot}5, \quad E = -1{\cdot}192$$

No.	Observed pressure p.	Calculated height in miles.	Temperature τ.		Density ϱ.
			Calculated.	Observed.	Calculated.
1	0·76568	0·000	+30°·75	+30°·75*	·99999
2	·5381	1·895	14·74	12·50	·74276
3	·5143	2·131	12·68	11·00	·71514
4	·4968	2·310	11·10	8·50	·69473
5	·4905	2·376	10·52	10·50*	·68736
6	·4666	2·632	8·24	12·00	·65928
7	·4626	2·676	7·85	11·00	·65457
8					
9	·4528	2·785	6·87	8·75	·64299
10	·4404	2·926	5·61	8·25	·62828
11	·4353	2·985	4·99	6·50	·62222
12	·4229	3·130	3·67	5·25	·60744
13	·4141	3·236	2·80	1·00	·59692
14	·4114	3·269	2·61	4·25	·59346
15	·3985	3·428	1·05	2·50	·57818
16	·3918	3·512	0·28	0·00	·57010
17	·3901	3·533	+ 0·08	+ 0·50	·56805
18	·3717	3·772	− 2·12	− 3·00	·54576
19	·3696	3·800	− 2·37	− 1·50	·54321
20	·3670	3·835	− 2·70	− 3·25	·54004
21	·3339	4·295	− 7·00	− 7·00*	·49947
22	·3288	4·369	− 7·70	− 9·50	·49317

The observations marked with an asterisk are those which were employed in deducing the constants.

The temperatures calculated by Mr. Russell from my formula may be considered as identical with the "température régularisée par la continuité", given by M. Biot in the *Conn. des Temps*, 1841, p. 13. The observations* of M. Gay Lussac of temperature are represented in the following diagram:

* The irregularities of the observations of temperature in any future ascent might perhaps be diminished if the ballast were suffered to escape gradually in a continued stream.

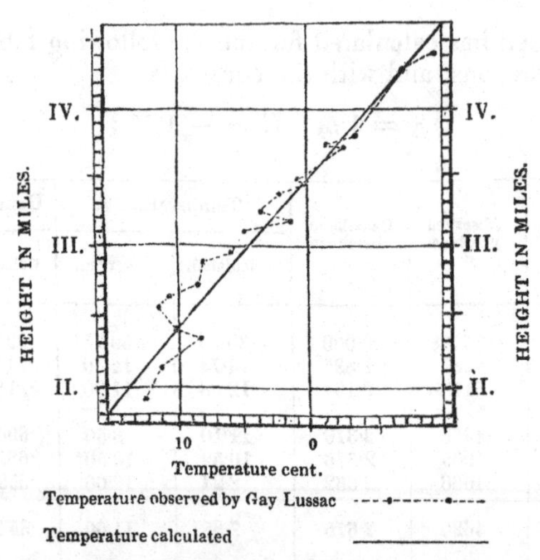

The abscissa represents the temperature in degrees of the centigrade thermometer, and the ordinate the height in miles.

At first the decrements of temperature are nearly equal for equal increments of altitude. These observations by no means furnish so good a criterion of the accuracy of my formula as the observations which have been made of the temperature of steam and other vapours. The determination of the constants γ and E for the atmosphere must be repeated at some future time; for it is obvious that no great reliance can be placed upon the extreme precision of the values now obtained * until other ascents have been made, and many similar observations have been compared together. We may then hope to obtain constants accurately appertaining to a mean state of the atmosphere, and the variations which take place in their values corresponding to fluctuations of the temperature; the pressure and the humidity of the atmosphere at the earth's surface may then be investigated. M. Biot has suggested that balloons furnished with self-registering instruments should be moored over each of the principal observatories of Europe. This plan appears to me subject to great difficulties. The weight of the line attached will diminish the buoyancy, so that I apprehend it will be found impossible to send up a balloon so fastened to any considerable altitude. The escape of gas will, I imagine, render it very difficult to maintain the balloon at the same height for any length of time. The height of the balloon will also be subject to great variations from the tension of the line changing with the force and direction of the wind. I am disposed to attach much greater

* Dulong found, for atmospheric air, perfectly dry, $\gamma = 1{\cdot}421$. See Poisson, *Méc.*, tom. ii. p. 646.

value to well-regulated ascents, in which every effort should be made to reach the highest possible altitude, and of course simultaneous observations should be made at the earth's surface at such short intervals of time that every observation of the aëronaut may be comparable with a similar observation at the surface of the earth. As, however, the density and temperature of the atmosphere above the height of five miles from the earth's surface can never be the subject of direct experiment and observation, the observations which can be made upon the conditions of steam and other vapours will always maintain an indirect importance from the light which they throw upon the conditions of the atmosphere. I do not think that an examination of observations made in aërostatic ascents will ever furnish a sure guide to the relation sought between the temperature and the pressure, although if such a relation is furnished by theory and corroborated by observations of other vapours, (which can be carried through a greater extent of the thermometric scale, and, above all, through the low pressures where the variations of temperature become more rapid,) the obervations of aëronauts may serve to determine with sufficient accuracy the constants involved in the formula for atmospheric air.

The following table,* calculated by Mr. Russell, shows the density and temperature of the air at different altitudes, calculated by means of my expressions and with the constants

$$\gamma = 1{\cdot}5, \quad E = -1{\cdot}192:$$

Height in miles.	Pressure p.		Temperature τ.		Density ϱ.
	Metres.	Inches.	Cent.	Fahr.	
0	0·76568	30·145	+ 30·75°	87·35°	1·00000
1	·63718	25·087	22·43	72·37	·85614
2	·52741	20·764	13·83	56·89	·73038
3	·43399	17·086	+ 4·94	40·89	·62066
4	·35480	13·968	− 4·23	24·39	·52515
5	·28799	11·336	13·71	+ 7·32	·44223
6	·23189	9·130	23·50	− 10·30	·37042
7	·18506	7·286	33·60	28·48	·30844
8	·14621	5·756	44·05	47·29	·25512
9	·11420	4·496	54·82	66·68	·20942
10	·08806	3·467	65·96	86·73	·17042
15	·01806	0·711	127·32	197·18	·05034
20	·00125	0·049	199·38	326·88	·00720
24	·00000	0·000	−265·95	−446·71	·00000

* In calculating this table, the law of Marriotte and Gay Lussac, expressed by the equation

$$p = k\varrho\,(1 + \alpha\,\theta),$$

has been implicitly supposed to hold good throughout: this of course is only conjectural, and it is not intended to attach precision to the temperatures assigned to the great altitudes.

As the expression which has served to calculate the temperatures evidently represents the state of the atmosphere far within the limits of the applicability of this or any other formula founded upon a state of repose to an atmosphere continually agitated by currents, it must of course serve to eliminate the density and to obtain an expression for the height in terms of the pressures and temperatures at the extremities of any atmospheric column.

If z be the altitude of the place above any fixed point, a the distance of the fixed point from the centre of the earth, g the force of gravity,

$$\frac{\mathrm{d}\,p}{\varrho} = -\frac{g\,a^2}{(a+z)^2}\,\mathrm{d}\,z,$$

and putting the expression for ϱ' at p. 3,

$$\frac{k\,(1+\alpha\,\theta)\,(p^{\beta}-E)\,\mathrm{d}\,p'}{p'\,(p'^{\beta}-E)} = -\frac{g\,a^2\,\mathrm{d}\,z'}{(a+z')^2}.$$

This expression can be integrated, and I find, supposing $z=0$, after a proper determination of the constants,

$$\frac{z'}{1+\frac{z'}{a}} =$$

$$\frac{k\,(1+\alpha\,\theta)}{g}\,\frac{\left\{p^{\beta}-p'^{\beta}\right\}\left\{\frac{1}{\alpha}+\theta'\right\}}{\beta\left\{p'^{\beta}\left(\frac{1}{\alpha}+\theta'\right)-p^{\beta}\left(\frac{1}{\alpha}+\theta\right)\right\}}\ \text{Nap. log}\left\{\frac{\left(\frac{1}{\alpha}+\theta'\right)}{\left(\frac{1}{\alpha}+\theta\right)}\left(\frac{p'}{p}\right)^{\beta}\right\}$$

If the variation of the force of gravity be neglected the pressures p, p' may be represented by the heights of the barometer h, h'. If M be the *modulus* or the quantity by which Naperian logarithms must be multiplied to give common logarithms, Laplace makes

$$\frac{k}{g\,M} = 18337^{\mathrm{m}}\cdot 46. \qquad \log M = 9\cdot 6377843.$$

In order to give an example of the use of this expression I take the 21st observation of Gay Lussac

$$h = \cdot 76568 \qquad \theta = \ 30\cdot 75$$
$$h' = \cdot 3339 \qquad \theta' = -\ 7\cdot 00$$

$$\log 18337{\cdot}46 = 4{\cdot}2633392$$

$$\log (1 + \alpha\theta) = 0{\cdot}0474015$$

$$\log \left\{ \frac{\left(h^{\beta} - h'^{\beta}\right)\left(\frac{1}{\alpha} + \theta'\right)}{h'^{\beta}\left(\frac{1}{\alpha} + \theta'\right) - h^{\beta}\left(\frac{1}{\alpha} + \theta\right)} \right\} = 0{\cdot}2696699$$

$$\log \left\{ \log \left\{ \frac{\left(\frac{1}{\alpha} + \theta'\right)}{\left(\frac{1}{\alpha} + \theta\right)} \left(\frac{h'}{h}\right)^{\beta} \right\} \right\} = 8{\cdot}7762776$$

$$3{\cdot}3566882$$

$$\log \beta \qquad 9{\cdot}5176049$$

$$3{\cdot}8390833$$

$$\frac{z'}{1 + \frac{z'}{a}} = 6903{\cdot}7$$

$\log a = 6{\cdot}8041168$ in metres.
$z' = 6921{\cdot}7$ metres.

If

$$\left(\frac{p'}{p}\right)^{\frac{\gamma-1}{\gamma}} = 1 - q \qquad \frac{E p^{\frac{1-\gamma}{\gamma}}}{1 - E p^{\frac{\gamma-1}{\gamma}}} = -H \text{ as before, p. 3.}$$

The expression for z' may be put into the form

$$\frac{z'}{1 + \frac{z'}{a}} = -\frac{k(1 + \alpha\theta)}{g H \beta} \text{ Nap. log } (1 - Hq).$$

If $\gamma = 1{\cdot}49138$ when $p' = 0$, $q = 1$, we get for the superior limit of the atmosphere an altitude of about 24 miles, or 38918 metres.

Ultimately the intensity of the cold deprives the air of its elasticity*. The density therefore requires in strictness to be re-

* See Poisson, *Théorie de la Chaleur*, p. 460, "On peut se représenter une colonne atmosphérique qui s'appuie sur la mer, par example, comme un fluide élastique terminé par deux liquides, dont l'un a une densité et une température ordinaires, et l'autre une température et une densité excessivement faibles." See also Biot, *Conn. de Temps*, 1841.

presented by a discontinuous function; for the formula suggested in this treatise is of course only applicable so long as the air exists in the state of an elastic vapour. The freezing point of air is unknown, and we cannot decide when this condition ceases to obtain.

Delambre estimates the height of the atmosphere as deduced from the phenomena of twilight* at 70,800 metres; but this calculation is open to objection. See *Conn. de Temps*, 1841, p. 58.

I have given the example of the calculation of a height by an observation of the barometer, in order to show how my formula for the density may be employed; but however inaccurate in principle the method in use may be, it is sufficiently exact for elevations accessible to man. In all inquiries, however, connected with the condition of the higher regions of the atmosphere, and in the various hypotheses which may be made respecting the decrement of temperature, the corresponding height must be calculated by an appropriate formula, procured agreeably to the hypothesis which may be adopted. Our information respecting the state of the higher regions of the atmosphere is I think more likely to be improved by observations made in aëronautic ascents than by those made on the sides of mountains.

Let $u = -$ Nap. log $(1 - Hq)$ $\qquad i = \dfrac{k(1+\alpha\theta)}{agH\beta}$

$$\frac{z'}{1+\dfrac{z'}{a}} = a\,i\,u.$$

At the summit of the atmosphere $q = 1$, if u'' be the corresponding value of u,

$$u'' = -\text{ Nap. log}(1 - H) \qquad c^{-u''} = 1 - H,$$

c being the number of which the hyperbolic logarithm is unity.

$$\frac{E\,p^{-\beta}}{1 - E\,p^{-\beta}} = -H. \quad \text{See p. 3.}$$

p being the pressure at the lower station; the pressure for $\cdot 76568^{\text{m}}$ or 30·14 inches of mercury in the barometer being unity.

I get, when

$$\gamma = 1{\cdot}5 \qquad \beta = -\frac{1}{3} \qquad E = -1{\cdot}192,$$

the following formula for calculating heights by observations of the barometer:

* See Delambre's *Astronomie*, vol. i. p. 337, and Lalande's *Ast.*, vol. ii. art. 2270.

$$\frac{z'}{1+\frac{z'}{a}} = [4{\cdot}7404605] \frac{(1+\alpha\theta)}{H} \log(1 - Hq) \text{ in French metres,}$$

$$= [5{\cdot}2564585] \frac{(1+\alpha\theta)}{H} \log(1 - Hq) \text{ in English feet,}$$

$$= [1{\cdot}5338195] \frac{(1+\alpha\theta)}{H} \log(1 - Hq) \text{ in English miles}$$

the temperature θ at the lower station being reckoned from the freezing point.

$$\text{Log}\,\alpha = 7{\cdot}3187588 \text{ for Fahrenheit's scale.}$$

If we assume the 21st observation of Gay Lussac, and suppose $\gamma = 1{\cdot}4$, I find

$$\beta = -\,{\cdot}2857 \qquad E = -\,{\cdot}8405 \qquad \log H = 9{\cdot}6596173$$

In Fahrenheit's scale

$$\tau = \frac{[2{\cdot}9935785]}{p^\beta + {\cdot}8405} - 448^\circ.$$

Height in miles $= [1{\cdot}9885722] \log(1 - Hq)$.

If we suppose $\gamma = 1{\cdot}5$, I find

$$\beta = -\,{\cdot}3333 \qquad E = -\,1{\cdot}1920 \qquad \log H = 9{\cdot}7354232$$

$$\tau = \frac{[30{\cdot}694832]}{p^\beta + 1{\cdot}1920} - 448^\circ.$$

Height in miles $= [1{\cdot}8457978] \log(1 - Hq)$.

If we suppose $\gamma = 1{\cdot}6$, I find

$$\beta = -\,{\cdot}375 \qquad E = -\,1{\cdot}5112 \qquad \log H = 9{\cdot}7794573$$

$$\tau = \frac{[3{\cdot}1285240]}{p^\beta + 1{\cdot}5112} - 448^\circ.$$

Height in miles $= [1{\cdot}7506111] \log(1 - Hq)$.

Mr. Russell has calculated for me the following table in order to show in what manner the density and temperature of the atmosphere vary in the higher regions under these three different suppositions.

Height in miles.	$\beta = -{\cdot}2857$.			$\beta = -\frac{1}{3}$.			$\beta = -{\cdot}375$.			Height in miles.
	p.	τ.	ϱ.	p.	τ.	ϱ.	p.	τ.	ϱ.	
0	1·0000	+87°	1·0000	1·0000	+87°	1·0000	1·0000	+87°	1·0000	0
4	·4628	+24	·5248	·4630	+24	·5248	·4631	+25	·5249	4
8	·1906	−45	·2534	·1902	−48	·2543	·1900	−50	·2553	8
12	·0656	−122	·1076	·0645	−130	·1086	·0635	−137	·1093	12
16	·0167	−206	·0367	·0153	−223	·0365	·0141	−240	·0362	16
20	·0022	−298	·0080	·0015	−330	·0068	·0009	−361	·0055	20
24	·0000	−399	·0003							24
	Limit 25·81 miles.			23·896 miles.			22·52 miles.			

By making $\gamma = 1{\cdot}5$, the expression for the density becomes simplified, $\frac{1}{1-\gamma} = -2$,

$$\varrho' = \frac{\varrho}{H^2} c^{-u} \left\{ c^{-u} - 1 + H \right\}^2 .$$

See p. 3.

If $\frac{\varrho'}{\varrho} = 1 - \omega$

$$1 - \omega = \frac{1}{H^2} \left\{ c^{-3u} - 2(1-H)c^{-2u} + (1-H)^2 c^{-u} \right\}$$

It must be recollected that the difficulty of determining the densities at different altitudes, and that of determining altitudes by observations of the barometer, rest in finding the accurate law of the temperature. So that if the expression which I have here suggested for the temperature be adopted, the expression for the density, and those for finding the elevation by observations of the barometer, follow as a matter of course, and their accuracy is unquestionable.

The employment of the formula in p. 26, for calculating heights, amounts to determining the constant E from the observations themselves, and not from previous observations. But if the constants are supposed to be known, as in calculating a series of observations made under the same circumstances, it is more simple to employ the expression

$$\frac{z'}{1 + \frac{z'}{a}} = a\,i\,u.$$

The day on which M. Gay Lussac made his ascent was very

warm, and the values of γ and H determined from his observations may differ slightly from those mean values which will be obtained hereafter from more complete data. The preceding theory supposes implicitly that a given temperature at the earth's surface always corresponds in any given place to a given pressure; this, owing to the currents, the winds, and to other causes, is not the case; for the atmosphere is never in a state of repose, and its temperature and density are in a continual state of oscillation about their mean values. The constants γ and E may also be subject to variations from fluctuations in the quantity of aqueous vapour diffused through the atmosphere.

If the decrements of temperature are the same for equal increments of altitude, which observation shows is nearly the case at small elevations,

$$\theta - \theta' = A\, z',$$

θ being the temperature at the lower station, θ' at the upper, and z' as before, the altitude of the latter reckoned from the former,

$$1 + \alpha\,\theta' = 1 + \alpha\,(\theta - A\, z'),$$

and if the variation of the force of gravity be neglected

$$\frac{d\,p'}{p'} = -\frac{g\,d\,z'}{k\,\{1 + \alpha\,(\theta - A z')\}}$$

$$z' = \frac{(1 + \alpha\,\theta)}{\alpha\,A}\left\{1 - \left(\frac{p'}{p}\right)^{\frac{k\,\alpha\,A}{g}}\right\}$$

$$\varrho' = \varrho\left\{\frac{1 + \alpha\,(\theta - A\,z')}{1 + \alpha\,\theta}\right\}^{\frac{g}{k\,\alpha\,A} - 1}$$

p' being the pressure at the upper station, and p at the lower.

Mr. Ivory assumes, Phil. Trans., 1838, p. 192,

$$\frac{1 + \alpha\,\theta'}{1 + \alpha\,\theta} = 1 - f \log\frac{\varrho}{\varrho'} + (f - f')\,\frac{\left(\log\frac{\varrho}{\varrho'}\right)^2}{1\,.\,2}$$

$$-\,(f - 2f + f'')\,\frac{\left(\log\frac{\varrho}{\varrho'}\right)^3}{1\,.\,2\,.\,3} + \&c.$$

But Mr. Ivory afterwards neglects the terms depending upon f', f'', &c., so that he virtually assumes

$$\frac{1+\alpha\,\theta'}{1+\alpha\,\theta} = 1 - f\left\{\log\frac{\varrho}{\varrho'} - \frac{\left(\log\frac{\varrho}{\varrho'}\right)^2}{1\,.\,2} + \&c.\right\}$$

$$= 1 - f\left\{1 - \frac{\varrho'}{\varrho}\right\} = \frac{p'\,\varrho}{p\,\varrho'}$$

$$\frac{p'}{p} = (1-f)\,\frac{\varrho'}{\varrho} + f\,\frac{\varrho'^2}{\varrho^2}$$

Mr. Ivory makes the constant $f = \frac{2}{9}$, p. 197, so that

$$\frac{p'}{p} = [9{\cdot}8908555]\,\frac{\varrho'}{\varrho} + [9{\cdot}3467875]\,\frac{\varrho'^2}{\varrho^2}$$

$$\mathrm{d}\,p' = p\,(1-f)\,\frac{\mathrm{d}\,\varrho'}{\varrho} + 2\,p\,f\,\varrho'\,\frac{\mathrm{d}\,\varrho'}{\varrho^2}$$

$$= -\,\frac{g\,\varrho'\,\mathrm{d}\,z'}{\left(1+\frac{z'}{a}\right)^2}$$

$$\frac{z'}{1+\frac{z'}{a}} = a\,i\,u^* = \frac{p\,(1-f)}{g\,\varrho}\,\log\frac{\varrho}{\varrho'} + \frac{2\,p\,f}{g\,\varrho}\left(1 - \frac{\varrho'}{\varrho}\right)$$

$$= \frac{k\,(1+\alpha\,\theta)\,(1-f)}{g}\,\log\frac{\varrho}{\varrho'} + \frac{2\,k\,(1+\alpha\,\theta)\,f}{g}\left(1-\frac{\varrho'}{\varrho}\right)$$

$$= [0{\cdot}9635418]\,\log\frac{\varrho}{\varrho'} + [0{\cdot}3582881]\left(1-\frac{\varrho'}{\varrho}\right)$$

for 50° Fahr. at the lower station.

As we cannot make direct observations of the temperature and density of the highest regions of the atmosphere, it becomes very important to avail of all indirect means of investigation. The problem of Astronomical Refractions furnishes us with valuable data in this respect, and any hypothesis relative to the state of the atmosphere which will not satisfy the known phenomena of refraction must of course be discarded. In any investigation of this kind it is indispensable to employ a formula for z in terms of the density consistent with the hypothesis, which may be made respecting the decrement of temperature; it is equally indispensable to carry the integral which affords the amount of refraction through limits which are in conformity with the same supposition.

---

* $a\,i\,u = \sigma$ in Mr. Ivory's notation. In this page p is the pressure and ϱ is the density at the earth's surface.

ON

ASTRONOMICAL REFRACTIONS.

IF the constitution of the atmosphere be such as I have concluded, by proper substitutions in the differential equations of refraction, an accurate table of refractions is to be procured, which may be compared with that of M. Bessel obtained empirically.

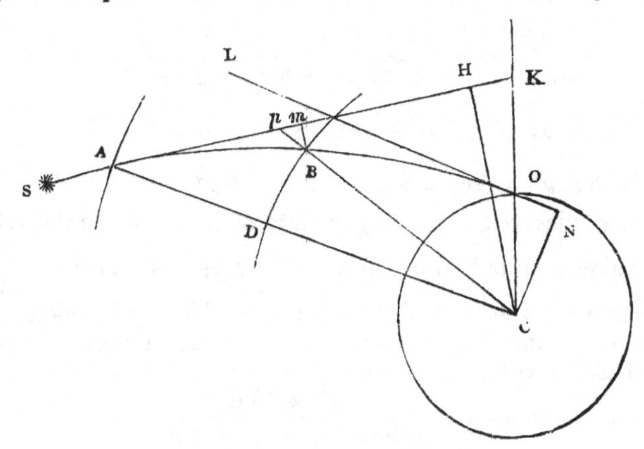

Let S A O N be the trajectory described by light emanating from the star S in its passage through the atmosphere to the earth's surface at O, θ the apparent zenith distance, or the angle which the tangent to the trajectory makes with the line C O K at O, C H perpendicular to S A K, the direction of the ray before it enters the atmosphere $= y$, $a =$ C O, then

$$\mathrm{d} . \delta \theta = \frac{\mathrm{d} y}{\sqrt{(a + z)^2 - y^2}}$$

$$y = a \sin \vartheta \sqrt{\frac{1 + 2 \mathrm{K} \rho}{1 + 2 \mathrm{K} \rho'}}$$

$$*\alpha = \frac{K\rho}{1 + 2K\rho} \qquad y = \frac{a \sin\theta}{\sqrt{1 - 2\alpha\omega}}$$

I assume these equations, which are proved by Mr. Ivory in the *Phil. Trans.*, 1838, and which are equivalent to similar equations given in the *Méc. Céleste.*

$$d.\delta\theta = \frac{\alpha \sin\theta \, d\omega}{(1 - 2\alpha\omega)\sqrt{\cos^2\theta + \left(\frac{z^2}{a} + \frac{z^2}{a^2}\right)(1 - 2\alpha\omega) - 2\alpha\omega}}$$

$\frac{z}{1 + \frac{z}{a}} = a\,i\,u.$ i being a constant and u a certain function of the density, which depends upon the constitution of the atmosphere, and which for the present may remain undefined.

$$\frac{z^2}{a} + \frac{z^2}{a^2} = 2iu + 3i^2u^2 + \&c.$$

$$d.\delta\theta = \frac{\alpha \sin\theta \, d\omega \{1 + 2\alpha\omega + \&c.\}}{\sqrt{\cos^2\theta + 2iu + 3i^2u^2 + \&c. - 2\alpha\omega}}$$

$$\text{if } x = u - \frac{\alpha}{i}\omega, \qquad iu - \alpha\omega = ix \qquad \omega = 1 - \frac{\varrho}{\varrho'}$$

$$2iu + 3i^2u^2 + \&c. - 2\alpha\omega = 2x + 3x^2.$$

"The quantities rejected being plainly of no account relatively to those retained. Further, because ω is always less than 1, $\frac{\alpha}{1 - 2\alpha\omega}$ is contained between α and $\alpha(1 + 2\alpha)$, and it may be taken equal to α, or to the mean value $\alpha(1 + \alpha)$†." Thus we have (See Phil. Trans. 1838. p. 205.)

$$d.\delta\theta = \sin\theta \times \frac{\alpha(1 + \alpha)\, d\omega}{\sqrt{\cos^2\theta + 2ix + 3i^2x^2}}$$

$$= \sin\theta \times \frac{\alpha(1 + \alpha)\, d\omega}{\sqrt{\cos^2\theta + 2ix}}$$

[1]

$$- \frac{3}{2}\,\frac{\sin\theta\,\alpha(1 + \alpha)\, i^2x^2\, d\omega}{(\cos^2\theta + 2ix)^{\frac{3}{2}}} + \&c.$$

[2]

The refraction will thus consist of two terms, which I proceed

---

* This quantity must not be confounded with the α which accompanies θ.

† Laplace introduces the same simplification. *Méc. Cél.*, vol. iv. p. 247.

to consider separately. The second term is minute, not amounting to 2″ sex. at the horizon.

Throughout this treatise on Astronomical Refractions one accent will be affixed to any symbol that it may denote the particular value of the variable which obtains at the surface of the earth, and two accents will be affixed when the particular value which obtains at the superior limit of the atmosphere is intended.

The limits of x, or x' and x'' corresponding to u' and u'', are

$x' = 0,\ x'' = u'' - \frac{\alpha}{i}$, because $\omega' = 0,\ \omega'' = 1$.

$x = u - \frac{\alpha}{i}\,\omega = u - \frac{\alpha}{i}\,\mathrm{f}\,u$; the function indicated by

the letter f for the present may remain undefined.

By Lagrange's theorem

$$u = x + \frac{\alpha}{i}\,\mathrm{f}\,x + \frac{\alpha^2}{2\,i^2}\,\frac{\mathrm{d}\,(\mathrm{f}\,x)^2}{\mathrm{d}\,x} + \frac{\alpha^3}{2\,.\,3\,i^3}\,\frac{\mathrm{d}^2\,(\mathrm{f}\,x)^3}{\mathrm{d}\,x^2} + \&c.$$

$$= x + \frac{\alpha}{i}\,\omega.$$

Hence

$$\omega = \mathrm{f}\,x + \frac{\alpha}{2\,i}\,\frac{\mathrm{d}\,(\mathrm{f}\,x)^2}{\mathrm{d}\,x} + \frac{\alpha^2}{2\,.\,3\,i^2}\,\frac{\mathrm{d}^2\,(\mathrm{f}\,x)^3}{\mathrm{d}\,x^2} + \&c.$$

Let $x = x'' - X,\quad u = u'' - U,\quad \omega = 1 - \mathrm{F}\,U$, then

$$x'' - X = u'' - U - \frac{\alpha}{i}\,\{1 - \mathrm{F}\,U\}$$

$$X'' = 0,\quad X' = x''$$

$$X = U - \frac{\alpha}{i}\,\mathrm{F}\,U$$

$$U = X + \frac{\alpha}{i}\,\mathrm{F}\,X + \frac{\alpha^2}{2\,i^2}\,\frac{\mathrm{d}(\mathrm{F}\,X)}{\mathrm{d}\,X} + \&c.$$

$$\omega = 1 - \mathrm{F}\,X - \frac{\alpha}{2\,i}\,\frac{\mathrm{d}\,(\mathrm{F}\,X)^2}{\mathrm{d}\,X} - \frac{\alpha^2}{2\,.\,3\,i^2}\,\frac{\mathrm{d}^2\,(\mathrm{F}\,X)^3}{\mathrm{d}\,X^2} - \&c.$$

This series may be used if the atmosphere extends only to a finite altitude.

Let

$$\sqrt{\frac{\cos^2\theta + 2\,i\,x}{i}} = z \qquad \cos^2\theta + 2\,i\,x = i\,z^2$$

$$\mathrm{d}\,x = z\,\mathrm{d}\,z \qquad x^2 = \frac{i^2\,z^4 - 2\,i\,z^2\cos^2\theta + \cos^4\theta}{4\,i^2}$$

The integral of $d\,.\,\delta\theta$ is to be taken from

$$z = \frac{\cos\theta}{\sqrt{i}} = z', \qquad \text{to } z = \frac{\sqrt{\cos^2\theta + 2\,i\,x''}}{\sqrt{i}} = z''.$$

Let $\left(\frac{d^n\,\omega}{d\,x^n}\right)'$ represent the value of $\left(\frac{d^n\,\omega}{d\,x^n}\right)$ at the former of these limits, and $\left(\frac{d^n\,\omega}{d\,x^n}\right)''$ its value at the latter, then integrating continually by parts,

$$\int \frac{\alpha \sin\theta\, d\,\omega}{\sqrt{\cos^2\theta + 2\,i\,x}} = \int \frac{\alpha \sin\theta \frac{d\,\omega}{d\,x} d\,x}{\sqrt{\cos^2\theta + 2\,i\,x}}$$

$$= \frac{\alpha \sin\theta}{\sqrt{i}} \left\{ \left(\frac{d\,\omega}{d\,x}\right)'' z'' - \left(\frac{d\,\omega}{d\,x}\right)' z' \right.$$

$$- \frac{1}{3}\left\{\left(\frac{d^2\,\omega}{d\,x^2}\right)'' z''^3 - \left(\frac{d^2\,\omega}{d\,x^2}\right)' z'^3\right\} \qquad [1]$$

$$\left. + \frac{1}{3\,.\,5}\left\{\left(\frac{d^3\,\omega}{d\,x^3}\right)'' z''^5 - \left(\frac{d^3\,\omega}{d\,x^3}\right)' z'^5\right\} \right\},$$

&c.

The second term is

$$- \frac{3\,\alpha \sin\theta\, i^2\, x^2\, d\,\omega}{2\,(\cos^2\theta + 2\,i\,x)^{\frac{5}{2}}}$$

$$= - \frac{3\,\alpha \sin\theta}{8\,i^{\frac{3}{2}}} \left\{ i^2 z^2 - 2\,i\cos^2\theta + \frac{\cos^4\theta}{z^2} \right\} \frac{d\,\omega}{d\,x} d\,\omega,$$

the integral of which is

$$- \frac{3\,\alpha \sin\theta}{8\,i^{\frac{3}{2}}} \left\{ \frac{d\,\omega}{d\,x}\left\{\frac{i^2 z^3}{3} - 2\,i\,z\cos^2\theta - \frac{\cos^4\theta}{z}\right\} \right.$$

$$- \frac{d^2\,\omega}{d\,x^2}\left\{\frac{i^2 z^5}{3\,.\,5} - \frac{2\,i\,z^3\cos^2\theta}{3} - z\cos^4\theta\right\}$$

$$\left. + \frac{d^3\,\omega}{d\,x^3}\left\{\frac{i^2 z^7}{3\,.\,5\,.\,7} - \frac{2\,i\,z^5}{3\,.\,5}\cos^2\theta - \frac{z^3}{3}\cos^4\theta\right\} \right\},$$

$- \text{\&c.}$

$$= \frac{3\,\alpha \sin\theta}{S\,\sqrt{i}} \left\{ \frac{1}{z} \left\{ \frac{1}{i}\frac{d\,\omega}{d\,x} \cos^4\theta \right\} \right.$$

$$+ z \left\{ -\frac{1}{i}\frac{d^2\,\omega}{d\,x^2} \cos^4\theta + z\frac{d\,\omega}{d\,x} \cos^2\theta \right\}$$

$$+ \frac{z^3}{3} \left\{ \frac{1}{i}\frac{d^3\,\omega}{d\,x^3} \cos^4\theta + z\frac{d^2\,\omega}{d\,x^2} \cos^2\theta - i\frac{d\,\omega}{d\,x} \right\} \qquad [2]$$

$$+ \frac{z^5}{3\,.\,5} \left\{ -\frac{1}{i}\frac{d^4\,\omega}{d\,x^4} \cos^4\theta + z\frac{d^3\,\omega}{d\,x^3} \cos^2\theta + i\frac{d^2\,\omega}{d\,x^2} \right\}$$

$$\left. + \frac{z^7}{3\,.\,5\,.\,7} \left\{ \frac{1}{i}\frac{d^5\,\omega}{d\,x^5} \cos^4\theta - z\frac{d^4\,\omega}{d\,x^4} \cos^2\theta - i\frac{d^3\,\omega}{d\,x^3} \right\} \right\}$$

In order to take this quantity between the proper limits, it is only necessary to write it first with two accents and then with one accent, and take the difference of the quantities so expressed.

Instead, however, of employing the preceding expressions, I shall now introduce the auxiliary quantity e employed by Mr. Ivory. Let

$$\tan\phi = \frac{\sqrt{2\,i\,x''}}{\cos\theta} \qquad\qquad e = \tan\frac{\phi}{2}$$

$$\tan\phi = \frac{2\,e}{(1 - e^2)^2}$$

$$\sqrt{\cos^2\theta + 2\,i\,x} = \frac{\sqrt{2\,i\,x''}}{2\,e} \sqrt{(1 - e^2)^2 + \frac{4\,e^2\,x}{x''}}$$

I assume with Mr. Ivory

$$\sqrt{(1 - e^2)^2 + \frac{4\,e^2\,x}{x''}} = 1 - e^2 + 2\,e^2\,z$$

$$\frac{d\,x}{\sqrt{\cos^2\theta + 2\,i\,x}} = \frac{2\,e\,x''\,d\,z}{\sqrt{2\,i\,x''}}$$

then $\qquad x = x''\,z - x''\,e^2\,(z - z^2).$

Suppose $d\,\omega$ contains any term of the form $A\,c^{-b\,x}\,d\,x$, then

$\dfrac{\alpha \sin\theta\,d\,\omega}{\sqrt{\cos^2\theta + 2\,i\,x}}$ will contain the term

$$\frac{2\,\alpha \sin\theta\,A\,e\,c^{-x''\,z} \times c^{x''\,e^2\,(z - z^2)}\,x''\,d\,z}{\sqrt{2\,i\,x''}},$$

which is to be integrated from $z = 0$ to $z = 1$. This integral may be exhibited under two aspects; in the first, which is that given by Mr. Ivory, the coefficients of the different powers of e consist of a finite number of terms. In other forms of the integral, which will be given here, applicable to all atmospheres of finite altitude, the coefficients are composed of an infinite number of terms, converging with rapidity and in a form suited for numerical computation.

$$c^{-b x'' z} = 1 - b x'' z + \frac{b^2 x''^2 z^2}{1 . 2} - \&c.$$

and the single term $A c^{-b x} \mathrm{d} x$ in $\mathrm{d} \omega$ will give in

$$\int \frac{\alpha (1 + \alpha) \sin \theta \,\mathrm{d} \omega}{\sqrt{\cos^2 \theta + 2 i x}} \text{ the term}$$

$$\frac{2 A \alpha (1 + \alpha) \sin \theta}{\sqrt{2 i x''}} \Big\{ \Big\{ x'' - b x''^2 z + \frac{b^2 x''^3}{2} z^2 - \frac{b^3 x''^3}{2 . 3} z^3 + \&c. \Big\} e \,\mathrm{d} z$$

$$+ b (1 - z) \Big\{ x''^2 z - b x''^3 z^2 + \frac{b^2 x''^4 z^3}{2} - \frac{b^3 x''^5}{2 . 3} z^4 + \&c. \Big\} e^3 \,\mathrm{d} z$$

$$+ b^2 (1 - z)^2 \Big\{ x''^3 z^2 - b x''^4 z^3 + \frac{b^2 x''^5 z^4}{2} - \frac{b^3 x''^6}{2 . 3} z^5 + \&c. \Big\} e^5 \,\mathrm{d} z$$

$$+ \&c. \Big\}$$

$$\int_0^1 z^m (1 - z)^n = \frac{n (n - 1) (n - 2) (n - 3) \ldots\ldots 1}{(m + 1) (m + 2) \ldots\ldots (m + n + 1)}.$$

Hence it will be seen that if $\mathrm{d} \omega$ contains any number of terms of the form $A c^{-b x} \mathrm{d} x$, the definite integral required

$$= \frac{2 (1 + \alpha) \alpha \sin \theta}{\sqrt{2 i x''}} \Big\{ \Big\{ x'' + \frac{x''^2}{2} \Big(\frac{\mathrm{d} \omega}{\mathrm{d} x}\Big)' + \frac{x''^3}{2 . 3} \Big(\frac{\mathrm{d} \omega^2}{\mathrm{d} x^2}\Big)' + \frac{x''^4}{2 . 3 . 4} \Big(\frac{\mathrm{d}^3 \omega}{\mathrm{d} x^3}\Big)' + \&c. \Big\} e$$

$$+ \Big\{ \frac{x''^2}{2 . 3} \Big(\frac{\mathrm{d} \omega}{\mathrm{d} x}\Big)' + \frac{x''^3}{3 . 4} \Big(\frac{\mathrm{d}^2 \omega}{\mathrm{d} x^2}\Big)' + \frac{x''^4}{2 . 4 . 5} \Big(\frac{\mathrm{d}^3 \omega}{\mathrm{d} x^3}\Big)' + \&c. \Big\} e^3 \qquad [1]$$

$$+ \Big\{ \frac{x''^3}{3 . 4 . 5} \Big(\frac{\mathrm{d}^2 \omega}{\mathrm{d} x^2}\Big)' + \frac{x''^4}{4 . 5 . 6} \Big(\frac{\mathrm{d}^3 \omega}{\mathrm{d} x^3}\Big)' + \frac{x''^5}{2 . 5 . 6 . 7} \Big(\frac{\mathrm{d}^4 \omega}{\mathrm{d} x^4}\Big)' + \&c. \Big\} e^5$$

$$+ \&c. \Big\}$$

Another expression for $\delta \theta$ may be obtained in the following manner. Suppose $\mathrm{d} \omega$ contains any term

$$A_n (x'' - x)^n \,\mathrm{d} x = A_n X^n \,\mathrm{d} x$$

$$x'' - x = x_{''}(1 - z)(1 + e^2 z)$$

$$\frac{A_n (x'' - x)^n \, \mathrm{d}\, x}{\sqrt{\cos^2\theta + 2\, i\, x}} = \frac{2\, A_n\, x''^{n+1} (1 - z)^n (1 + e^2 z)^n\, e\, \mathrm{d}\, z}{\sqrt{2\, i\, x''}}$$

$$\int_0^1 (1 - z)^n z^m = \frac{m(m-1)(m-2)\ldots\ldots 1}{(n+1)(n+2)\ldots\ldots(n+m+1)}$$

n and m being whole numbers. Hence

$$\delta\,\theta = \frac{2\,\alpha(1+\alpha)\sin\theta}{\sqrt{2\,i\,x''}} \left\{ \left\{ A_0\, x'' + \frac{A_1\, x''^2}{2} + \frac{A_2\, x''^3}{3} + \frac{A_3\, x''^4}{4} + \frac{A_4\, x''^5}{5} + \&\text{c.} \right\} e \right.$$

$$+ \left\{ \frac{A_1\, x''^2}{2.3} + \frac{2\, A_2\, x''^3}{3.4} + \frac{3\, A_3\, x''^4}{4.5} + \frac{4\, A_4\, x''^5}{5.6} + \&\text{c.} \right\} e^3$$

$$+ \left\{ \frac{2.1\, A_2\, x''^3}{3.4.5} + \frac{3.2\, A_3\, x''^4}{4.5.6} + \frac{4.3\, A_4\, x''^5}{5.6.7} + \frac{5.4\, A_5\, x''^6}{6.7.8} + \&\text{c.} \right\} e^5 \qquad [1]$$

$$+ \left\{ \frac{3.2.1\, A_3\, x''^4}{4.5.6.7} + \frac{4.3.2\, A_4\, x''^5}{5.6.7.8} + \frac{5.4.3\, A_5\, x''^6}{6.7.8.9} + \frac{6.5.4\, A_6\, x''^7}{7.8.9.10} + \&\text{c.} \right\} e^7$$

$$\left. + \&\text{c.} \right\}$$

A_0, A_1, A_2, &c. are constants, the numerical value of which depends upon the constitution of the atmosphere.

$$\omega = 1 - A_0\, X - \frac{A_1\, X^2}{2} - \frac{A_2\, X^3}{3} - \&\text{c.}$$

the first term is necessarily equal to unity, because when $X = 0$, $\omega = 1$, when $\omega = 0$, $X = X' = x''$, therefore generally

$$A_0\, x'' + \frac{A_1\, x''^2}{2} + \frac{A_2\, x''^3}{3} + \&\text{c.} = 1.$$

Let ω_1 be written for brevity instead of $\left(\frac{\mathrm{d}\,\omega}{\mathrm{d}\,u}\right)$,

ω_2 . . . . $\left(\frac{\mathrm{d}^2\,\omega}{\mathrm{d}\,u^2}\right)$,

ω_3 . . . . $\left(\frac{\mathrm{d}^3\,\omega}{\mathrm{d}\,u^3}\right)$,

then the quantities $\left(\frac{\mathrm{d}\,\omega}{\mathrm{d}\,x}\right)$, $\left(\frac{\mathrm{d}\,\omega^2}{\mathrm{d}\,x^2}\right)$ might be deduced from ω_1, ω_2, &c., in the following manner, without having recourse to the series

$$\omega = \text{f}\,x + \frac{\alpha}{2\,i}\,\frac{\text{d}\,.\,(\text{f}\,x)^2}{\text{d}\,x} + \frac{\alpha^2}{2\,.\,3\,i^2}\,\frac{\text{d}^2\,.\,(\text{f}\,x)^3}{\text{d}\,x^2} + \&\text{c}.$$

Since

$$u - \frac{\alpha}{i}\,\omega = x \qquad \frac{\text{d}\,x}{\text{d}\,u} = 1 - \frac{\alpha}{i}\,\frac{\text{d}\,\omega}{\text{d}\,u} = 1 - \frac{\alpha}{i}\,\omega_1$$

$$\frac{\text{d}\,\omega}{\text{d}\,u} = \frac{\text{d}\,\omega}{\text{d}\,x}\,\frac{\text{d}\,x}{\text{d}\,u}$$

therefore

$$\frac{\text{d}\,\omega}{\text{d}\,x} = \frac{\omega_1}{1 - \frac{\alpha}{i}\,\omega_1},$$

similarly

$$\frac{\text{d}^2\,\omega}{\text{d}\,x^2} = \frac{\omega_2}{\left(1 - \frac{\alpha}{i}\,\omega_1\right)^3}$$

$$\frac{\text{d}^3\,\omega}{\text{d}\,x^3} = \frac{\omega_3}{\left(1 - \frac{\alpha}{i}\,\omega_1\right)^4} + \frac{3\,\frac{\alpha}{i}\,(\omega_2)^2}{\left(1 - \frac{\alpha}{i}\,\omega_1\right)^5}$$

$$\frac{\text{d}^4\,\omega}{\text{d}\,x^4} = \frac{\omega_4}{\left(1 - \frac{\alpha}{i}\,\omega_1\right)^5} + \frac{10\,\frac{\alpha}{i}\,\omega_3\,\omega_2}{\left(1 - \frac{\alpha}{i}\,\omega_1\right)^6} + \frac{15\,\frac{\alpha^2}{i^2}\,(\omega_2)^3}{\left(1 - \frac{\alpha}{i}\,\omega_1\right)^7}$$

The quantities $\frac{\text{d}\,\omega}{\text{d}\,X}$, $\frac{\text{d}^2\,\omega}{\text{d}\,X^2}$, &c., might also be deduced from $\frac{\text{d}\,\omega}{\text{d}\,U}$, $\frac{\text{d}^2\,\omega}{\text{d}\,U^2}$, &c., by similar expressions, only changing the signs of those terms which are multiplied by uneven powers of $\frac{\alpha}{i}$. I have not, however, found it convenient to have recourse to this method of obtaining the development of ω in terms of X. I have employed the series

$$\omega = 1 - \text{F}\,X - \frac{\alpha}{2\,i}\,\frac{\text{d}(\text{F}X)^2}{\text{d}\,X} - \frac{\alpha^2}{2\,.\,3i^2}\,\frac{\text{d}^2(\text{F}X)^3}{\text{d}\,X^2} - \&\text{c.},$$

and I have found $\frac{\alpha}{2\,i}\,(\text{F}\;X)^2$ by actual multiplication, $\frac{\alpha^2}{2\,.\,3\,i^2}(\text{F}\,X)^3$ by multiplying $\frac{\alpha}{2\,i}(\text{F}\,X)^2$ by $\frac{\alpha}{3\,i}$ $(\text{F}\;X)$, &c. This process, though somewhat tedious, is extremely easy. As it may be carried on systematically, and the numbers follow each other, it is not liable to error.

So far all is general; it now remains to make some supposition with regard to the function $f\,u$, upon which the constitution of the atmosphere depends. If we take, as in p. 3, see also p. 24,

$$\omega = 1 - H^{\frac{1}{1-\gamma}} c^{-u} \left\{ c^{-u} - 1 + H \right\}^{\frac{1}{\gamma-1}}$$

$$u = u'' - U. \quad \text{See p. 35.}$$

$$\omega = 1 - H^{\frac{1}{1-\gamma}} c^{-u''} c^{U} \left\{ c^{-u''} c^{U} - 1 + H \right\}^{\frac{1}{\gamma-1}}$$

$$c^{-u''} = 1 - H, \quad \text{therefore}$$

$$\omega = 1 - \frac{(1-H)^{\frac{\gamma}{\gamma-1}}}{H^{\frac{1}{\gamma-1}}} c^{U} \left\{ c^{U} - 1 \right\}^{\frac{1}{\gamma-1}}$$

If we take, as in p.19, $\gamma = 1{\cdot}5$,

$$1 - \omega = \frac{c^{-u}}{H^2} \left\{ c^{-u} - 1 + H \right\}^2 = (1-q)^2 (1 - H q)$$

$$f\,x = 1 - \frac{c^{-x}}{H^2} \left\{ c^{-x} - 1 + H \right\} \qquad p = p' (1-q)^3$$

$$\omega = 1 - \frac{(1-H)^3}{H^2} \left\{ c^{3U} - 2 c^{2U} + c^{U} \right\},$$

$$i = \frac{k(1 + \alpha \theta')}{a g H \beta}. \quad \text{See p. 25,} \qquad \theta = \frac{\varrho'}{p'} \left\{ \frac{1}{\alpha} + \theta' \right\} \frac{p}{\varrho} - \frac{1}{\alpha}$$

$$\frac{E p^{\frac{\gamma-1}{\gamma}}}{1 - E p^{\frac{\gamma-1}{\gamma}}} = -H. \quad \text{See p. 3.}$$

In page 29 I found $H = {\cdot}54378$ (from the observations of M. Gay Lussac) corresponding to the temperature $87{\cdot}^\circ 35$ of Fahrenheit, and to $30{\cdot}145$ inches of mercury in the barometer. As the uncertainty with respect to the values of γ and E appertaining to the mean state of the atmosphere makes it useless to have recourse to greater refinement, I shall now suppose that this value of H will be sufficiently exact for the temperature 50° of Fahrenheit and for 30 inches of mercury in the barometer at the earth's surface; the sequel will show that this hypothesis is admissible, and the calculation of i will stand thus: when $\gamma = 1{\cdot}5$

$$\log \frac{k}{g\,M} = 4{\cdot}2633392 \qquad \log \beta = 9{\cdot}5228787$$

$$\log M = 9{\cdot}6377843 \qquad \log H = 9{\cdot}7354232$$

$$\log(1 + \alpha\,\theta') = 0{\cdot}0159881 \qquad a = 6{\cdot}8041168$$

$$3{\cdot}9171116 \qquad 6{\cdot}0624187$$

$$6{\cdot}0624187$$

$$\log i = 7{\cdot}8546929 \qquad i = {\cdot}0071564$$

$$u'' = \text{Nap. log}\,\frac{1}{1-H} = {\cdot}78478.$$

The following table shows the constitution of the atmosphere with this system of constants. It should be recollected that in calculating this table, as well as those in p. 22 and p. 27, the law of Mariotte and Gay Lussac,

$$p = k\,\varrho\,(1 + \alpha\,\theta),$$

is implicitly supposed to hold good at very low temperatures, which is to a certain extent conjectural. For this reason, and for the reason that we have not at present sufficient data for determining with great precision the constants γ and E, it is not intended to attach precision to the temperatures, densities and pressures given in the following table for the altitudes beyond 5 miles. The following example will serve to show how the table was calculated:

Calculation of the Pressure, Temperature, and Density for the height of 10 miles.

$$\log 10 = 1{\cdot}0000000 \qquad i = 7{\cdot}8546929$$

$$\log a = 3{\cdot}5974758 \text{ in miles} \qquad \log M = 9{\cdot}6377843$$

$$7{\cdot}4025242 \qquad 8{\cdot}2169086$$

$${\cdot}002526$$

$$7{\cdot}4025242$$

$$\log 1{\cdot}002526 = 0{\cdot}0011364 \qquad {\cdot}152925$$

$$9{\cdot}8470750 = \log(1 - H\,q)$$

$$7{\cdot}4013878 \qquad {\cdot}703194$$

$$8{\cdot}2169086 \qquad {\cdot}296806 = H\,q$$

$$9{\cdot}1844792$$

$$\log Hq = 9{\cdot}4725517$$
$$\log H = 9{\cdot}7354232$$
$$9{\cdot}7371285$$
$$\cdot 54592 = q$$
$$\cdot 45408 = 1 - q$$

$$\log(1-q) = 9{\cdot}6571324 \qquad 9{\cdot}6571324 \qquad 1{\cdot}2201080$$
$$2 \qquad 3 \qquad \log p = 0{\cdot}4485185$$
$$9{\cdot}3142648 \qquad 8{\cdot}9713972 \qquad 1{\cdot}6686265$$
$$9{\cdot}8470750 \qquad 1{\cdot}4771213 \qquad \log \varrho = 9{\cdot}1613398$$
$$9{\cdot}1613398 \qquad 0{\cdot}4485185 \qquad 2{\cdot}5072867$$
$$\varrho = \cdot 14499 \qquad p = 2{\cdot}81 \qquad 321{\cdot}6$$
$$448{\cdot}0$$

$$\tau = [1{\cdot}2201080]\,\frac{p}{\varrho} - 448^\circ \qquad \tau = -\,126{\cdot}4$$

Table showing the constitution of the Atmosphere.

Height in miles.	Pressure p.	Temp. τ.	Density ϱ.
	Inch.	Fahr.	
0	30·00	+50·0	1·00000
1	24·61	35·0	·84611
2	20·07	19·5	·71294
3	16·25	+ 3·4	·59798
4	13·06	−13·3	·49903
5	10·41	30·6	·41403
10	2·81	126·4	·14499
15	·45	240·6	·03573
22·35	......	−448·0	....

According to this system of constants, the ascent for depressing Fahrenheit's thermometer 1° is about 352 feet.

If $\log \alpha$ (in sex. sec.) $= 1{\cdot}7669538$ $\qquad \alpha$* $= \cdot 00028348$

$$x'' = u'' - \frac{\alpha}{i} = \cdot 74514 \qquad \log\left\{\frac{(1-H)^3}{H^2}\right\} = 9{\cdot}5066765.$$

* I have adjusted the value of α so that the mean refraction at 45° might exactly agree with that of M. Bessel.

$$\text{F } U = \frac{(1-H)^3}{H^2}\left\{c^{3U} - 2c^{2U} + c^{U}\right\}.$$ See p. 32.

$$= [9{\cdot}5066765]\ U^2 + [9{\cdot}8077064]\ U^3 + [9{\cdot}8254352]\ U^4$$
$$+ [9{\cdot}6827677]\ U^5 + [9{\cdot}4289405]\ U^6 + [9{\cdot}0902531]\ U^7$$
$$+ [8{\cdot}6829116]\ U^8 + [8{\cdot}2178251]\ U^9 + [7{\cdot}7028036]\ U^{10}$$
$$+ [7{\cdot}1436673]\ U^{11} + [6{\cdot}5450488]\ U^{12} + [5{\cdot}9104758]\ U^{13}$$
$$+ [5{\cdot}2429802]\ U^{14} + [4{\cdot}5449962]\ U^{15} + [3{\cdot}8186763]\ U^{16} + \&c.$$

F X is found by changing U into X in the above series. Although the development of ω might be obtained by procuring the quantities $\frac{d\,\omega}{d\,X}$, $\frac{d^2\,\omega}{d\,X^2}$, &c., from $\frac{d\,\omega}{d\,U}$, $\frac{d\,\omega^2}{d\,U^2}$, through the expressions given in p. 37. I have preferred employing the series

$$d\,\omega = 1 - \text{F}\,X - \frac{\alpha}{2\,i}\,\frac{d(\text{F}\,X)^2}{d\,X} - \frac{\alpha^2}{2\,.\,3\,i^2}\,\frac{d^2(\text{F}\,X)^3}{d\,X^2}$$

$$\frac{d\,\omega}{d\,x} = \frac{d\,\text{F}\,X}{d\,X} + \frac{\alpha}{2\,i}\,\frac{d^2(\text{F}\,X)^2}{d\,X^2} + \frac{\alpha^2}{2\,.\,3\,i^2}\,\frac{d^3(\text{F}\,X)^3}{d\,X^3} + \&c.$$

By involution from the expression for F X the following were obtained :

$$\frac{\alpha}{2\,i}(\text{F}\,X)^2 = [7{\cdot}3105250]\ X^4 + [7{\cdot}9125849]\ X^5 + [8{\cdot}2225672]\ X^6$$
$$+ [8{\cdot}3653575]\ X^7 + [8{\cdot}3901249]\ X^8 + [8{\cdot}3259417]\ X^9$$
$$+ [8{\cdot}1894155]\ X^{10} + [7{\cdot}9925434]\ X^{11} + [7{\cdot}7440827]\ X^{12}$$
$$+ [7{\cdot}4492548]\ X^{13} + [7{\cdot}1140008]\ X^{14} + [6{\cdot}7436508]\ X^{15}$$
$$+ [6{\cdot}3433889]\ X^{16} + [5{\cdot}9048724]\ X^{17} + \&c.$$

$$\frac{\alpha^2}{2\,.\,3\,i^2}(\text{F}\,X^3) = [4{\cdot}9382822]\ X^6 + [5{\cdot}7162512]\ X^7 + [6{\cdot}1990788]\ X^8$$
$$+ [6{\cdot}5124208]\ X^9 + [6{\cdot}7058125]\ X^{10} + [6{\cdot}8073789]\ X^{11}$$
$$+ [6{\cdot}8341245]\ X^{12} + [6{\cdot}7797954]\ X^{13} + [6{\cdot}7124763]\ X^{14}$$
$$+ [6{\cdot}5779048]\ X^{15} + [6{\cdot}3758464]\ X^{16} + [6{\cdot}1424833]\ X^{17}$$
$$+ [5{\cdot}9091202]\ X^{18} + \&c.$$

$$\frac{\alpha^3}{2.3.4\,i^3}(\text{F}\,X^4) = [2{\cdot}4411007]\ X^8 + [3{\cdot}3458338]\ X^9 + [3{\cdot}9500667]\ X^{10}$$
$$+ [4{\cdot}3853642]\ X^{11} + [4{\cdot}6996664]\ X^{12} + [4{\cdot}9204991]\ X^{13}$$
$$+ [5{\cdot}0668412]\ X^{14} + [5{\cdot}1470479]\ X^{15} + [5{\cdot}1817987]\ X^{16}$$
$$+ [5{\cdot}1502344]\ X^{17} + [5{\cdot}0588813]\ X^{18} + [4{\cdot}9792296]\ X^{19} + \&c.$$

$$\frac{a^4}{2.3.4.5 i^4}(\mathrm{F} X^5) = [9 \cdot 8470092]\, X^{10} + [0 \cdot 1480392]\, X^{11} + [1 \cdot 5496223]\, X^{12}$$
$$+ [2 \cdot 0850470]\, X^{13} + [2 \cdot 4878504]\, X^{14} + [2 \cdot 8026972]\, X^{15}$$
$$+ [3 \cdot 0171128]\, X^{16} + [3 \cdot 1689304]\, X^{17} + [3 \cdot 2999429]\, X^{18}$$
$$+ [3 \cdot 3830969]\, X^{19} + [3 \cdot 4680509]\, X^{20} + \&c.$$

The coefficients of the different powers of X in these series become very small, but they acquire large multipliers from the successive differentiations which are required to give the corresponding terms in the expression for ω.

I find with this constitution of the atmosphere, A_n being the coefficient of X^n in the expression for $\frac{d\omega}{dx}$.

$$A_1 = \cdot 6422$$
$$A_2 = 1 \cdot 9268 + \cdot 0245 = 1 \cdot 9513$$
$$A_3 = 2 \cdot 6761 + \cdot 1635 + \cdot 0010 = 2 \cdot 8406$$
$$A_4 = 2 \cdot 4085 + \cdot 5008 + \cdot 0109 + \cdot 0001 - 2 \cdot 9203$$
$$A_5 = 1 \cdot 6110 + \cdot 9741 + \cdot 0531 + \cdot 0007 = 2 \cdot 6389$$
$$A_6 = \cdot 8617 + 1 \cdot 3750 + \cdot 1640 + \cdot 0045 = 2 \cdot 4052$$
$$A_7 = \cdot 3854 + 1 \cdot 5250 + \cdot 3657 + \cdot 0192 + \cdot 0003 = 2 \cdot 2956$$
$$A_8 = \cdot 1486 + 1 \cdot 3920 + \cdot 6353 + \cdot 0595 + \cdot 0019 = 2 \cdot 2373$$
$$A_9 = \cdot 0504 + 1 \cdot 0812 + \cdot 9009 + \cdot 1428 + \cdot 0074 \ldots\ldots = 2 \cdot 1827$$
$$A_{10} = \cdot 0153 + \cdot 7322 + 1 \cdot 0335 + \cdot 2802 + \cdot 0229 + \cdot 0007 \ldots\ldots = 1 \cdot 0848$$
$$A_{11} = \cdot 0042 + \cdot 4389 + 1 \cdot 1265 + \cdot 4596 + \cdot 0561 + \cdot 0029 \ldots\ldots = 2 \cdot 6882$$
$$A_{12} = \cdot 0010 + \cdot 2366 + 1 \cdot 0329 + \cdot 6852 + \cdot 1094 + \cdot 0090 \ldots\ldots = 2 \cdot 0741$$
$$A_{13} = \cdot 0002 + \cdot 1163 + \cdot 8410 + \cdot 8072 + \cdot 2051 + \cdot 0226 \ldots\ldots = 1 \cdot 9924$$
$$A_{14} = \ldots\ldots + \cdot 0529 + \cdot 5644 + \cdot 8410 + \cdot 3371 + \cdot 0492 \ldots\ldots = 1 \cdot 8466.$$

$$\frac{d\omega}{dx} = \cdot 6422\, X + 1 \cdot 9513\, X^2 + 2 \cdot 8406\, X^3 + 2 \cdot 9203\, X^4$$
$$+ 2 \cdot 6389\, X^5 + 2 \cdot 4052\, X^6 + \&c.$$

Hence by substituting these values of A_1, A_2, &c., in the expression for $\delta\theta$ given in p. 36, I find the first term in the refraction

$$= \sin\theta \, \{1132'' \cdot 8\, e + 639'' \cdot 9\, e^3 + 220'' \cdot 4\, e^5 + 60'' \cdot 5\, e^7 + 17'' \cdot 8\, e^9 + 5'' \cdot 5\, e^{11} + \&c.\} \qquad [1]$$

At the horizon $e = 1$, and this portion of the horizontal refraction $= 2076'' \cdot 9$.

The second term in the refraction is

$$-\frac{3}{2}\frac{\sin\theta\,\alpha\,i^2\,x^2\,d\,\omega}{(\cos^2\theta+2\,i\,x)^{\frac{5}{2}}}$$

$$=-\frac{3.4\sqrt{i}\,\alpha\sin\theta\,x''^2\,z^2\,\{1-e^2+e^2\,z\}^2\,e^3\frac{d\,\omega}{d\,x}\,d\,z}{2\sqrt{2\,x''}\,\{1-e^2+2\,e^2\,z\}^2}$$

$$=-\frac{3.4\sqrt{i}\,\alpha\sin\theta\,x''^2\,z^2\,e^3}{2\sqrt{2\,x''}}\frac{d\,\omega}{d\,x}\,d\,z\left\{1-2\,z\,e^2\right.$$

$$\left.+\{5\,z^2-2\,z\}\,e^4-\&c.\right\}$$

Suppose $d\,\omega$ contains any term

$$A_n\,(x''-x)^n\,d\,x$$

$$x''-x=x''\,(1-z)\,(1+e^2\,z)$$

$$d.\,\delta\,\theta=-\frac{3.4\sqrt{i}\,\alpha\sin\theta\,x''^{n+2}}{2\sqrt{2\,x''}}\,A_n(1-z)^n\,(1+e^2\,z)^n\,z^2\,e^3\,d\,z$$

$$\left\{1-2\,z\,e^2+\{5\,z^2-2\,z\}\,e^4-\&c.\right\}$$

Neglecting the higher powers of e

$$\delta\,\theta=-\frac{3.4}{2}\frac{\sqrt{i}\,\alpha\sin\theta\,e^3}{\sqrt{2\,x''}}\left\{\frac{2.1\,A_1\,x''^3}{2.3.4}+\frac{2.1\,A_2\,x''^4}{3.4.5}+\frac{2.1\,A_3\,x''^5}{4.5.6}\right.$$

$$\left.+\&c.\right\} \qquad [2]$$

With the same constants as before, $\gamma=1{\cdot}5$, $H=\cdot54378$

$$\delta\,\theta=-[1{\cdot}3861838]\sin\theta\,e^3\left\{\frac{2.1\,A_1\,x''^3}{2.3.4}+\frac{2.1\,A_2\,x''^4}{3.4.5}+\frac{2.1\,A_3\,x''^5}{4.5.6}\right.$$

$$\left.+\&c.\right\} \qquad [2]$$

$$=-1''{\cdot}5\sin\theta\,e^3.$$

This term thus amounts to only $1''{\cdot}5$ at the horizon; according to Mr. Ivory it does not amount to more than $1''$.

Hence, finally, the refraction is expressed by the following series :—

$$\text{Ref.}=\sin\theta\,\{1132''{\cdot}8\,e+638''{\cdot}4\,e^3+220''{\cdot}4\,e^5$$

$$+60''{\cdot}5\,e^7+17''{\cdot}8\,e^9+5''{\cdot}5\,e^{11}+\&c.\}$$

$$= \sin\theta \{[3\cdot0541728]\, e + [2\cdot8051475]\, e^3 + [2\cdot3443834]\, e^5 + [1\cdot7821564]\, e^7 + [1\cdot2501754]\, e^9 + [0\cdot7409070]\, e^{11} + \&c.\}$$

$$\tan\phi = \frac{[9\cdot0139814]}{\cos\theta} \qquad e = \tan\frac{\phi}{2}.$$

Mr. Russell has calculated a table of refractions from the above formula; and the following comparative view has been drawn up, with Bessel's table in the *Tabulæ Regiomontanæ* (which may be considered as the result of observations), with the table published annually in the *Conn. des Temps*, and with Mr. Ivory's table, recently published in the Phil. Trans. 1838, p. 224.

Tables of Mean Refractions.

Bar. 30 inch. Therm., Fahr., 50°.

App. Zenith Dist.	Mean Refractions.				App. Zenith Dist.
	Calculated.			Observed.	
	Conn. des Temps.	Ivory.	New Table*.	Tab. Reg.	
10°	10″30	10″30	10″30	10″30	10°
20	21·26	21·26	21·26	21·26	20
30	33·72	33·72	33·72	33·72	30
40	48·99	48·99	48·99	48·99	40
45	58·36	58·36	58·36	58·36	45
50	69·52	69·52	69·51	69·52	50
55	83·25	83·25	83·24	83·24	55
60	100·86	100·85	100·85	100·85	60
65	124·65	124·65	124·63	124·62	65
70	159·22	159·16	159·16	159·11	70
75	214·83	214·70	214·68	214·58	75
80	320·63	320·19	320·08	319·88	80
81	354·33	353·79	353·64	353·38	81
82	395·37	394·68	394·47	394·20	82
83	445·87	445·42	445·11	444·86	83
84	511·22	509·86	509·34	509·23	84
85	595·80	593·96	593·13	593·38	85
85½	648·34	646·21	645·15	647·10	85½
86	710·07	707·43	706·04	707·15	86
86½	783·07	779·92	778·08	777·36	86½
87	870·37	866·76	864·30	864·59	87
87½	975·89	971·93	968·84	972·21	87½
88	1105·1	1101·35	1097·26	1101·40	88
88½	1265·0	1262·6	1257·66	1265·5	88½
89	1464·9	1466·8	1461·49	1481·8	89
89½	1716·4	1729·5	1725·70	1764·9	89½
90		2072·6	2075·4		90

* The constitution of the atmosphere is shown by the table in p. 40.

The following table shows the errors of the table of the *Conn. des Temps*, of Mr. Ivory's table, and of my table, assuming Bessel's to be correct.

Zenith Dist.	Error of Table of Conn. des Temps.	Error of Mr. Ivory's Table.	Error of New Table.	Zenith Dist.	Error of Table of Conn. des Temps.	Error of Mr. Ivory's Table.	Error of New Table.
70°	+ ·11″	+ ·05″	+ ·05″	86°	+ 3·12″	+ ·28″	− 1·11″
75	+ ·25	+ ·12	+ ·10	86½	+ 5·71	+ 2·56	+ ·72
80	+ ·75	+ ·31	+ ·20	87	+ 5·78	+ 2·17	− ·29
81	+ ·95	+ ·41	+ ·26	874	+ 3·68	− ·28	− 3·37
82	+ 1·17	+ ·48	+ ·27	88	+ 3·70	− ·05	− 4·14
83	+ 1·01	+ ·56	+ ·25	88½	− 0·50	− 2·90	− 7·80
84	+ 1·99	+ ·63	+ ·11	89	− 16·90	− 15·00	− 20·30
85	+ 2·42	+ ·58	− ·25	89½	− 48·50	− 35·40	− 39·20
85½	+ 1·24	− ·89	− 1·95				

I think that the discrepancies about 85½, 86, 86½, are caused by irregularities in the refractions of the *Tab. Reg.* Groombridge, who made many observations for the purpose of determining the amount of the refraction near the horizon, makes the horizontal refraction, for barometer 30 inch, and therm. Fahr. 50°, 2075″·4.* There is, however, some uncertainty respecting this quantity, and generally respecting the amount of refractions near the horizon. Upon this point see Delambre, *Ast.*, vol. i. p. 319. Mr. Ivory says " There is great probability that the horizontal refraction is very near 2070″, and does not exceed that quantity."

But for the irregularity in Bessel's table, which is clearly seen in the diagram inserted in the annexed plate, my table of mean refractions would be identical with the table of that distinguished astronomer to within 3 degrees of the horizon. It may therefore be safely concluded that the refractions, which belong to the atmosphere, constituted as I have supposed, in conformity with my theory of the heat of steam and other vapours, are consistent with observation.

The quantities denoted by u, (ϱ), s in the *Mécanique Céleste* correspond to the quantities, $i\,x$, ϱ', and $\frac{z}{a}$ of this treatise. The equation

$$s - \alpha\left[1 - \frac{\varrho}{(\varrho)}\right] = u, \text{ Méc. Cél., vol. iv. p. 262,}$$

corresponds to the equation

* This curious coincidence, with my value of the horizontal refraction, is of course partly accidental.

$$u - \frac{\alpha}{i}\,\omega = x \text{ of p. 34.}$$

Laplace assumes the relation between ω and x.

$$\varrho = (\varrho)\left[1 + \frac{f\,u}{l'}\right] c^{-\frac{u}{l'}}$$

or in the notation of this treatise

$$\varrho = \varrho'\left\{1 + \frac{i\,f}{l'}\,x\right\} c^{-\frac{i\,x}{l'}}$$

$$\omega = -\frac{i\,f}{l'}\,x\,c^{-\frac{i\,x}{l'}}$$

f and l' being arbitrary quantities, such that

$$f = \cdot 49042 \qquad\qquad l' = \cdot 000741816.$$

A table, similar to that which I have given in p. 43, showing the constitution of the atmosphere, which Laplace has assumed, would be instructive, and would enable us to judge of the admissibility of the conditions attributed to the higher regions of the atmosphere by that great philosopher.

In this treatise I have obtained an expression* for the altitude in terms of the pressure, founded upon the conditions of elastic vapours generally; this gives the relation between u and ω (see p. 30), from which a relation between x and u must afterwards be sought. When on the contrary the relation between ω and x is assumed (as was done by Laplace) an advantage may be gained in the calculation of the refraction, at the expense, however, of a simple and intelligible definition of the constitution of the atmosphere; and such a relation is of course also unconnected with any considerations founded upon the nature of caloric.

Mr. Ivory assumes the relation

$$\frac{p}{p'} = \frac{7}{9}\,\frac{\varrho}{\varrho'} + \frac{2}{9}\,\frac{\varrho^2}{\varrho'^2}$$

p' denoting the pressure, and ϱ' the density of the atmosphere at the earth's surface. From this relation it follows that (see p. 32)

$$* \quad \frac{z}{1 + \frac{z}{a}} = -\,a\,i \log(1 - H\,q).$$

$$a\,i\,u = \frac{7\,k\,(1+\alpha\,\theta')}{9g}\log\frac{\varrho'}{\varrho} + \frac{4\,k\,(1+\alpha\,\theta')}{9\,g}\left(1-\frac{\varrho}{\varrho'}\right)$$

k, α, θ' are L, β, τ', in Mr. Ivory's notation.

When u is a simple function of ω, this value of u may be substituted in the equation

$$x = u - \frac{\alpha}{i}\,\omega^{*}, \qquad \text{(p. 34)}$$

and the value of ω in terms of x may be found at once by the reversion of the series.

Mr. Ivory makes $\dfrac{k\,(1+\alpha\,\theta')}{a\,g} = i$, so that

$$x = -\log(1-\omega) + f\log(1-\omega) + \left\{2f - \frac{\alpha}{i}\right\}\omega$$

$$= -\log(1-\omega) + f\log(1-\omega) + h\,\omega.$$

This equation corresponds to the equation of Mr. Ivory

$$x = u - \lambda\,(1 - c^{-u}) - f\,\frac{\mathrm{d}\,c^{-u}\,R_2}{c^{-u}\,\mathrm{d}\,u} - f'\,\frac{\mathrm{d}^2\,c^{-u}\,R_4}{c^{-u}\,\mathrm{d}\,u^2} - \&\text{c.}$$

p. 203, when $f' = 0$. $R_2 = 1 - u - c^{-u}$ $\omega = 1 - c^{-u}$.

The table of mean refractions given by Mr. Ivory is founded upon the supposition that f', f'', &c. $= 0$.

Let $i'\,x' = -\dfrac{k\,(1+\alpha\,\theta)\,(1-f)}{a\,g}\log(1-\omega)$

$$+\left\{\frac{2\,k\,(1+\alpha\,\theta)}{a\,g}\,f - \alpha\right\}\omega$$

and let $i' = \dfrac{k\,(1+\alpha\,\theta)\,(1-f)}{a\,g} = (1-f)\,i$

$$h' = \frac{h}{1-f} \qquad \lambda = \frac{\alpha}{i} \qquad h = 2f - \lambda$$

$$x' = -\log(1-\omega) + h'\,\omega$$

i and h are identical with the quantities represented by those letters by Mr. Ivory, if

* Mr. Ivory has the equivalent equation $\frac{\sigma}{a} = \frac{s}{a} + \alpha\,\omega = i\,x + \alpha\,\omega$, p. 203. Mr. Ivory's σ is $a\,i\,u$ in the notation of this treatise.

$\alpha = \cdot 0002835, \qquad i = \cdot 0012958, \qquad h = \cdot 22566, \; f = \frac{2}{9}$

then $\qquad i' = \cdot 0010078, \qquad h' = \cdot 29012.$

By Lagrange's theorem I find

$$\omega = 1 - c^{h'} c^{-x'} + h' c^{2h'} c^{-2x'} - \frac{3\, h'^2}{1 \,.\, 2} c^{3h'} c^{-3x'}$$

$$+ \frac{4^2\, h'^3}{1 \,.\, 2 \,.\, 3} c^{4h'} c^{4x'} - \&c.$$

$$\frac{d\,\omega}{d\,x'} = c^{h'} c^{-x'} - 2\, h' c^{2h'} c^{-2x'} + \frac{3^2\, h'^2}{1 \,.\, 2} c^{3h'} c^{-3x'}$$

$$+ \frac{4^3\, h'^3}{1 \,.\, 2 \,.\, 3} c^{4h'} c^{-4x'} - \&c.$$

The first part of the refraction is given by the expression

$$\alpha\,(1 + \alpha) \sin\theta \int_0^{\infty} \frac{\frac{d\,\omega}{d\,x'}\, d\,x'}{\sqrt{\cos^2\theta + 2\, i'\, x'}}$$

Let

$$n \left\{ \frac{\cos^2\theta}{2\, i'} + x' \right\} = z^2$$

$$n\, x' = z^2 - \frac{n \cos^2\theta}{2\, i'} \qquad\qquad d\,x = \frac{2\, z\, d\, z}{n}$$

$$\int_0^{\infty} \frac{c^{-n\,x'}\, d\,x'}{\sqrt{\cos^2\theta + 2\, i'\, x'}} = \frac{2\, c^{z'}}{\sqrt{2\, i'}\,\sqrt{n}} \int_{z'}^{\infty} c^{-z^2}\, d\, z$$

At the horizon $\cos\theta = 0, \quad z' = 0$

$$\int_0^{\infty} \frac{c^{-n\,x'}\, d\,x'}{\sqrt{2\, i'\, x'}} = \frac{2}{\sqrt{2\, i'}\,\sqrt{n}} \int_0^{\infty} c^{-z^2}\, d\, z = \frac{\sqrt{\pi}}{\sqrt{2\, i'}\,\sqrt{n}}$$

This part of the horizontal refraction

$$= \frac{\alpha\,(1 + \alpha)\sqrt{\pi}}{\sqrt{2\, i'}} \left\{ c^{h'} - \frac{2\, h'\, c^{2h'}}{\sqrt{2}} + \frac{3^2\, h'^2\, c^{3h'}}{1 \,.\, 2 \sqrt{3}} - \frac{4^3\, h'^3\, c^{4h'}}{1 .2 .3 \sqrt{4}} + \&c. \right\}$$

$$= \frac{\alpha\,(1 + \alpha)\sqrt{\pi}}{\sqrt{2\, i}} \left\{ 1 + \frac{1}{2} f + h - \frac{2\, h}{\sqrt{2}} \right.$$

$$+\frac{1.3}{2.4}f^2+\frac{3}{2}hf+\frac{h^2}{1.2}-\frac{2.3}{\sqrt{2}\,2}hf\cdot\frac{2^2h^2}{\sqrt{2}}+\frac{3^2h^2}{\sqrt{3}\,1.2}$$

$$+\frac{1.3.5}{2.4.6}f^3+\frac{3.5}{2.4}hf^2+\frac{5}{1.2.2}h^2f+\frac{h^3}{1.2.3}$$

$$-\frac{2.3.5}{\sqrt{2}\,2.4}hf^2\quad\frac{2^2.5}{\sqrt{2}\,2}h^2f\quad\frac{2^3h^3}{\sqrt{2}\,1.2}$$

$$+\frac{3^2.5}{\sqrt{3}\,2.2}h^2f+\frac{3^3}{1.2\sqrt{3}}h^3-\frac{4^3}{1.2.3\sqrt{4}}h^3$$

$$+\frac{1.3.5.7}{2.4.6.8}f^4+\frac{3.5.7}{2.4.6}hf^3+\frac{5.7}{1.2.2.4}h^2f^2+\frac{7h^3f}{1.2.3.2}+\frac{h^4}{1.2.3.4}$$

$$-\frac{2.3.5.7}{\sqrt{2}\,2.4.6}hf^3-\frac{2^2.5.7}{\sqrt{2}\,2.4}h^2f^2-\frac{2^3.7}{\sqrt{2}\,1.2.2}h^3f$$

$$-\frac{2^4h^4}{\sqrt{2}\,1.2.3}+\frac{3^2.5.7}{1.2\sqrt{3}.2.4}h^2f^2+\frac{3^3.7}{1.2\sqrt{3}\,2}h^3f+\frac{3^4h^4}{1.2\sqrt{3}\,1.2}$$

$$-\frac{4^3.7}{1.2.3\sqrt{4}.2}h^3f-\frac{4^4}{1.2.3.4\sqrt{4}}h^4+\frac{4^5}{1.2.3.4.5\sqrt{5}}h^4+\&c.$$

$$=\frac{\alpha(1+\alpha)\sqrt{\pi}}{\sqrt{2i}}\left\{1+\lambda(\sqrt{2}-1)-f\left(2\sqrt{2}-\frac{5}{2}\right)\right.$$

$$\left.+h^2\left\{\frac{1}{2}-2\sqrt{2}+\frac{3}{2}\sqrt{3}\right\}-\frac{3}{2}hf(\sqrt{2}-1)+\frac{3}{8}f^2\right\}$$

when the higher powers of f and h are rejected, and this expression agrees with that given by Mr. Ivory, Phil. Trans., 1838, p. 207.

$$\frac{\alpha(1+\alpha)\sqrt{\pi}}{\sqrt{2i}}=2036''\cdot5$$

In atmospheres which extend to an infinite distance m (or x'' in the notation of this treatise) is infinite and e always $=1$, so that in this case the method employed by Mr. Ivory in p. 211 of his memoir, Phil. Trans., 1838, would seem at least to require further elucidation. Mr. Ivory has avoided this consideration, which would otherwise arise with the atmosphere which he has assumed, by imposing an arbitrary limit to the altitude of his atmosphere, while, however, if I am not mistaken, upon his own assumption, the density and the pressure are still finite. When n is large the numerators of the separate quantities of which the quantity A_{2n+1} in p. 211 is composed become large also.

I do not find in Mr. Ivory's paper any remarks tending to prove that the quantities which he has discarded depending upon the higher powers of f and h are incapable of producing any sensible effect; taken separately they are by no means insignificant. Nor do I think it follows as a matter of course, even if the positive and negative terms are numerically of equal value at the horizon, and so fortunately cut one another out, that the same thing will happen necessarily at all other altitudes. Unless the approximation is pushed so far as to secure the retention of all the sensible terms, or those which fairly come within the limits of the errors of observation, any comparison of the result with the valuable table of M. Bessel is illusory and only calculated to lead to incorrect conclusions. It is also indispensable that the relation implied or expressed between z and ω should be in exact conformity with the conditions attributed to the atmosphere, and in this respect the table of mean refractions of the late Mr. Atkinson in the Memoirs of the Astronomical Society appears to me not to rest upon a solid foundation.

Mr. Ivory connects the pressure and the density by the relation

$$\frac{p}{p'} = \cdot 77777 \frac{\varrho}{\varrho'} + \cdot 22222 \frac{\varrho^2}{\varrho'^2}.$$

M. Biot finds

$$\frac{p}{p'} = \cdot 761909002718 \frac{\varrho}{\varrho'} + \cdot 238167190564 \frac{\varrho^2}{\varrho'^2}$$
$$- \cdot 000076193282,$$

when the coefficients are so taken as to apply as nearly as the question will admit of throughout the whole extent of the atmosphere. But, by a careful examination of the data, M. Biot finds that at the earth's surface the following relation is more accurate.

$$\frac{p}{p'} = \cdot 956643870584 \frac{\varrho}{\varrho'} + \cdot 120146052460 \frac{\rho^2}{\rho'^2}$$
$$- \cdot 076789923044 \qquad \text{(p. 69.)}$$

and at the upper limit of the atmosphere

$$\frac{p}{p'} = \cdot 6604978157646 \frac{\rho}{\rho'} + \cdot 4159581823536 \frac{\rho^2}{\rho'^2}$$
$$- \cdot 00006605394115.$$

According to my view this equation does not contain the true mathematical law which connects the density and pressure, but of course a parabola of this kind may always be found which will osculate the true curve at any given point.

In the *Comptes Rendus des Séances de l'Académie des Sciences*, tom. viii. p. 95, M. Biot verified and adopted a calculation of Lambert, who found from the phenomena of twilight the altitude of the atmosphere (hauteur des dernières particules d'air réfléchissantes) to be 29·115 metres.

It is unnecessary to dwell any further at present upon this subject, because if my theory of the Heat of Vapours be correct, the calculation of Astronomical Refractions, founded upon conditions which are not in conformity with that theory, becomes a problem of mere curiosity.

THE END.

PRINTED BY RICHARD AND JOHN E. TAYLOR,
RED LION COURT, FLEET STREET.

ON

THE DISCOVERY OF THE PLANET NEPTUNE.

LECTURE DELIVERED AT THE BROMLEY LITERARY INSTITUTE.

BY

SIR J. W. LUBBOCK, BART., F.R.S., F.G.S., F.L.S.

LONDON:
GOODHALL & CO., 9, PANCRAS LANE. E.C.

1861.

LONDON:
PRINTED BY GOODHALL AND CO.
9, PANCRAS LANE, E.C.

ON THE DISCOVERY OF THE PLANET NEPTUNE.

I have chosen this subject for my lecture this evening, in the first place, because it is generally considered as the most interesting event which has happened in the science of Astronomy since the discovery of the law of gravitation by Newton, and because the circumstances which attended it are not so generally known as they deserve to be; yet, those circumstances are not involved in any ambiguity or obscurity. I shall place before you the clearest documentary evidence, proving that Mr. Adams had ascertained and *published* the existence of the planet, and its exact place in the heavens, many months before this had been done by the great French Astronomer, M. Le Verrier. And the only question which can possibly arise is, what constitutes publication. Some rule certainly ought to obtain, for the history of science would be thrown into confusion; and great injustice would be the consequence, if, when one philosopher has proclaimed to the world some important truth, another is to be allowed to step in and share in the reward, upon his simple assertion that he also had arrived at the same conclusion. In this case, I shall show you that Mr. Adams communicated his discovery to the Astronomer Royal; and I maintain that that mode of publication is as honest, as complete, and as indisputable as if Mr. Adams had communicated his discovery to the public in the columns of the *Times'* Newspaper; unfortunately, it did not become so generally known as it would have been in the latter case. But, suppose any one were to make a discovery in physiology, and were to communicate it in a letter to that eminent physiologist now occupying the high position of the President of the Royal Society, and suppose Sir B. Brodie were to throw away the letter into his waste-basket, would that be a reason that the author of the letter should lose his claim to be considered as the discoverer? It is a matter very greatly to be regretted that Mr. Adams did not take steps to give greater publicity to the marvellous result he had obtained, the knowledge of which was,

I believe, for some time confined to the Astronomer Royal and to Prof. Challis. It is also much to be regretted that the Astronomer Royal did not institute a search for the planet, at least, in the vicinity of the place in the heavens assigned to it by Mr. Adams. In this case it would probably have been seen in this country before it had been heard of in France. But, before we blame Mr. Airy, or impute neglect to him, or to Prof. Challis, we must bear in mind all the circumstances of the case. In fact, it is clear from letters of Mr. Airy, published by himself, that he believed it was impossible to arrive at a knowledge of the place of the planet, if it existed, from the data which Mr. Adams possessed; he, therefore, did not feel justified in employing the resources of the Observatory under his control, in searching for an object, the existence of which was a matter of speculation, and the place of which, if it existed, might be far remote from that assigned to it by Mr. Adams. The sequel shows that Mr. Airy was mistaken; but it is the misfortune of Mr. Adams, that the very magnitude of the difficulties which he had successfully overcome, operated to his disadvantage, by creating a mistrust in the minds of Mr. Airy and Prof. Challis, which led them to delay the examination of the heavens with the hope of discovering the planet, until after Le Verrier had announced his discovery of the place of the planet, and had assigned to it a place identical with that which Mr. Adams many months previously had communicated to Mr. Airy. This mistrust served still further to delay the actual confirmation of the existence of the planet by observation; for it induced the Astronomer Royal to suggest to Prof. Challis, to whom the search was eventually entrusted, to examine, not only the vicinity of the place assigned to the planet by Adams and Le Verrier, but a very much larger area. The area which Prof. Challis undertook to examine was, in fact, so large, that although he actually saw the planet three times before it was found at Berlin, he failed to recognise it, as he would have done, if he had confined his attention to a smaller area. In the sequel, I shall place before you the documents to which I have briefly alluded in these prefatory remarks; and I shall explain how it happened that the planet was discovered by the astronomers at Berlin, the same night that the communication reached them from Le Verrier, requesting them to search for it.

There is no event in the history of Astronomy which can be compared with the discovery of the planet Neptune, many other heavenly bodies have, indeed, been found which were not known to the ancients, but not under circumstances which present the same degree of interest.

I will briefly advert to some of the more remarkable discoveries of this kind, in order to contrast them hereafter with the discovery of Neptune.

On the 7th of January, 1610, Galileo examined Jupiter with a telescope, which he had recently succeeded in constructing, and his attention was arrested by three small but bright stars which appeared in his vicinity. He imagined them to be three fixed stars. Happening, by mere accident, to examine Jupiter again on the 8th of January, he was surprised to find the stars arranged quite differently from what they were when he first saw them. They were now all on the west side of the planet. The following night was cloudy, and Jupiter could not be observed. On the 10th, and on the 11th, Galileo again saw two stars, and they were both on the east of Jupiter. He now came to the conclusion that they were three stars revolving round Jupiter in the same manner as Venus and Mercury revolved round the Sun. On the 13th, he finally saw all the four satellites.

Galileo's discovery excited, as may be imagined, great interest and great amazement. Some denied their existence, as Sizzi, a Florentine Astronomer, who maintained that as there were only seven days of the week, and seven metals, there could be only seven planets. Galileo's friend Dini rejected them as needless and superfluous, unless Galileo could declare their effect, that is, their influence upon the events of life, and in what manner they were to be taken into account in casting nativities. Even the illustrious Kepler wondered how there could be any addition to the number of the planets, as he had found, thirteen years previously, and published in his Mysterium Cosmographicum, the notion that the number of the planets depended upon the number of the regular solids, the dodecahedron, the cube, etc., and that the planets could be neither more nor less than six in number. Galileo afterwards discovered the spots on the Sun, and the phases of the planet Venus.

Galileo stumbled, as it were, upon these phenomena; but the discovery of the satellites of Jupiter, although not wearing the distinctive characters which render so interesting the discovery of Neptune, was of great importance; and it influence on the progress of the science of Astronomy, which was very great, is thus described by Sir J. Herschel:—

* "The discovery of these bodies was one of the first brilliant results of the invention of the telescope; one of the first great facts

* Herschel's Address to the Astronomical Society, 1827.

which opened the eyes of mankind to the system of the universe, which taught them the comparative insignificance of their own planet, and the superior vastness and nicer mechanism of those other bodies, which had before been distinguished from the stars only by their motion, and wherein none but the boldest thinkers had ventured to suspect a community of nature with our own globe. This discovery gave the holding turn to the opinions of mankind respecting the Copernican system; the analogy presented by these little bodies (little, however, only in comparison with the great central body about which they revolve) performing their beautiful revolutions in perfect harmony and order about it, being too strong to be resisted. This elegant system was watched with all the curiosity and interest the subject naturally inspired. The eclipses of the satellites speedily attracted attention; and the more when it was discerned, as it speedily was, by Galileo himself, that they afforded a ready method of determining the difference of longitudes of distant places on the earth's surface, by observations of the instants of their disappearances and reappearances, simultaneously made. Thus, the first astronomical solution of the great problem of the longitude, the first mighty step which pointed out a connection between speculative astronomy and practical utility, and which, replacing the fast dissipating dreams of astrology by nobler visions, showed how the stars might really, and without fiction, be called arbiters of the destinies of empires, we owe to the satellites of Jupiter, those atoms imperceptible to the naked eye, and floating like motes in the beam of their primary—itself an atom to our sight, noticed only by the careless vulgar as a large stars, and by the philosophers of former ages as something moving among the star, they knew not what, nor why: perhaps only to perplex the wise with fruitless conjectures, and harrass the weak with fears as idle as their theories."

In the year 1781, Herschel, then Mr. Herschel, and residing at Bath, discovered the planet Uranus. His attention was arrested by the appearance in the field of the telescope of a star having a disc, different in appearance from the spurious disc of a fixed star caused by irradiation. His attention being called to it, he soon found that it had a proper motion; he therefore presumed it to be a comet. It was not till some time after that it was found to move round the sun in a circular orbit, and to be, in fact, a planet.

The talent of Galileo and of Herschel was, in fact, displayed in the construction of the telescope, when Galileo, upon imperfect indications of what had been done in Holland, succeeded in constructing an instrument which increased the optical power of the human eye, and directed it to the heavens, he could not fail to detect appearances which could never have been otherwise observed.

The discovery of the numerous small planets which lie between Mars and Jupiter, was due to other methods; these I shall describe when I come to relate the mode in which the planet Neptune was first recognised in the field of the telescope by Galle and Encke.

When it was found that the satellites of Jupiter were continually seen at about the same distance from that planet, in the same way as the Moon accompanies the earth, it was impossible that the idea of some link or *force* connecting them with the planet could escape notice. Long before the time of Newton, vague and confused notions of *gravitation* were given out by Kepler and others. Not being sufficiently conversant with the resolution of forces, they were unable to explain why, if the Earth attracted the Moon, the Moon could be maintained at the same distance from it. Kepler gets over the difficulty by attributing an animal force to the Earth and Moon. In fact, Kepler literally believed that the Earth was an enormous living animal. "The Earth is not an animal like a dog, ready at every nod; but more like a bull, or an elephant." And, again, "Although in the preface to my 'Commentary on Mars,' I have mentioned it as probable that the waters are attracted by the Moon as iron is by a loadstone; yet, if any one uphold that the Earth regulates its breathing according to the motion of the Sea and Moon, as animals have daily and nightly alternations of sleep and waking, I shall not think his philosophy unworthy of being listened to."

Kepler derived his notion of a magnetic attraction among the planetary bodies from the writings of Gilbert, an English philosopher, who died in 1603. The following is an extract from his work, which was not published until the middle of the 17th century, but a knowledge of its contents may, in several instances, be traced back to the period at which it was written :—

"There are two primary causes of the motion of the seas—the moon, and the diurnal revolution. The moon does not act on the seas by its rays or its light. How then? Certainly by the common effort of the bodies, and (to explain it by something similar) by their magnetic attraction. It should be known, in the first place, that the whole quantity of water is not contained in the sea and rivers, but that the mass of earth (I mean this globe) contains moisture and spirit much deeper even than the sea. The moon draws this out by sympathy, so that they burst forth on the arrival of the moon, in consequence of the attraction of that star; and for the same reason, the quicksands which are in the sea open themselves more, and perspire their moisture and spirits during the flow of the tide, and the whirlpools in the sea disgorge copious waters; and as the star retires, they devour the same again, and attract the spirits and moisture of the terrestrial globe. Hence the moon attracts, not so much the sea as the subterranean spirits and humours; and the interposed earth has no more power of resistance than a table or any other dense body has to resist the force of a magnet. The sea rises from the greatest depths, in consequence of the ascending humours and spirits; and when it is raised up, it necessarily flows on to the shores, and from the shores it enters the rivers."

The most remarkable paragraph which occurs in Kepler, relating to gravitation, appears to me to be the following, which is probably the earliest indication of any correct notion with respect to the figure of the Earth and the force of gravity: "If the Earth were not round, heavy bodies would not tend from every side in a straight line towards the centre of the earth, but to different points from different sides." In fact, the Earth not being a perfect sphere, the force of gravity does not act in a straight line drawn to the centre, and the deviation is sensible in delicate astronomical observations.

Kepler speculated upon a real attractive force residing in the Sun; but he considered the planets as quiet and unwilling to move when left alone; and that this virtue, supposed by him to proceed in every direction of the Sun, swept them round, just as the sails of a windmill would carry round anything which became entangled in them. Hence, Ross, a cotemporary writer, asserted that, "Kepler's opinion, that the planets are moved round by the sunne, and that this is done by sending forth a magnetic virtue, and that the sunbeams are like the teeth of a wheel taking hold of the planets, are senseless crotchets fitter for a wheeler or a miller, than a philosopher."* Kepler also had a correct idea with respect to the diminution of the force of attraction: "Something in the nature of light which is observed to diminish similarly at increased distances."

With indomitable perseverence, Kepler speculated on the results of

* The New Planet, no Planet, or the Earth no Wandering Star.

a few principles taken for granted by him, from very precarious analogies, as the causes of the phenomena observed in nature. He was disturbed by the manifold inconveniences of the common theory of the universe, and embraced that of Copernicus, which he learned from his tutor Māstlin. He pertinaciously sought the causes, which make the number, the size, and the motion of the orbits such as they are. He observed that the motions in every case seemed to be connected with the distance; and he argued that if God had adapted motions to the orbits in some relation to the distances, it was probable that he had also arranged the distances themselves in relation to something else.

After an appalling quantity of labour, for we must remember that logarithms were not then invented, after many trials and many failures, Kepler at last succeeded in discovering two of the celebrated theorems called Kepler's laws: the first is, that the planets move in ellipses round the Sun, placed in one of the foci; the second, that the time of describing any arc is proportioned in the same orbit to the area included between the arc and the two bounding distances from the Sun. The third, which was not discovered till twelve years later, is that the squares of the periodic times of the planets are as the cubes of their mean distances from the Sun; or, as Kepler expressed it, "The periodic times of the planets are exactly in the sesquiplicate proportion of their orbits or circles."

The following table is intended to illustrate the third law of Kepler:—

	Mean Distance.		Period in Years.	
Mercury.	·3871	·0580	·2408	·0580
Venus.	·7233	·3784	·6152	·3784
The Earth.	1·000	1·000	1·000	1·000
Mars.	1·524	3·537	1·881	3·537
Jupiter.	5·203	140·7	11·86	140·7
Saturn.	9·539	867·6	29·46	867·6
Uranus.	19·18	7059	84·02	7059
Neptune.	30·04	27097	164·6	27097

The second column contains the cubes of the numbers in the first column; the fourth column contains the squares of the numbers in the third column; and the numbers in the second and the fourth columns are identical. The mean distance and the periods are taken from Sir J. Herschel's treatise on Astronomy. Kepler also discovered that the plane of the orbit of a planet passed through the sun, a proposition of equal importance with the three others that are called his laws. The remarks which I have addressed you upon the subject of the discoveries of Galileo and Kepler, are taken from the lives of those Astronomers written by my lamented friend Drinkwater Bethune, and published in the library of the Society for the Diffusion of useful Knowledge.

Newton deduced these laws from the law of gravitation; that is, he showed that if the law of attraction varied inversely as the square of the distance from the Sun, these laws were a necessary consequence. About the same time,* Huyghens, Wren, Hook, and Halley had obtained theorems relating to circular motion; but Newton obtained the conditions of a body moving in an ellipse which was a much more difficult problem. Newton afterwards made other important applications of the theory of gravitation to the Moon and to the tides, and he showed that if the planets move in ellipses† round the Sun, placed in the focus, the law of attraction must vary inversely as the square of the distance.

Ever since the time of Newton, the efforts of the greatest mathematicians have been and are still directed to carrying out into all their details the views of Newton, details which neither the inaccurate observations which he possessed or the methods which he made use of were adequate to reach. They gave rules for preparing tables for finding, at any future time, the places of the Moon and planets, as well as the eclipses of Jupiter's satellites. These tables have progressively improved; but, probably, many years will yet elapse before they can be pronounced to have arrived at the utmost limit of human ingenuity.

This is called the problem of three bodies; suppose two planets revolving round the Sun approximately in circles or ellipses of small eccentricity, to determine the irregularities, or perturbations, as they are called, produced by their mutual attractions upon each other. The problem which Adams and Le Verrier solved was the converse of this. It had not been before attempted; and it presented still greater difficulties.

Before the discovery of Uranus by Sir W. Herschel, in 1781, astronomers had frequently observed it, supposing it to be a fixed star.

---

* Casus corollarii sexti obtinet in corporibus cœlestibus (ut seorsum collegerunt etiam nostrates Wrennius, Hookius, at Halleius).

† Revolvatur corpus in ellipsi: requiritur lex vis centripetœ tendentis ad umbilicum ellipseos.—*Newton.*

In 1690, Flamsteed had designated it as a star of the sixth magnitude; and from that year, and before 1781, its position had been determined no less than nineteen different times. Delambre was the first who calculated tables of the planet, which, for a few years, represented its motion with tolerable accuracy. With time, however, the discordance between the tables and observations constantly increased. This led to the publication of new Tables by Mr. Alexis Bouvard, in 1821. It was fully shown, in the introduction to these tables, that, when every correction for perturbation was applied, it was still impossible to reconcile the observations of Flamsteed, Lemonnier, Bradley, and Mayer, made before 1781, with those made after. The orbit adopted in these tables represented the observations made in the years immediately following the publication of the tables. But, in five or six years, the discordance, again growing up, became so great, that it could not escape notice. The following letter was addressed by the Rev. T. J. Hussey, who had recently passed through Paris, to the Astronomer Royal:—

"Hayes, Kent, 17th November, 1834.

"With M. Alexis Bouvard I had some conversation upon a subject I had often meditated, which will probably interest you, and your opinion may determine mine. Having taken great pains last year with some observations of Uranus, I was led to examine closely Bouvard's tables of that planet. The apparently inexplicable discrepancies between the ancient and modern observations suggested to me the possibility of some disturbing body beyond Uranus, not taken into account because unknown. My first idea was to ascertain some approximate place of this supposed body empirically, and then with my large reflector set to work to examine all the minute stars thereabouts: but I found myself totally inadequate to the former part of the task. If I could have done it formerly, it was beyond me now, even supposing I had the time, which was not the case. I therefore relinquished the matter altogether; but, subsequently, in conversation with Bouvard, I inquired if the above might not be the case; his answer was, that, as might have been expected, it had occurred to him, and some correspondence had taken place between Hansen and himself respecting it. Hansen's opinion was, that one disturbing body would not satisfy the phenomena; but that he conjectured there were two planets beyond Uranus. Upon my speaking of obtaining the places empirically, and then sweeping closely for the bodies, he fully acquiesced in the propriety of it, intimating that the previous calculations would be more laborious than difficult; that, if he had leisure, he would undertake them and transmit the results to me, as the basis of a very close and accurate sweep. I have not heard from him since on the subject, and have been too ill to write. What is your opinion on the subject? If you consider the idea as possible, can you give me the limits, roughly, between which this body, or those bodies, may probably be found during the ensuing winter? As we might expect an excentricity [inclination?] approaching rather to that of the old planets than of the new, the breadth of the Zone to be examined will be comparatively inconsiderable. I may be wrong; but I am disposed to think that, such is the perfection of my equatoreal's object glass, I could distinguish, almost at once, the difference of light of a small planet and a star. My plan of proceeding, however, would be very different: I should accurately map the whole space within the required limits, down to the minutest star I could discern; the interval of a single week would then enable me to ascertain any change. If the whole of this matter do not appear to you a chimæra, which, until my conversation with Bouvard, I was afraid it might, I shall be very glad of any sort of hint respecting it."

To this letter the Astronomer Royal returned the following answer:—

"Observatory, Cambridge, 1834, Nov. 23.

"I have often thought of the irregularity of Uranus, and, since the receipt of your letter, have looked more carefully to it. It is a puzzling subject; but I give it as my opinion, without hesitation, that it is not yet in such a state as to give the smallest hope of making out the nature of any external action on the planet. Flamsteed's observations I reject (for the present) without ceremony: but the two observations by Bradley and Mayer cannot be rejected. Thus the state of things is this,— the mean motion and other elements derived from the observations between 1781 and 1825 give considerable errors in 1750, and give *nearly the same errors* in 1834, *when the planet is at nearly the same part of its orbit.* If the mean motion had been determined by 1750 and 1834, this would have indicated nothing: but the fact is, that the mean motions were determined (as I have said) independently. This does not look like irregular perturbation. The observations would be well reconciled if we could from theory bring in two terms; one a small error in Bouvard's excentricity and perihelion, the other a term depending on twice the longitude. The former, of course, we could do; of the latter there are two, viz., a term in the equation of the centre, and a term in the perturbations by Saturn. The first I have verified completely (formula and numbers); the second I have verified generally, but not completely: I shall, when I have an opportunity, look at it thoroughly. So much for my doubts as to the certainty of any extraneous action. But if it were certain that there were any extraneous action, I doubt much the possibility of determining the place of a planet which produced it. I am sure it could not be done till the nature of the irregularity was well determined from several successive revolutions."

It will be seen from this letter that what Adams and Le Verrier actually performed was believed by the Astronomer Royal to be impossible. The same opinion was again expressed in a letter to M. Eugene Bouvard, dated Oct. 12, 1837 :—

"Royal Observatory, Greenwich, 1837, Oct. 12.

"You will see by this statement that the errors of longitude are increasing with fearful rapidity, while those of latitude are nearly stationary. . . . I cannot conjecture what is the cause of these errors, but I am inclined, in the first instance, to ascribe them to some error in the perturbations. There is no error in the pure elliptic theory (as I found by examination some time ago). If it be the effect of any unseen body, it will be nearly impossible ever to find out its place."

The first notice of Mr. Adams occurs in a letter from Prof. Challis to the Astronomer Royal, dated Cambridge, Feb. 13, 1844:—

"Cambridge Observatory, Feb. 13, 1844.

"A young friend of mine, Mr. Adams, of St. John's College, is working at the theory of Uranus, and is desirous of obtaining errors of the tabular geocentric longitudes of this planet, when near opposition, in the years 1846—1826, with the factors for reducing them to errors of heliocentric longitude. Are your reductions of the planetary observations so far aduanced that you could furnish these data? and is the request one which you have any objection to comply with? If Mr. Adams may be favoured in this respect, he is further desirous of knowing, whether, in the calculation of the tabular errors, any alterations have been made in Bouvard's Tables of Uranus besides that of Jupiter's mass."

The Astronomer Royal immediately sent Mr. Adams the observations he required. It appears that by the following autumn, that is, the autumn of 1845, not 1846, as printed in Mr. Grant's history, p. 172. Mr. Adams had obtained the place of the planet, for he called on one of the last days of October, 1845, at the Royal Observatory, Greenwich, in the absence of the Astronomer Royal, and left the following important paper:—

"According to my calculations, the observed irregularities in the motion of Uranus may be accounted for by supposing the existence of an exterior planet, the mass and orbit of which are as follows:—

Mean Distance (assumed nearly in accordance with Bode's law)	38,4
Mean Sidereal Motion in 365,25 days	1°30′ 9
Mean Longitude, 1st October, 1845	323 34
Longitude of Perihelion	315 55
Excentricity	0,1610
Mass (that of the Sun being unity)	0,0001556

He left also a statement of the residual errors which remained in the positions of Uranus, after the perturbations caused by the new planet were taken into account, showing that the original errors had almost entirely disappeared.

The Astronomer Royal wrote a letter to Mr. Adams, acknowledging the receipt of this important communication, to which Mr. Adams did not reply. Thus, apparently, the matter would have dropped, had not M. Le Verrier appeared upon the scene.

I have thus endeavoured to place before you the conditions of the problem, and the motives which induced Mr. Adams to attempt to discover the place of the unseen planet; but, as I think it will be more satisfactory to you to have the account in his own words, I will quote Mr. Adams's own statement contained in the preamble to his Memoir "On the observed irregularities in the motion of Uranus on the hypothesis of disturbances caused by a more distant planet; with a determination of the mass, orbit, and position of the disturbing body":—

"The irregularities in the motions of Uranus have for a long time engaged the attention of Astronomers. When the path of the planet became approximately known, it was found that, previously to its discovery by Sir W. Herschel, in 1781, it had several times been observed as a fixed star by Flamsteed, Bradley, Mayer, and Lemonnier. Although these observations are doubtless very far inferior in accuracy to the modern ones, they must be considered valuable, in consequence of the great extension which they give to the observed arc of the planet's orbit. Bouvard, however, to whom we owe the Tables of Uranus at present in use, found that it was impossible to satisfy these observations, without attributing much larger errors to the modern observations that they admit of, and consequently founded his Tables exclusively on the latter. But, in a very few years, sensible errors began again to show themselves, and though the Tables were formed so recently as 1821, their error at the present time exceeds two minutes of space, and is still rapidly increasing. There appeared, therefore, no longer any sufficient reason for rejecting the ancient observations, especially since, with the exception of Flamsteed's first observation, which is more than twenty years anterior to any of the others, they are mutual confirmatory of each other.

"Now that the discovery of another planet has confirmed, in the most brilliant manner, the conclusions of analysis, and enabled us with certainty to refer these irregularities to their true cause, it is unnecessary for me to enter at length upon the reasons which led me to reject the various other hypotheses which had been formed to account for them.

It is sufficient to say, that they all appeared to be very improbable in themselves, and incapable of being tested by any exact calculation. Some had even supposed that at the great distance of Uranus from the Sun, the law of attraction becomes different from that of the inverse square of the distance. But the law of gravitation was too firmly established for this to be admitted, till every other hypothesis had failed, and I felt convinced that in this, as in every previous instance of the kind, the discrepancies which had for a time thrown doubts on the truth of the law, which eventually afford the most striking confirmation of it.

My attention was first directed to this subject several years since, by reading Mr. Airy's valuable Report on the recent progress of Astronomy. I find among my papers the following memorandum, dated July 3, 1841:—'Formed a design, in the beginning of this week, of investigating, as soon as possible after taking my degree, the irregularities in the motion of Uranus which are yet unaccounted for; in order to find whether they may be attributed to the action of an undiscovered planet beyond it, and, if possible, thence to determine approximately the elements of its orbit, &c., which would probably lead to its discovery.'

Accordingly, in 1843, I attempted a first solution of the problem, assuming the orbit to be a circle, with a radius equal to twice the mean distance of Uranus from the Sun. Some assumption as to the mean distance was clearly necessary in the first instance, and Bode's law appeared to render it probable that the above would not be far from the truth. This investigation was founded exclusively on the modern observations, and the errors of the Tables were taken from those given in the Equations of Condition of Bouvard's Tables as far as the year 1821, and subsequently from the observations given in the *Astronomische Nachrichten*, and from the Cambridge and Greenwich Observations. The result showed that a good general agreement between theory and observation might be obtained; but the larger differences occurring in years where the observations used were deficient in number, and the Greenwich Planetary Observations being then in process of reduction, I applied to Mr. Airy, through the kind intervention of Professor Challis, for the observations of some years, in which the agreement appeared least satisfactory. The Astronomer Royal, in the kindest possible manner, sent me, in February, 1854, the results of all the Greenwich Observations of Uranus.

" Meanwhile, the Royal Academy of Sciences of Göttingen had proposed the Theory of Uranus as the subject of their mathematical prize, and although the little time which I could spare from important duties in my college prevented me from attempting the complete examination of the theory, which a competition for the prize would have required, yet this fact, together with the possession of such a valuable series of observations, induced me to undertake a new solution of the problem. I now took into account the most important terms depending on the first power of the eccentricity of the disturbing planet, retaining the same assumption as before with respect to the mean distance. For the modern observations, the errors of the Tables were taken exclusively from the Greenwich Observations as far as the year, 1830, with the exception of an observation by Bessel, 1823; and subsequently from the Cambridge and Greenwich Observations, and those given in various numbers of the *Astronomische Nachrichten*. The errors of the Tables for the ancient Observations were taken from those given in the Equations of Condition of Bouvard's Tables. After obtaining several solutions differing little from each other, by gradually taking into account more and more terms of the series expressing the Perturbations, I communicated to Professor Challis, in September, 1845, the final values which I had obtained for the mass, heliocentric longitude, and elements of the orbit of the assumed planet. The same results, slightly corrected, I communicated in the following month to the Astronomer Royal. The eccentricity coming out much larger than was probable, and later observations showing that the theory founded on the first hypothesis as to the mean distance, was still sensibly in error, I afterwards repeated my investigation, supposing the mean distance to be about $\frac{1}{30}$th part less than before. The result, which I communicated to Mr. Airy, in the beginning of September of the present year, appeared more satisfactory than my former one, the eccentricity being smaller, and the errors of theory, compared with late observations, being less, and led me to infer that the distance should be still further diminished."

The motives which induced M. Le Verrier to devote his mind to this subject were, of course, the same that had influenced Mr. Adams; they are explained with great clearness in M. Le Verrier's memoir; but, as they would not add anything to the information which is afforded by the long quotation from Mr. Adams, it is unnecessary for me to do more than refer to them.

In the autumn of the year 1845, M. Le Verrier, at the urgent request of M. Arago, undertook a most careful and elaborate revision of the Tables of Uranus. He recalculated the perturbations produced by Jupiter and Saturn, a work of immense labour. The result obtained was that he generally established the correctness of former theories; but he added many small terms.

The second memoir of M. Le Verrier was presented to the Academy on the 1st June, 1846. In this paper, with an enormous amount of labour, and with wonderful mathematical skill, M. Le Verrier demonstrated that it was impossible to account for the movements of Uranus, without the introduction of a new planet, and he assigned 325° for its heliocentric longitude on the 1st January, 1847.

This determination of Le Verrier, and its coincidence with the determination of Mr. Adams, in the preceding year, recalled the attention of the Astronomer Royal to the question; and the subject appears to have come under discussion at a meeting of the Board of Visitors of the Observatory of Greenwich, held on the 29th of June, 1846.

In consequence, Mr. Airy wrote a letter to Prof. Challis, dated July 9th, 1846, of which the following is an extract:—

"You know that I attach importance to the examination of that part of the heavens in which there is . . . reason for suspecting the existence of a planet exterior to Uranus. I have thought about the way of making such examination, but I am convinced that (for various reasons, of declination, latitude of place, feebleness of light, and regularity of superintendence) there is no prospect whatever of its being made with any chance of success, except with the Northumberland Telescope."

On the 13th of July, Mr. Airy transmitted to Prof. Challis "suggestions for the examination of a portion of the heavens in search of the external planet, which is presumed to exist, and to produce disturbance in the motion of Uranus." These "suggestions" contemplated the examination of a part of the heavens 30° long, in the direction of the ecliptic and 10° broad.

Prof. Challis commenced the search on July 29th, and had actually observed the planet on August 4, 1846. The portion of the heavens which Prof. Challis proposed to examine was so large, that, to scrutinize it thoroughly, would have required many more observations than he could have made in six months.

On the 23rd September, Dr. Galle, at Berlin, received a letter from M. Le Verrier, asking him to search for the planet, and stating that probably the planet would present a disc.

The same evening, Dr. Galle compared a map of that portion of the heavens which had been constructed by Dr. Bremiker, and immediately found that a star of the eighth magnitude was missing in the map near the place which M. Le Verrier had assigned to the planet. On that and on the following night, by comparing this with the neighbouring stars, a proper motion was detected in conformity with Le Verrier's elements by Galle and Encke.

The exact place of the planet was found to be—

Sept. 23·5 Longitude, 325°52′

Le Verrier's elements gave Longitude 324°58′, that is within 1° of the truth. Moreover, the diameter of the planet was ″3, as Le Verrier had predicted.

The *optical* discovery of Neptune by Galle and Encke was made under precisely the same circumstances as the discovery of the small planets by Mr. Hind and others, which now amount to sixty-three in number. Mr. Hind, at Mr. Bishop's observatory in the Regent's Park, has discovered no less than ten. Mr. Hind's method, which has since been followed by others, consisted in mapping most carefully every object he could discover in a certain portion of the heavens successively, near the ecliptic, and, after a certain lapse of time, when favourable circumstances presented themselves, again of re-examining the same portion of the heavens, and comparing it with the map. If a star appeared in the heavens which could not be found in the map, only two hypotheses were admissible, either that it had at first been omitted by mistake, or that, being a *planet*, it had since moved into its present position. In the latter case, Mr. Hind's instruments enabled him to detect a proper motion in a few hours. By the kindness of Mr. Hind, I am enabled to place before you one of these maps, with the positions marked upon it, in which the planets Euterpe and Thalia were discovered by him.

On the 12th October, 1846, Prof. Challis addressed the following letter to Mr. Airy :—

" Cambridge Observatory, October 12, 1846.

" I had heard of the discovery [of the new planet] on October 1. . . . I find that my observations would have shewn me the planet in the early part of August, if I had only discussed them. I commenced observing on July 29, attacking first of all, as it was prudent to do, the position which Mr. Adams's calculations assigned as the most probable place of the planet. On July 30, I adopted the method of observing which I spoke of to you. . . . In this way I took all the stars to the 11th magnitude in a zone of 9' in breadth, and was sure that none brighter than the 11th escaped me. My next observations were on August 4. On this day . . . I took stars here and there in a zone of about 70' in breadth, purposely selecting the brighter, as I intended to make them reference-points for the observations in zones of 9' breadth. Among these stars was the planet. A comparison of this day's observations, with a good star-map, would most probably have detected it. On account of moonlight, I did not observe again till August 12. On that day I went over again the zone of 9 breadth, which I examined on July 30. . . . The space gone over on August 12 exceeded in length that of July 30, but included the whole of it. On comparing [at a later time] the observations of these two days, I found that the zone of July 30 contained every star in the responding portion of the zone of August 12, except one star of the 8th magnitude. This, according to the principle of search, which in the want of a good star-map I had adopted, must have been a planet. It had wandered into the latter zone in the interval between July 30 and August 12. By this statement you will see, that, after four days of observing, the planet was in my grasp, if only I had examined or mapped the observations. I delayed doing this, partly because I thought the probability of discovery was small till a much larger portion of the heavens was scrutinised; but chiefly because I was making a grand effort to reduce the vast number of comet observations which I have accumulated; and this occupied the whole of my time when I was not engaged in observing. I actually compared, to a certain extent, the observations of July 30 and August 12, soon after taking them, more for the sake of testing the two methods of observing adopted on those days than for any other purpose; and I stopped short within a very few stars of the planet. After August 12, I continued my observations with great diligence, recording the positions of, I believe, some thousands of stars; but I did not again fall in with the planet, as I took positions too early in right ascension. . . . On Sept. 29, however, I saw, for the first time, Le Verrier's last results, and on the evening of that day I observed strictly according to his suggestions, and within the limits he recommended; and I was also on the look-out for a disk. Among 300 stars which I took that night, I singled out one, against which I directed my assistant to note 'seems to have a disc,' which proved to be the planet. I used on this, as on all other, occasions, a power of 160. This was the third time I obtained an approximate place of the planet before I heard of its discovery."

Thus was the new planet detected, and the triumph of Adams and Le Verrier was complete. An object shining by the light reflected from the Sun at that enormous distance was necessarily very faint. It might, indeed, have been so faint as to be invisible with the most powerful telescopes. But, nevertheless, the memoirs of Adams and Le Verrier would have remained monuments of their unrivalled ability, and would ever have continued to be admired as models of clearness and perspicuity.

I wish I may have been so far successful as to have communicated to you any portion of my enthusiastic admiration of the discovery of the planet Neptune, and of the respect which I entertain towards ADAMS and LE VERRIER, its illustrious authors. Let us not forget that the respect which is due towards a great country like England is not due to the soil we tread upon, or to our buildings, or to anything material, but to the nation, that is, to the individuals of which the nation is composed. Newton raised the glory of England in the seventeenth century, and we are grateful to him for it. Adams has raised the glory of England in the nineteenth century; may he live to make other discoveries, shedding lustre upon his country.

I have endeavoured to place before you some of the circumstances connected with the discovery of the planet Neptune. I am quite aware that, in so doing, I am far from being able to do justice to the merits of M. M. Adams and Le Verrier. My aim has not been to offer to you a technical treatise, in which the student may find a minute description of the methods employed by the great astronomers. I have avoided, as much as possible, whatever would tend to encumber my lecture with algebraic or geometrical symbols; and I have merely endeavoured to place before you that *"central thread of common sense on which the pearls of analytical research are invariably strung." My "utmost pretension has been to place you on the threshold of this particular wing of the temple of Science, or rather on an eminence exterior to it. Admission to its sanctuary, and to the privileges and feelings of a votary, is only to be gained by one means, — *a sound and sufficient knowledge of mathematics, the great instrument of all exact enquiry, without which no man can ever make such advances in this or any other of the higher departments of science, as can entitle him to form an independent opinion on any subject of discussion within their range.*"

---

* Herschel's Treatise on Astronomy.

Printed in the United States
By Bookmasters